PENGUIN BOOKS

A History of the World in Twelve Maps

Jerry Brotton is Professor of Renaissance Studies at Queen Mary University of London, and a leading expert in the history of maps and Renaissance cartography. His most recent book, *The Sale of the Late King's Goods: Charles I and his Art Collection* (2006), was short-listed for the Samuel Johnson Prize as well as the Hessell-Tiltman History Prize. In 2010, he was the presenter of the BBC4 series 'Maps: Power, Plunder and Possession'.

D0256214

JERRY BROTTON

# A History of the World in Twelve Maps

PENGUIN BOOKS

PENGUIN BOOKS

Published by the Penguin Group
Penguin Books Ltd, 80 Strand, London WC2R ORL, England
Penguin Group (USA) Inc., 375 Hudson Street, New York, New York 10014, USA
Penguin Group (Canada), 90 Eglinton Avenue East, Suite 700, Toronto, Ontario, Canada M4P 2Y3
(a division of Pearson Penguin Canada Inc.)
Penguin Ireland, 25 St Stephen's Green, Dublin 2, Ireland (a division of Penguin Books Ltd)
Penguin Group (Australia), 707 Collins Street, Melbourne, Victoria 3008, Australia
(a division of Pearson Australia Group Pty Ltd)
Penguin Books India Pvt Ltd, 11 Community Centre, Panchsheel Park, New Delhi – 110 017, India
Penguin Group (NZ), 67 Apollo Drive, Rosedale, Auckland 0632, New Zealand
(a division of Pearson New Zealand Ltd)
Penguin Books (South Africa) (Pty) Ltd, Block D, Rosebank Office Park,
181 Jan Smuts Avenue, Parktown North, Gauteng 2193, South Africa

Penguin Books Ltd, Registered Offices: 80 Strand, London WC2R ORL, England

www.penguin.com

First published by Allen Lane 2012
Published in Penguin Books 2013
004

Copyright © Jerry Brotton, 2012

The moral right of the author has been asserted

Typeset by Jouve (UK), Milton Keynes
Printed in Great Britain by Clays Ltd, St Ives plc

A CIP catalogue record for this book is available from the British Library

ISBN: 978-0-141-03493-5

www.greenpenguin.co.uk

MIX
Paper from
responsible sources
FSC™ C018179

Penguin Books is committed to a sustainable
future for our business, our readers and our planet.
This book is made from Forest Stewardship
Council™ certified paper.

*For my wife, Charlotte*

# Contents

# List of Figures

# List of Illustrations

# Introduction

In 1881, the Iraqi-born archaeologist Hormuzd Rassam discovered a small fragment of a 2,500-year-old cuneiform clay tablet in the ruins of the ancient Babylonian city of Sippar, today known as Tell Abu Habbah, on the south-west outskirts of modern-day Baghdad. The tablet was just one of nearly 70,000 excavated by Rassam over a period of eighteen months and shipped back to the British Museum in London. Rassam's mission, inspired by a group of English Assyriologists who were struggling to decipher cuneiform script, was to discover a tablet which it was hoped would provide a historical account of the biblical Flood.[1] At first, the tablet was overlooked in favour of more impressive, complete examples. This was partly because Rassam, who could not read cuneiform, was unaware of its significance, which was appreciated only at the end of the nineteenth century when the script was successfully translated. Today, the tablet is on public display at the British Museum, labelled as 'The Babylonian Map of the World'. It is the first known map of the world.

The tablet discovered by Rassam is the earliest surviving object that represents the whole world in plan from a bird's-eye view, looking down on the earth from above. The map is composed of two concentric rings, within which are a series of apparently random circles, oblongs and curves, all of which are centred on a hole apparently made by an early pair of compasses. Evenly distributed around the outer circle are eight triangles, only five of which remain legible. Only when the cuneiform text is deciphered does the tablet begin to make sense as a map.

The outer circle is labelled 'marratu', or 'salt sea', and represents an ocean encircling the inhabited world. Within the inner ring the most prominent curved oblong running through the central hole depicts the Euphrates River, flowing from a semicircle in the north labelled 'mountain', and ending in the southern horizontal rectangle described as 'channel' and 'swamp'. The rectangle bisecting the Euphrates is labelled 'Babylon', surrounded by an arc of circles representing cities and regions including Susa (in southern Iraq), Bit Yakin (a district of Chaldea, near where Rassam himself was born), Habban (home of the ancient Kassite tribe), Urartu (Armenia), Der and Assyria. The triangles emanating outwards from the outer circle of sea are labelled 'nagû', which can be translated as 'region' or 'province'. Alongside them are cryptic legends describing distances (such as 'six leagues between where the sun is not seen'),[2] and exotic animals – chameleons, ibexes, zebus, monkeys, ostriches, lions and wolves. These are uncharted spaces, the mythical, faraway places beyond the circular limits of the known Babylonian world.

The cuneiform text at the top of the tablet and on its reverse reveals that this is more than just a map of the earth's surface: it is a comprehensive diagram of Babylonian cosmology, with the inhabited world as its manifestation. The tantalizing fragments speak of the creation myth of the battle between the Babylonian gods Marduk and Ti'amat. In Babylonian mythology, Marduk's victory over what the tablet calls the 'ruined gods' led to the foundation of heaven and earth, humanity and language, all centred on Babylon, created 'on top of the restless sea'. The tablet, made from the earth's clay, is a physical expression of Marduk's mythical accomplishments, the creation of the earth and subsequent achievements of human civilization, fashioned out of the watery primal chaos.

The circumstances of the tablet's creation remain obscure. The text on the back of the tablet identifies its scribe as a descendant of someone called 'Ea-bēl-ilī' from the ancient city of Borsippa (Birs Nimrud), to the south of Sippar, but why it was made and for whom remains a mystery. Nevertheless, we can tell that this is an early example of one of the most basic objectives of human understanding: to impose some kind of order and structure onto the vast, apparently limitless space of the known world. Alongside its symbolic and mythic description of the world's origins, the tablet's map presents an abstraction of terrestrial reality. It comprehends the earth by categorizing it in circles, triangles, oblongs and dots, unifying writing and image in a world picture at the centre of which lies Babylon. More than two millennia before the dream of looking at the earth from deep space became a reality, the Babylonian world map offers its viewers the chance to look down on the world from above, and adopt a god-like perspective on earthly creation.

Even today, the most committed traveller can never hope to experience more than a fraction of the earth's surface area of more than 510 million square kilometres. In the ancient world, even short-distance travel was a rare and difficult activity, generally undertaken with reluctance and positively feared by those who did so.[3] To 'see' the world's dimensions reproduced on a clay tablet measuring just 12 by 8 centimetres must have been awe-inspiring, even magical. This is the world, the tablet says, and Babylon is the world. To those who saw themselves as part of Babylon, it was a reassuring message. To those who saw it and were not, the tablet's description of Babylonian power and dominion was unmistakable. No wonder that from ancient times, the kind of geographical information relayed by objects like the Babylonian tablet was the preserve of the mystical or ruling elite. As we shall see throughout this book, for shamans, savants, rulers and religious leaders, maps of the world conferred arcane, magical authority on their makers and owners. If such people understood the secrets of creation and the extent of humanity, then surely they must know how to master the terrestrial world in all its terrifying and unpredictable diversity.

Although the Babylonian world map represents the first known attempt to map the whole known world, it is a relatively late example of human mapmaking. The earliest known examples of prehistoric art showing the landscape in plan are inscribed on rock or clay and predate the Babylonian world map by more than 25,000 years; they stretch back to the Upper Palaeolithic period of 30,000 BC. These early inscriptions, much debated by archaeologists as to their date and meaning, seem to represent huts with human figures, livestock enclosures, divisions between basic dwellings, depictions of hunting grounds, even rivers and mountains. Most are so stark that they might easily be mistaken for abstract, geometrical attempts to represent the spatial distribution of objects or events when they are in fact probably more symbolic marks, connected to indecipherable mythic, sacred and cosmological references for ever lost to us. Today, archaeologists are more cautious than their nineteenth-century predecessors in ascribing the term 'map' to these early pieces of rock art; establishing a clear date for the emergence of prehistoric rock art seems to be as futile as defining when a baby first learns to differentiate itself spatially from its immediate environment.[4]

The urge to map is a basic, enduring human instinct.[5] Where would we be without maps? The obvious answer is, of course, 'lost', but maps provide answers to many more questions than simply how to get from one place to another. From early childhood onwards, we make sense of ourselves in relation to the wider physical world by processing information spatially. Psychologists call this activity 'cognitive mapping', the mental device by which individuals acquire, order and recall information about their spatial environment, in the process of which they distinguish and define themselves spatially in relation to a vast, terrifying, unknowable world 'out there'.[6] Mapping of this kind is not unique to humans. Animals also use mapping procedures, such as the scent-marking of territory performed by dogs or wolves, or the location of nectar from a hive defined by the 'dance' of the honey bee.[7] But only humans have made the crucial leap from mapping to map*making*.[8] With the appearance of permanent graphic methods of communication more than 40,000 years ago, humans developed the ability to translate ephemeral spatial information into permanent and reproducible form.

So what is a map? The English word 'map' (and its derivatives) is used in a variety of modern European vernaculars such as Spanish, Portuguese and Polish, and comes from the Latin term *mappa*, meaning a tablecloth, or napkin. The French word for map – *carte* – originates in a different Latin word, *carta*, which also provides the root for the Italian and Russian words for map (*carta* and *karta*) and refers to a formal document, which in turn is derived from the Greek word for papyrus. The ancient Greek term for map – *pinax* – suggests a different kind of object. A *pinax* is a tablet made of wood, metal or stone, on which words or images were drawn or incised. Arabic takes the term in a more visual direction: it uses two words, *ṣūrah*, translated as 'figure', and *naqshah*, or 'painting', while Chinese has adopted a similar word, *tu*, meaning a drawing or a diagram.[9] The term 'map' (or 'mappe') only enters the English language in the sixteenth century, and between then and the 1990s more than 300 competing definitions of it have been proposed.[10]

Today, scholars generally accept the definition provided in the ongoing multi-volume *History of Cartography*, published since 1987 under the general editorship of J. B. Harley and David Woodward. In their preface to the first volume, Harley and Woodward proposed a new English definition of the word. 'Maps', they said, 'are graphic representations that facilitate a spatial understanding of things, concepts, conditions, processes, or events in the human world.'[11] This definition (which will be adopted throughout this book) 'naturally extends to celestial cartography and to the maps of imagined cosmographies', and frees them from more restricted geometrical definitions of the term. By including cosmography – which describes the universe by analysing the earth and the heavens – Harley and Woodward's definition of maps enables us to see archaic artefacts like the Babylonian world map as both a cosmic diagram and a map of the world.

Self-conscious perceptions of maps, and the science of their creation, are relatively recent inventions. For thousands of years what different cultures have called 'maps' were made by people who did not think of them as being in a category separate from the writing of formal documents, painting, drawing or inscribing diagrams on a range of different media from rock to paper. The relationship between maps and what we call geography is even more subtle. Since the Greeks, geography has been defined as the graphic (*graphein*) study of the earth (*gē*), of which mapping represents a vital part. But as an intellectual discipline geography was not properly formalized as either a profession or a subject of academic study in the West until the nineteenth century.

It is in this disparate variety of maps – as cloths, tablets, drawings or prints – that much of their remarkable power and enduring fascination lies. A map is simultaneously both a physical object and a graphic document, and it is both written and visual: you cannot understand a map without writing, but a map without a visual element is simply a collection of place names. A map draws on artistic methods of execution to create an ultimately imaginative representation of an unknowable object (the world); but it is also shaped by scientific principles, and abstracts the earth according to a series of geometrical lines and shapes. A map is concerned with space as its ultimate aim, according to Harley and Woodward's definition. It offers a spatial understanding of events in the human world; but, as we shall see in this book, it is often also about time, as it asks the viewer to observe how these events unfold one after another. We of course look at maps visually, but we can also read them as a series of different stories.

All these strands meet in the type of map that is the subject of this book: maps of the world. But just as much as the term 'map' has its own elusive and shifting qualities, so too does the concept of 'the world'. 'World' is a man-made, social idea. It refers to the complete physical space of the planet but can also mean a collection of ideas and beliefs that constitute a cultural or individual 'world view'. For many cultures throughout history, a map has been the perfect vehicle to express both these ideas of 'world'. Centres, boundaries and all the other paraphernalia included in any map of the world are defined as much by these 'world views' as they are by the mapmaker's physical observation of the earth, which is never made from a neutral cultural standpoint anyway. The twelve maps in this book all present visions of the physical space of the whole world which result from the ideas and beliefs that inform them. A world view gives rise to a world map; but the world map in turn defines its culture's view of the world. It is an exceptional act of symbiotic alchemy.[12]

World maps pose challenges and opportunities for the mapmaker different from those involved in mapping local areas. To begin with, their scale means they are never seriously used as route-finding devices to enable their users to get from one location on the earth's surface to another. But the most significant difference between local and world mapping is one of perception, and presents a serious problem in making any map of the world. Unlike a local area, the world can never be apprehended in a single synoptic gaze of the mapmaker's eye. Even in ancient times, it was possible to locate natural or man-made features from which to look down on a small area at an oblique angle (a 'bird's-eye' perspective) and see its basic elements. Until the advent of photography from space, no such perspective was available to perceive the earth.

Before that momentous innovation, the mapmaker creating a world map drew on two resources in particular, neither of which was physically part of the earth: the sky above and his own imagination. Astronomy enabled him to observe the movement of the sun and the stars and to estimate the size and shape of the earth. Connected to such observations were the more imaginative assumptions based on personal prejudice and popular myths and beliefs, which indeed still exert their power over any world map, as we shall see. The use of photographic satellite imagery is a relatively recent phenomenon that allows people to believe they see the earth floating in space; for three millennia before that, such a perspective always required an imaginative act (nevertheless, a photograph from space is not a map, and it is also subject to conventions and manipulations, as I point out in this book's final chapter on online mapping and its use of satellite imagery).

Further challenges and opportunities beyond perception affect all world maps, including those chosen in this book, and each one can be seen in embryo by looking again at the Babylonian world map. An overriding challenge is abstraction. Any map is a substitute for the physical space it claims to show, constructing what it represents, and organizing the infinite, sensuous variety of the earth's surface according to a series of abstract marks, the beginnings of borders and boundaries, centres and margins. Such markers can be seen in the rudimentary lines of topographical rock art, or the increasingly regular geometrical shapes of the kind on the Babylonian tablet. When these lines are applied to the whole earth, a map not only represents the world, but imaginatively produces it. For centuries the only way of comprehending the world was through the mind's eye, and world maps showed, imaginatively, what the physically unknowable world might look like. Mapmakers do not just reproduce the world, they construct it.[13]

A logical consequence of mapping as a powerful imaginative act is that, in the dictum coined by the Polish-American philosopher Alfred Korzybski in the 1940s, 'the map is not the territory'.[14] Rather like the relation between language and the objects it denotes, the map can never consist of the territory it purports to represent. 'What is on the paper map', argued the English anthropologist Gregory Bateson, 'is a representation of what was in the retinal representation of the man who made the map; and as you push the question back, what you find is an infinite regress, an infinite series of maps. The territory never gets in at all.'[15] A map always manages the reality it tries to show. It works through analogy: on a map a road is represented by a particular symbol which bears little resemblance to the road itself, but viewers come to accept that the symbol is *like* a road. Rather than imitating the world, maps develop conventional signs which we come to accept as standing in for what they can never truly show. The only map that can ever completely represent the territory it depicts would be on the effectively redundant scale of 1 : 1. Indeed, the selection of scale, a proportional method of determining a consistent relationship between the size of the map and the space it represents, is closely related to the problem of abstraction, and has been a rich source of pleasure and comedy for many writers. In Lewis Carroll's *Sylvie and Bruno Concluded* (1893), the other-worldly character Mein Herr announces that '[w]e actually made a map of the country, on a scale of *a mile to the mile!*' When asked if the map has been used much, Mein Herr admits, 'It has never been spread out', and that 'the farmers objected: they said it would cover the whole country, and shut out the sunlight! So we now use the county itself, as its own map, and I assure you it does nearly as well.'[17] The conceit was taken a stage further by Jorge Luis Borges, who, in his one-paragraph short story 'On Rigour in Science' (1946), recast Carroll's account in a darker key. Borges describes a mythical empire where the art of mapmaking had reached such a level of detail that

> the Colleges of Cartographers set up a Map of the Empire which had the size of the Empire itself and coincided with it point by point. Less Addicted to the Study of Cartography, Succeeding Generations understood that this widespread Map was Useless and not without Impiety they abandoned it to the Inclemencies of the Sun and of the Winters. In the deserts of the West some mangled Ruins of the Map lasted on, inhabited by Animals and Beggars; in the whole Country there are no other relics of the Disciplines of Geography.[18]

Borges understood both the timeless quandary and potential hubris of the mapmaker: in an attempt to produce a comprehensive map of their world, a process of reduction and selection must take place. But if his 1 : 1 scale map is an impossible dream, what scale should a mapmaker choose to ensure their world map does not endure the fate he described? Many of the world maps described in this book offer an answer, but none of their chosen scales (or indeed anything else about them) has ever been universally accepted as definitive.

A further problem that presents itself is one of perspective. At what imaginary location does the mapmaker stand before beginning to map the world? The answer, as we have already seen, invariably depends upon the mapmaker's prevailing world view. In the case of the Babylonian world map, Babylon lies at the centre of the universe, or what the historian Mircea Eliade has called the '*axis mundi*'.[19] According to

Eliade, all archaic societies use rites and myths to create what he describes as a 'boundary situation', at which point 'man discovers himself becoming conscious of his place in the universe'. This discovery creates an absolute distinction between a sacred, carefully demarcated realm of orderly existence, and a profane realm which is unknown, formless and hence dangerous. On the Babylonian world map, such sacred space circumscribed by its inner ring is contrasted with the profane space defined by the outer triangles, which represent chaotic, undifferentiated places antithetical to the sacred centre. Orienting and constructing space from this perspective repeats the divine act of creation, shaping form out of chaos, and placing the mapmaker (and his patron) on a par with the gods. Eliade argues that such images involve the creation of a centre that establishes a vertical conduit between the terrestrial and divine worlds, and which structures human beliefs and actions. Perhaps the hole at the centre of the Babylonian world map, usually regarded as the result of a pair of compasses marking out the map's circular parameters, is rather a channel between one world and the next.

The kind of perspective adopted by the Babylonian world map could also be called egocentric mapping. Throughout most of recorded history, the overwhelming majority of maps put the culture that produced them at their centre, as many of the world maps discussed in this book show. Even today's online mapping is partly driven by the user's desire to first locate him- or herself on the digital map, by typing in their home address before anywhere else, and zooming in to see that location. It is a timeless act of personal reassurance, locating our selves as individuals in relation to a larger world that we suspect is supremely indifferent to our existence. But if such a perspective literally centres individuals, it also elevates them like gods, inviting them to take flight and look down upon the earth from a divine viewpoint, surveying the whole world in one look, calmly detached, gazing upon what can only be imagined by earthbound mortals.[20] The map's dissimulating brilliance is to make viewers believe, just for a moment, that such a perspective *is* real, that they are not still tethered to the earth, looking at a map. And here is one of the map's most important characteristics: the viewer is positioned simultaneously inside *and* outside it. In the act of locating themselves on it, the viewer is at the same moment imaginatively rising above (and outside) it in a transcendent moment of contemplation, beyond time and space, seeing everything from nowhere. If the map offers its viewer an answer to the enduring existential question 'Where am I?', it does so through a magical splitting which situates him or her in two places at the same time.[21]

This problem of defining where the viewer stands in relation to a map of the world is one geographers have struggled with for centuries. For Renaissance geographers, one solution was to compare the viewer of a map to a theatre-goer. In 1570 the Flemish mapmaker Abraham Ortelius published a book containing maps of the world and its regions entitled *Theatrum orbis terrarum* – the 'Theatre of the World'. Ortelius used the Greek definition of theatre – *theatron* – as 'a place for viewing a spectacle'. As in a theatre, the maps that unfold before our eyes present a creative version of a reality we think we know, but in the process transform it into something very different. For Ortelius, as for many other Renaissance mapmakers, geography is 'the eye of history', a theatre of memory, because, as he put it, 'the map being laid before our eyes, we may behold things done or places where they were done, as if they were at this time present'. The map acts like a mirror, or 'glass', because 'the charts being placed, as it were certain glasses before our eyes, will the longer be kept in memory, and make the deeper impression in us'. But, like all the best dramatists, Ortelius concedes that his 'glasses' are a process of creative negotiation, because on certain maps 'in some places, at our discretion, where we thought good, we have altered some things, some things we have put out, and otherwhere, if it seemed to be necessary, we have put in' different features and places.[22]

Ortelius describes the position from which a viewer looks at a world map, which is closely related to orientation – the location from which we take our bearings. Strictly speaking, orientation usually refers to *relative* position or direction; in modern times it has become established as fixing location relative to the points on a magnetic compass. But long before the invention of the compass in China by the second century AD, world maps were oriented according to one of the four cardinal directions: north, south, east and west. The decision to orientate maps according to one prime direction varies from one culture to another (as will be seen from the twelve maps discussed in this book), but there is no purely geographical reason why one direction is better than any other, or why modern Western maps have naturalized the assumption that north should be at the top of all world maps.

Why north ultimately triumphed as the prime direction in the Western geographical tradition, especially considering its initially negative connotations for Christianity (discussed in Chapter 2), has never been fully explained. Later Greek maps and early medieval sailing charts, or portolans, were drawn using magnetic compasses, which probably established the navigational superiority of the north–south axis over an east–west one; but even so there is little reason why south could not have been adopted as the simplest point of cardinal orientation instead, and indeed Muslim mapmakers continued to draw maps with south at the top long after the adoption of the compass. Whatever the reasons for the ultimate establishment of the north as the prime direction on world maps, it is quite clear that, as subsequent chapters will show, there are no compelling grounds for choosing one direction over another.

Perhaps the most complex problem of all that confronts the mapmaker is one of projection. For modern cartographers, 'projection' refers to a two-dimensional drawing on a plane surface of a three-dimensional object, namely the globe, using a system of mathematical principles. It was only consciously formulated as a method in the second century AD by the Greek geographer Ptolemy, who employed a grid of geometrical lines of latitude and longitude (called a graticule) to project the earth onto a flat surface. Prior to this, maps like the Babylonian example provided no apparent projection (or scale) to structure their representation of the world (though of course they still projected a geometrical image of the world based on their cultural assumptions about its shape and size). Over the centuries, circles, squares, rectangles, ovals, hearts, even trapezoids and a variety of other shapes have been used to project the globe onto a plane, each one based on a particular set of cultural beliefs. Some of these assumed a spherical earth, some of them did not: on the Babylonian world map the world is represented as a flat disc, with its inhabited dimensions encircled by sea, beyond which are its literally shapeless edges. Early Chinese maps also appear to accept the belief in a flat earth, although as we shall see this is partly based on their own particular fascination with the square as a defining cosmological principle. By at least the fourth century BC the Greeks had shown that the earth was a sphere, and produced a series of circular maps projected onto a plane surface.

All these projections struggled with an enduring geographical and mathematical conundrum: how is the whole earth reduced to a single flat image? Once the earth's sphericity was scientifically proved, the problem was compounded: how was it possible to project the sphere accurately onto a plane surface?[23] The answer, as the German mathematician Carl Friedrich Gauss conclusively proved in his work on projections in the 1820s, was that it was not possible. Gauss showed that a curved sphere and a plane were not isometric: in other words, the terrestrial globe could never be mapped onto the plane surface of a map using a fixed scale without some form of distortion of shape or angularity; we shall see some of the many distortions which have been adopted in the course of this book.[24] Despite Gauss's insight, the search for 'better', more accurate, projections only intensified (even Gauss went on to offer his own method of projection). Even today, the problem remains hidden though in plain view, invariably acknowledged on world maps and atlases, but buried in the technical detail of their construction.

One of the many paradoxes of maps is that, although mapmakers have been creating them for thousands of years, our study and understanding of them is still in its relative infancy. It was only in the nineteenth century in Europe that the academic discipline of geography came into existence, coinciding with the professionalization of the mapmaker, who was redesignated with the more scientific title of 'cartographer'. As a result, geography has only recently begun a systematic attempt to understand the history of maps and their role in different societies. In 1935 Leo Bagrow (1881–1957), a Russian naval officer trained in archaeology, founded *Imago Mundi*, the first journal dedicated to the study of the history of cartography, followed in 1944 by the completion of his *Die Geschichte der Kartographie* (*History of Cartography*), the first comprehensive study of its subject.[25] Since then, only a handful of popular books on the subject have been published by experts in the field, and the multi-volume *History of Cartography* edited by Harley and Woodward (who have both tragically died since the project's inception) will not be brought up to the present for years to come. Cartography remains a subject in need of a discipline, its study generally undertaken by scholars trained (like myself) in a variety of other fields, its future even more uncertain than the maps it seeks to interpret.

This book tells a story which shows that, despite the strenuous efforts of generations of cartographers, the ultimate claims of scientific cartography have never been realized. The first great national survey of a nation based on Enlightenment principles of science, the *Carte de Cassini*, discussed in Chapter 9, was never really finished, and its global equivalent, the International Map of the World, conceived at the end of the nineteenth century, and whose story is told in the Conclusion, was abandoned towards the end of the twentieth. Geography's erratic development as an academic and professional discipline over the last two centuries has meant that it has been relatively slow to question its intellectual assumptions. In recent years, geographers have developed serious reservations about their involvement in the political partition of the earth. Belief in the objectivity of maps has found itself subject to profound revision, and it is now recognized that they are intimately connected to prevailing systems of power and authority. Their creation is not an objective science but a realist endeavour, and aspires to a particular way of depicting reality. Realism is a stylistic representation of the world, just like naturalism, classicism or romanticism, and it is no coincidence that the claims for cartography's objectivity reached their height at the same moment as the ascendancy of the realist novel in Europe in the nineteenth century. Instead of arguing that mapmaking follows an inexorable progress towards scientific accuracy and objectivity, this book will argue that it is a 'cartography without progress', which provides different cultures with particular visions of the world at specific points in time.[26]

The book takes twelve world maps from cultures and moments in world history, and examines the creative processes though which they tried to resolve the problems faced by their makers, from perception and abstraction to scale, perspective, orientation and projection. The problems are constant, but the responses are specific to the mapmaker's particular culture, and we discover that what drove them was as much personal, emotional, religious, political and financial as geographical, technical and mathematical. Each map either shaped people's attitudes to the worlds in which they lived, or crystallized a particular world view at specific moments in global history – often both. These twelve maps were created at particularly crucial moments, when their makers took bold decisions about how and what to represent. In the process they created new visions of the world that aimed not only to explain to their audiences that this was what the world looked like, but to convince them of why it existed, and to show them their own place within it. Each map also encapsulates a particular idea or issue that both motivated its creation and captured its contemporaries' understanding of the world, from science, politics, religion and empire to nationalism, trade and globalization. But maps are not always shaped exclusively by ideology, conscious or unconscious. Inchoate emotional forces have also played their part in making them. The examples here range from the pursuit of intellectual exchange in an Islamic map from the twelfth century, to global conceptions of toleration and equality in Arno Peters's controversial world map published in 1973.

Although this book makes no claim to provide anything approaching a comprehensive story of the history of cartography, it does offer several challenges to prevailing assumptions about the subject. The first is that, however we interpret the history of maps, it is not an exclusively Western activity. Current research is revealing just how far pre-modern, non-Western cultures are part of the story, from the Babylonian world map to Indian, Chinese and Muslim contributions. Secondly, there is also no hidden agenda of evolution or progress in the historical mapping of the world. The maps examined are the creation of cultures which perceive physical, terrestrial space in different ways, and these perceptions inform the maps they make. This leads to the third argument, that each map is as comprehensible and as logical to their users as the other, be it the medieval Hereford *mappamundi* or Google's geospatial applications. The story told here is therefore a discontinuous one, marked by breaks and sudden shifts, rather than the relentless accumulation of increasingly accurate geographical data.

The map, whatever its medium or its message, is always a creative interpretation of the space it claims to represent. The critical 'deconstruction' of maps as objective representations of reality by writers like Korzybski, Bateson and others has left them looking like malevolent tools of ideology, weaving a conspiratorial web of deceit and dissimulation wherever they are to be found. Instead, the maps in this book are interpreted more as a series of ingenious arguments, creative propositions, highly selective guides to the worlds they have created. Maps allow us to dream and fantasize about places we shall never see, either in this world or in other, as yet unknown worlds. Perhaps the best metaphorical description of maps was graffitied in 45-centimetre letters on a wall next to the railway line approaching Paddington Station in London: 'Far away is close at hand in images of elsewhere.' A metaphor, like a map, involves carrying something across from one place to another.

Maps are always images of elsewhere, imaginatively transporting their viewers to faraway, unknown places, recreating distance in the palm of your hand. Consulting a world map ensures that faraway is always close at hand.

'How valuable a good map is,' wrote the seventeenth-century painter Samuel van Hoogstraten in a similar vein, 'wherein one views the world as from another world.'[27] Oscar Wilde developed Hoogstraten's transcendental sentiment when he famously remarked that a 'map of the world that does not include Utopia is not even worth glancing at, for it leaves out the one country at which Humanity is always landing. And when Humanity lands there, it looks out, and seeing a better country, sets sail.'[28] Maps always make choices about what they include and what they omit, but it is at the moment such decisions are made that Wilde dreams of the possibility of creating a different world – or even new worlds beyond our knowledge (which is one of the reasons that science fiction writers have been drawn irresistibly to maps). As Ortelius admitted, every map shows one thing, but therefore not another, and represents the world in one way, and as a consequence not in another.[29] Such decisions might often be political, but they are always creative. The ability expressed by all the mapmakers in this book to rise above the earth and look down on it from a divine perspective represents an idealistic leap of imaginative faith in humanity, but so powerful is this vision that various political ideologies have sought to appropriate it for their own ends.

This legacy brings the discussion right up to the present day, and the ongoing controversy surrounding the increasing domination of digital online mapping applications, exemplified by the subject of my final chapter, Google Earth. After nearly two millennia of being made on stone, animal skins and paper, maps are now changing in ways unknown since the invention of print in the fifteenth century and are facing imminent obsolescence as the world and its maps become digitized and virtual. Perhaps these new applications will create an unprecedented democratization of maps, allowing greatly increased public access, even giving people the ability to build their own maps. But it seems more likely that the corporate interests of multinational companies will bring a new world of online maps in which access is prescribed by financial imperatives, subject to political censorship and indifferent to personal privacy. One of the arguments of this book is that anyone who wants to understand the consequences of online mapping and why the virtual, online map of the world looks like it does today needs a longer perspective, one that reaches back as far as the first Greek attempts to map the known world and beyond.

The world is always changing, and so are maps. But this book is not about maps that have changed the world. From the Greeks to Google Earth, it is not in the nature of maps meaningfully to change anything. Instead, maps offer arguments and propositions; they define, recreate, shape and mediate. Invariably, they also fail to reach their objectives. Many of the maps chosen were heavily criticized at the moment of their completion, or were quickly superseded. Others were neglected at the time, or subsequently dismissed as outdated or 'inaccurate', falling into obscurity. But they all bear witness that one way of trying to understand the histories of our world is by exploring how the spaces within it are mapped. Space has a history, and I hope this book goes a little way towards telling that history through maps.

# I

# Science

*Ptolemy's* Geography, c. *AD 150*

*Alexandria, Egypt,* C. *AD 150*

Sailing to Alexandria by sea from the east, the first thing a classical travel-ler saw on the horizon was the colossal stone tower of the Pharos, on a small island at the entrance to the city's port. At more than 100 metres high, the tower acted as a landmark for sailors along the largely feature-less Egyptian coastline. During the day a mirror, positioned at its apex, beckoned sailors, and at night fires were lit to guide pilots into shore. But the tower was more than just a navigational landmark. It announced to travellers that they were arriving in one of the great cities of the ancient world. Alexandria was founded by Alexander the Great in 334 BC, who named the city after himself. Following his death it became the capital of the Ptolemaic dynasty (named after one of Alexander's gen-erals) that would rule Egypt for more then 300 years, and spread Greek ideas and culture throughout the Mediterranean and the Middle East.[1] Gliding past the stone Pharos, a traveller entering the port in the third century BC was confronted by a city laid out in the shape of a chlamys, the rectangular woollen cloak worn by Alexander and his soldiers, an iconic image of Greek military might. Alexandria, like the rest of the civilized world at the time, was wrapped in the mantle of Greek influ-ence, the 'umbilicus' of the classical world. It was a living example of a Greek polis transplanted onto Egyptian soil.

The city's rise represented a decisive shift in the political geography of the classical world. Alexander's military conquests had transformed the Greek world from a group of small, insular Greek city states into a series of imperial dynasties spread across the Mediterranean and Asia. This concentration of wealth and power within empires like the

Ptolemaic dynasty brought with it changes to warfare, technology, science, trade, art and culture. It led to new ways of people interacting, doing business, swapping ideas and learning from each other. At the centre of this evolving Hellenistic world, stretching from Athens to India between *c.* 330 BC and *c.* 30 BC, stood Alexandria. From the west it welcomed the merchants and traders from the great Mediterranean ports and cities as distant as Sicily and southern Italy, and grew rich from its trade with the rising power of Rome. From the north, it took its cultural influences from Athens and the Greek city-states. It acknowledged the influence of the great Persian kingdoms to the east, and from the south it absorbed the wealth of the fertile Nile Delta and the vast trading routes and ancient kingdoms of the sub-Saharan world.[2]

Like most great cities that stand at a crossroads of people, empires and trade, Alexandria also became a nucleus for learning and scholarship. Of all the great monuments that define Alexandria, none is more potent in the Western imagination than its ancient library. Founded by the Ptolemies *c.* 300 BC, the Alexandria library was one of the first public libraries, designed to hold a copy of every known manuscript written in Greek, as well as translations of books from other ancient languages, particularly Hebrew. The library held thousands of books, written on papyrus rolls, and all catalogued and available for consultation. At the heart of their network of royal palaces, the Ptolemies established a 'Mouseion', or museum, originally a shrine dedicated to the nine Muses (or goddesses), but which the Ptolemies redefined as a place for the worship of the muses of learning and scholarship. Here, scholars were invited to study, with promises of lodging, a pension and, best of all, access to the library. From across Greece some of the period's greatest minds were lured to work in the museum and its library. Euclid (*c.* 325–265 BC), the great mathematician, came from Athens; the poet Callimachus (*c.* 310–240 BC) and the astronomer Eratosthenes (*c.* 275–195 BC) both came from Libya; Archimedes (*c.* 287–212 BC), the mathematician, physicist and engineer, travelled from Syracuse.

The Alexandria library was one of the first systematic attempts to gather, classify and catalogue the knowledge of the ancient world. The Ptolemies decreed that any books entering the city were to be seized by the authorities and copied by the library's scribes (although their owners sometimes discovered that only a copy of their original book was returned). Estimates of the number of books held in the library have proved notoriously difficult to make due to wildly contradictory claims by classical sources, but even conservative assessments put the number at more than 100,000 texts. One classical commentator gave up trying to count. 'Concerning the number of books and the establishment of libraries,' he wrote, 'why need I even speak when they are all the memory of men?'[3] The library was indeed a vast repository for the collective memory of a classical world contained within the books it catalogued. It was, to borrow a phrase from the history of science, a 'centre of calculation', an institution with the resources to gather and process diverse information on a range of subjects, where 'charts, tables and trajectories are commonly at hand and combinable at will', and from which scholars could synthesize such information in the search for more general, universal truths.[4]

It was here, in one of the great centres of calculation and knowledge, that modern mapmaking was born. Around AD 150 the astronomer Claudius Ptolemaeus wrote a treatise entitled *Geōgraphikē hyphēgēsis*, or 'Guide to Geography', which would become known simply as the *Geography*. Sitting in the ruins of the once great library, Ptolemy compiled a text that claimed to describe the known world and which would come to define mapmaking for the next two millennia. Written in Greek on a papyrus roll over eight sections, or 'books', the *Geography* summarized a thousand years of Greek thinking on the size, shape and scope of the inhabited world. Ptolemy defined his task as a geographer as being to 'show the known world as a single and continuous entity, its nature and how it is situated, by taking account only of the things that are associated with it in its broader, general outlines', which he listed as 'gulfs, great cities, the more notable peoples and rivers, and the more noteworthy things of each kind'. His method was simple: 'The first thing one has to investigate is the earth's shape, size, and position with respect to its surroundings, so that it will be possible to speak of its known part, how large it is and what it is like,' and 'under which parallels of the celestial sphere each of the localities is known'.[5] The *Geography* that resulted was many things simultaneously: a topographical account of the latitude and longitude of more than 8,000 locations in Europe, Asia and Africa; an explanation of the role of astronomy in geography; a detailed mathematical guide for making maps of the earth and its regions; and the treatise that provided the Western geographical tradition with an enduring definition of geography – in short, a complete mapmaking kit as conceived by the ancient world.[6]

No text before or since Ptolemy's would provide such a comprehensive account of the earth and how to describe it. After its completion, Ptolemy's *Geography* disappeared for a thousand years. No original copies from Ptolemy's own time have survived, and it only reappeared in thirteenth-century Byzantium, with maps, drawn by Byzantine scribes, which were clearly based on Ptolemy's description of the earth and the position of its 8,000 locations, and which show the classical world as it appeared to him in second-century Alexandria. In ascending order, the Mediterranean, Europe, North Africa, the Middle East and parts of Asia look relatively familiar. The Americas and Australasia, southern Africa and the Far East, unknown to Ptolemy, are all missing, as is the Pacific and most of the Atlantic Ocean. The Indian Ocean is shown as an enormous lake, with southern Africa running right round the bottom half of the map to join an increasingly speculative Asia east of the Malaysian peninsula. Nevertheless, this is a map that we seem to understand: oriented with north at the top, it has place names marking key regions, and is constructed using a graticule. Like most of his Greek forebears as far back as Plato, Ptolemy understood that the earth was round, and used this grid to address the difficulty of projecting a spherical earth onto a plane, or flat surface. He acknowledged that drawing a rectangular map required a graticule 'to achieve a resemblance to a picture of a globe, so that on a flattened surface, too, the intervals established on it will be in as good proportion as possible to the true intervals'.[7]

All this makes it tempting to see Ptolemy's *Geography* as a remarkably early harbinger of modern mapmaking. Unfortunately, it is not that simple. Scholarly opinion remains divided as to whether or not Ptolemy ever himself drew maps to accompany the *Geography*: many historians argue that the thirteenth-century Byzantine copies contain the first maps to illustrate his text. Unlike disciplines such as medicine, there was no field or 'school' of Greek geography. There are virtually no recorded examples of the practical use of maps in classical Greece, and certainly no instances of Ptolemy's book being used in this way.

Turning to Ptolemy's biography to try to understand the significance of his book offers little help. Virtually nothing is known about his life. There is no autobiography, no statue, not even an account written by a contemporary. Many of his other scientific treatises remain lost. Even the *Geography* itself was scattered across the Christian and Muslim communities that emerged to fill the void left by the collapse of the Roman Empire. The early Byzantine manuscripts bear few clues as to how much the text had changed since Ptolemy first wrote it. The little we do know about Ptolemy is based on his surviving scientific works, and vague descriptions of him written by much later Byzantine sources. His taking the name 'Ptolemaeus' indicates that he was probably a native and inhabitant of Ptolemaic Egypt, which, during his lifetime, was already under the control of the Roman Empire. 'Ptolemaeus' also suggests, although does not prove, descent from Greek ancestors. 'Claudius' indicates that he possessed Roman citizenship, possibly granted to a forebear by the emperor Claudius. The astronomical observations recorded in his earliest scientific works suggest that he flourished during the reigns of the emperors Hadrian and Marcus Aurelius, giving approximate dates of a birth around AD 100, and death no later than AD 170.[8] This is all that we know about Ptolemy's life.

The creation of Ptolemy's *Geography* is in some respects a paradox. Though the book is arguably the most influential in the history of map-making, it is uncertain, as we have seen, if it even contained maps. Its author, a mathematician and astronomer, did not regard himself as a geographer, and his life is a virtual blank. He lived in one of the great cities of late-Hellenistic learning, but at a time when its power and influence were already past their apogee. Rome had overthrown the Ptolemies in 30 BC, and oversaw the gradual decline and dispersal of the once great library. But Ptolemy was fortunate. It was only as the great flowering of the Hellenistic world began its slow decline that conditions were conducive for the creation of the book that would define both geography and mapmaking; the world had to reach its nadir before it was possible to describe its geography. If the Alexandria library assembled and then lost the 'memory of men', Ptolemy's *Geography* represented the memory of a significant part of man's world. But such a text still required its author's immersion in nearly a millennium of Greek literary, philosophical and scientific speculation on the heavens and the earth before it could be written.

Although archaic Greece had no word for 'geography', from at least the third century BC the early Greeks referred to what we would call a 'map' as *pinax*. The other term often used was *periodos gēs*, literally 'circuit of the earth' (a phrase that would form the basis of many subsequent treatises on geography). Although both these terms for maps would eventually be superseded by the Latin term *mappa*, the later classical Greek formulation of geography has endured, formed by the compound of the noun *gē*, or earth, with the verb *graphein*, to draw or write.[9] These terms offer some insight into the ways in which the Greeks approached maps and geography. A *pinax* is a physical medium on which images or words are inscribed, and *periodos gēs* implies a physical activity, specifically 'going round' the earth in a circular fashion. The etymology of *geo-graphy* also suggests that it was both a visual (drawn) activity and a linguistic (written) statement. Although all these terms were increasingly used from the third century BC, they were subsumed within the more recognizable branches of Greek learning, namely *mythos* (myth), *historia* (history), or *physiologia* (natural science).

From its earliest beginnings, Greek geography emerged from philosophical and scientific speculations on the origins and creation of the universe, rather than any specifically practical need. Looking back on its origins while writing his own seventeen-book *Geography* at around the time of the birth of Christ, the Greek historian and self-styled geographer Strabo (*c.* 64 BC–AD 21) claimed that 'the science of Geography' was 'a concern of the philosopher'. The knowledge needed to practise geography was, for Strabo, 'possessed solely by the man who has investigated things both human and divine'.[10] For the Greeks, maps and geography were part of a wider speculative enquiry into the order of things: explanations, both written and visual, of the origins of the cosmos and mankind's place within it.

The earliest account of what we would call Greek geography appears in the work of the poet Strabo calls 'the first geographer': Homer, whose epic poem the *Iliad* is usually dated to the eighth century BC. At the end of book 18, as the war between the Greeks and Trojans reaches its climax, Thetis, mother of the Greek warrior Achilles, asks Hephaestus, the god of fire, to provide her son with armour in which to fight his Trojan adversary, Hector. Homer's description of the 'huge and mighty shield' that Hephaestus fashions for Achilles is one of the earliest literary examples of *ekphrasis*, a vivid description of a work of art. But it can also be seen as a cosmological 'map', or what one Greek geographer would call a '*kosmou mimēma*', or 'image of the world',[11] a moral and symbolic

Fig. 1 The shield of Achilles, bronze cast designed by John Flaxman, 1824.

depiction of the Greek universe, in this case composed of five layers or concentric circles. At its centre were 'the earth, and sky, and sea, the weariless sun and the moon waxing full, and all the constellations that crown the heavens'. Moving outwards, the shield portrayed 'two fine cities of mortal men', one at peace, one at war; agricultural life showing the practice of ploughing, reaping and vintage; the pastoral world of 'straight-horned cattle', 'white-woolled sheep'; and finally 'the mighty river of Ocean, running on the rim round the edge of the strong-built shield'.[12]

Although Homer's description of the shield of Achilles may not immediately strike the modern reader as either a map or an example of geography, the Greek definitions of both terms suggest otherwise.

Strictly speaking, Homer provides a *geo-graphy* – a graphic account of the earth – which gives a representation, symbolic in this instance, of the origins of the universe and mankind's place within it. It also adheres to Greek definitions of a map as *pinax* or *periodos gēs*: the shield is both a physical object on which words are inscribed, as well as a circuit of the earth, circumscribed within the limits of 'the mighty river of Ocean', which defines the boundary (*peirata*) of a potentially boundless (*apei-ron*) world. Later Greek commentators saw Homer's description as providing not just a geography, but also a story of creation itself: a cosmogony. Hephaestus, god of fire, represents the basic element of creation, and the construction of the circular shield is an allegory of the formation of a spherical universe. The shield's four metals (gold, silver, bronze and tin) represent the four elements, while its five layers correspond to the earth's five zones.[13]

As well as a cosmogony, the shield of Achilles is also a description of the known world as it appears to anyone who looks up from the horizon and gazes at the sky. The earth is a flat disc, encircled all around by sea, with the sky and stars above, and the sun rising in the east and setting in the west. This was the shape and scope of the *oikoumenē*, the Greek term for the inhabited world. Its root lay in the Greek *oikos*, 'house' or 'dwelling space'. As the word tells us, the early Greek perception of the known world, like that of most archaic communities, was primarily ego-centric, emanating outwards from the body and its sustaining domestic space. The world began with the body, was defined by the hearth, and ended at the horizon. Anything beyond this was boundless chaos.

For the Greeks, geography was intimately connected to an understanding of cosmogony, because to understand the origins of the earth (*Gē*) was to understand creation. For poets like Homer and more explicitly Hesiod in his *Theogony* (*c.* 700 BC), creation begins with Chaos, the formless mass that precedes the three other entities, Tartaros (the primordial god of the gloomy pit beneath the earth), Eros (the god of love and procreation) and, most importantly, Gaia (the female personification of Earth). Both Chaos and Gaia produce children, Nyx (Night) and Uranus (Sky). From her subsequent union with Uranus, Gaia produces the twelve Titan deities: six sons – Oceanus, Hyperion, Coeus, Cronus, Iapetus and Crius – six daughters – Mnemosyne, Phoebe, Rhea, Tethys, Theia and Themis – who in turn are defeated by the Olympian gods led by Zeus. Unlike the Christian tradition, human creation in the earliest Greek accounts is contradictory and often secondary to the struggles of the deities. Homer never provides an explicit account of mortal creation, in contrast with Hesiod, who claims that mankind is created by the Titan Cronus, but with little explanation as to why. In other versions of the myth, mortals are created by the Titan Prometheus, who incurs the wrath of Zeus by providing humans with the gift of 'fire', or spirit of self-conscious knowledge. In other versions of the creation myth, in Hesiod and others, mankind is denied any explicitly divine identity, and is born from the soil or earth.[14]

These ambiguous explanations of the birth of humanity in early Greek mythical accounts of creation contrast with emerging scientific and naturalist accounts of 'the order of things' which began to appear in the sixth century BC in the Ionian city of Miletus (in modern-day Turkey), among a group of thinkers who offered a recognizably scientific argument to explain creation. Miletus was well positioned to absorb the influence of Babylonian theories of creation and astronomical observations on the movement of the stars that stretched back as far as 1800 BC, represented, as we saw at the beginning of this book, on clay tablets which showed the earth encircled by water and with Babylon near its centre. The Milesian philosopher Anaximander (c. 610–546 BC) was, according to Diogenes Laertius, a third-century AD biographer, 'the first to draw the outline of the sea and the land', and who 'published the first geographical map [geographikon pinaka]'.[15]

Like most of the Greek writers who discussed geography before Ptolemy, very little of Anaximander's writings or maps survives; when trying to piece together a coherent story of Greek geography we have to rely on their memorial reconstruction and reportage by later Greek writers, the so-called doxographers. These include figures like Plutarch, Hippolytus and Diogenes Laertius, who all recount the lives and key doctrines of earlier writers. It is often difficult to assess the significance of much later writers dealing with geography, including Strabo and his Geography, which is disproportionately influential simply because it has survived. Nevertheless, virtually all Greek writers point to Anaximander as the first thinker to provide a compelling account of what he himself is believed to have called 'the order of things'. Anaximander offered a variation on Hesiod's originating Chaos by proposing that in the beginning was eternal boundlessness, or *apeiron*. The boundlessness somehow secreted a 'seed' which then produced flame, 'which grew around the air about the earth like bark around a tree'.[16] As the earth began to form, the enveloping 'flame' broke away to create 'rings' of planets, stars, the moon and the sun (in ascending order). These rings encircled the earth, but were only visible because of 'vents' through which the heavenly bodies can be seen from earth as circular objects. Anaximander argued that human life came from primeval moisture (in some versions, mankind is born of thorny bark, in others it evolves from fish). As a naturalistic explanation of the creation of the universe and humanity, this was a significant development on earlier accounts based on gods and myth, but it is Anaximander's explanation of the earth's place in this cosmogony which is particularly original. The doxographers tell us that Anaximander argued 'the earth is aloft, not dominated by anything; it remains in place because of the similar distance from all points [of the celestial circumference]', and that its shape 'is cylindrical, with a depth one third of its width'.[17] From this cosmogony came a new cosmology – the study of the physical universe. Abandoning Babylonian and earlier Greek beliefs that the earth floated on water or air, Anaximander introduced a purely geometrical and mathematical cosmology, in which the earth sits at the centre of a symmetrical cosmos in perfect equilibrium. It is the earliest known scientifically argued concept of a geocentric universe.

Anaximander's rational claims for the physical origins of creation defined all subsequent Greek metaphysical speculation. His impact on Greek geography was also profound. Although no description of his world map remains, doxography provides some idea of how it might have looked. Imagine the earth as a circular drum, around which the heavenly rings circle: on one side of the drum lies an uninhabited world and on its other side the *oikoumenē*, encircled by the ocean. At its centre stood either Anaximander's homeland of Miletus, or the stone *omphalos*, the 'navel' of the world, recently established at Apollo's temple in Delphi, and from where most subsequent Greek maps would take their bearings. Written descriptions probably supplemented Anaximander's description: the mythical travels of the Argonauts and Odysseus; *periploi*, or nautical descriptions of seaborne travel across the Mediterranean; and accounts of the early colonization of regions in the Black Sea, Italy and the eastern Mediterranean.[18] The resulting map probably contained a rudimentary outline of Europe, Asia and Libya (or Africa) as vast islands, separated by the Mediterranean, Black Sea and the Nile.

Subsequent writers on geography would refine and develop Anaximander's map, but few could match his compelling cosmology. The Milesian statesman and historian Hecataeus (*fl.* 500 BC) wrote the first explicitly geographical treatise with the title *Periodos gēs*, or 'Circuit of the Earth', complete with a world map. The map is lost and only fragments of the *Periodos* remain, but they provide some indication of how far Hecataeus built on Anaximander's earlier geography. Hecataeus' *Periodos* describes Europe, Asia and Libya, beginning at the westernmost point of the known world, the Columns of Hercules (or Straits of Gibraltar), and moving eastwards round the Mediterranean, through the Black Sea, Scythia, Persia, India and Sudan, and ending on the Atlantic coast of Morocco. As well as writing about physical geography, Hecataeus was involved in the Ionian Revolt (*c.* 500–493 BC), in which several Ionian cities unsuccessfully rebelled against their Persian rulers.

Hecataeus' map remained tied to a perception of the world shaped as either a disc (as in Homer) or a cylinder (as in Anaximander). Such mythical and mathematical assumptions came under sustained attack from the first and arguably greatest of all Greek historians, Herodotus of Halicarnassus (*c.* 484–425 BC). In book 4 of his vast *History*, Herodotus breaks off from discussing the might of Persia and the northernmost limits of the known world in Scythia, to chastise geographers like Hecataeus. 'I cannot help but laugh', he writes, 'at the absurdity of all the map-makers – there are plenty of them – who show Ocean running like a river round a perfectly circular earth, with Asia and Europe of the same size.'[19] As a traveller and a historian, Herodotus had little time for the neat geographical symmetry of either Homer's myth or Anaximander's science. Although he reiterated Hecataeus' tripartite division of the world between Europe, Asia and Libya (Africa), Herodotus also carefully listed the people, empires and territories known to his contemporaries, before concluding that 'I cannot but be surprised at the method of mapping Libya, Asia, and Europe. The three continents do, in fact, differ very greatly in size. Europe is as long as the other two put together, and for breadth is not, in my opinion, to be compared with them.'[20] He

dismissed the assumption that the inhabited world was completely surrounded by water, and questioned why it was that 'three distinct
women's names should have been given to what is really a single
land-mass' – Europe (a Lebanese princess abducted by Zeus), Asia (the
wife of Prometheus – although in other traditions the son of the Thra-
cian king Cotys) and Libya (the daughter of Epaphus, son of Jupiter).[21]
Herodotus had little interest in the geometry or nomenclature of the
flat, disc-shaped world maps he describes (none of which survive). As
far as he was concerned, such abstract idealizations should be replaced
by the verifiable reality of empirical travel and personal encounters.

Herodotus implicitly raised questions about mapmaking that would
define it – and at times divide it – for centuries. Are claims to objectivity
of science, and in particular geometry, sufficient to make accurate maps
of the world? Or should mapmaking rely more on the noisy, often contradictory and unreliable reports of travellers to develop a more
comprehensive picture of the known world? One consequence of such
distinctions was to ask if mapping was a science or an art: was it primarily spatial or temporal, a visual or a written act? Although Greek
mapmaking remained based on mathematical and astronomical calculations, Herodotus raised the issue of how it gathered, assessed and
incorporated the raw data gathered by travellers in the creation of a
more comprehensive map of the world.

Herodotus' concerns found little immediate resonance among his
contemporaries, who continued to pursue mathematical and philosophical questions relating to the nature of the earth. Anaximander's belief
in a geometrically symmetrical universe was developed by Pythagoras
(fl. 530 BC) and his disciples, as well as Parmenides (fl. 480 BC), who is
credited with taking the logical step of suggesting that if the universe
was spherical, then so was the earth. But the first recorded statement on
the earth's sphericity comes towards the end of Phaedo (c. 380 BC),
Plato's celebrated dialogue on the last days of Socrates. The dialogue is
best known for its philosophical explanation of the Platonic ideas of the
immortality of the soul and the theory of ideal forms, but towards its
end Socrates offers an image of what he calls 'the wondrous regions in
the earth', as seen by the virtuous soul after death. 'I have been convinced', says Socrates, 'that if it is round and in the centre of the heaven,
it needs neither air nor any other such force to prevent its falling, but
the uniformity of the heaven in every direction with itself is enough to

support it, together with the equilibrium of the earth itself.'[22] What follows is a uniquely Platonic vision of the earth. Socrates explains that humanity inhabits only a fraction of its surface, dwelling in a series of hollows, 'varying in their shapes and sizes, into which water and mist and air have flowed together; and the earth itself is set in the heaven, a pure thing in pure surroundings, in which the stars are situated'. Socrates explains that 'this earth of ours' is a poor, 'corrupted' copy of 'the true earth', an ideal world which is only visible to the immortal soul.[23] Finally, in a remarkable description of global transcendence, he anticipates his own death, as he describes rising up and looking down on the spherical world:

> First of all the true earth, if one views it from above, is said to look like those twelve-piece leather balls, variegated, a patchwork of colours, of which our colours here are, as it were, samples that painters use. There the whole earth is of such colours, indeed of colours far brighter still and purer than these: one portion is purple, marvellous for its beauty, another is golden, and all that is white is whiter than chalk or snow; and the earth is composed of the other colours likewise, indeed of colours more numerous and beautiful than any we have seen.[24]

This unprecedented spectre of a spherical, glittering, ideal world, viewed by the immortal soul in a moment of spiritual transcendence, would be adopted in a range of subsequent global geographical imaginings, particularly within the Christian tradition of salvation and spiritual ascendancy. It would also define Plato's belief in the world's creation by a divine demiurge, or 'craftsman', put forward in *Timaeus*. This vision of the earth is central to Plato's argument regarding the theory of forms and the soul's immortality. Only the immortal soul can apprehend the ideal form of the world; but the mortal human intellect and imagination, in the shape of painters, mapmakers or mathematicians, are able to represent its divine, celestial order, albeit through poor reproductions. Even the mathematicians could only offer pale approximations of the ideal earth: Plato's allusion to the twelve-piece leather ball is a reference to Pythagoras' theory of the dodecahedron, the solid closest to the sphere. Plato's vision – more than two millennia before the dream of rising up above the earth and seeing it in its full glory became a reality in the age of extraterrestrial space travel – would prove a compelling, if elusive ideal for generations of geographers.

Having defined the earth within the wider context of creation, late classical Greek thinkers began to speculate on the relationship between the celestial and the terrestrial spheres, and how the former could help to measure the shape and extent of the latter. One of Plato's pupils, the mathematician and astronomer Eudoxus of Cnidus (*c.* 408–355 BC), produced a model of concentric celestial spheres rotating around an axis through the centre of the earth. Eudoxus made the intellectual leap of stepping outside the limits of the terrestrial world to imagine the universe (and the earth at its centre) beyond space and time by drawing a celestial globe seen from the 'outside' looking inwards, viewing the stars and the earth from a god-like perspective. This allowed him to plot the movements of the heavens on a terrestrial globe, showing how the main celestial circles (created by imaging the extension of the earth's axis out into space, around which the stars appear to circle), including the equator and the tropics, criss-crossed the earth's surface.

Eudoxus' geocentric universe was a major development in celestial mapmaking. It allowed him to develop a personified version of the zodiac (*zodiakos kuklos*, or 'circle of animals') which would shape all subsequent celestial mapmaking and astrology, and which still influences the language of modern-day geography, including the tropics of Cancer and Capricorn. As well as his astronomical calculations, Eudoxus wrote a lost text, the *Circuit of the Earth*, which is said to have made one of the first estimates of the earth's circumference, 400,000 stades (the notoriously difficult Greek method of measurement, defined as the distance covered by a plough in a single draft, and estimated anywhere between 148 and 185 metres).[25] By joining empirical observation of the heavens and the earth to the philosophical speculations of Anaximander and Plato, Eudoxus' calculations influenced the works of the most important of all ancient philosophers and his perceptions of the known world: Aristotle (384–322 BC).

Several of Aristotle's works contain detailed accounts of the shape and size of the earth, including his cosmographical treatise *On the Heavens*, and *Meteorology* (which strictly translated means 'the study of things aloft'), both written approximately 350 BC. In *On the Heavens*, Aristotle provided what we would regard as proper evidence that the earth is spherical. Drawing on Anaximander's cosmogony, he believed that the earth's 'mass will be everywhere equidistant from its centre', in other words, spherical. 'The evidence of the senses', Aristotle went on, 'further corroborates this.' 'How else', he asks, 'would eclipses of the moon show [curved] segments as we see them?' And why else does 'quite a small change of position to south or north cause a manifest alteration of the horizon', unless the earth is round?[26]

The *Meteorologica* took these arguments even further. Aristotle defined his subject as 'everything which happens naturally', and 'which takes place in the region which borders most nearly on the movements of the stars', and was 'closest to the earth.[27] Although the book now reads like an esoteric account of comets, shooting stars, earthquakes, thunder and lightning, it was part of an attempt by Aristotle to give shape and meaning to a geocentric universe. In the second book of the *Meteorologica*, Aristotle describes the inhabited world. 'For there are two habitable sectors of the earth's surface', 'one, in which we live, towards the upper pole, the other towards the other, that is the south pole ... these sectors are drum-shaped'. He concluded that 'present maps of the world', showing the *oikoumenē* as a circular, flat disc were 'absurd' for philosophical and empirical reasons:

> For theoretical calculation shows that it is limited in breadth and could, as far as climate is concerned, extend round the earth in a continuous belt: for it is not difference of longitude but latitude that brings great variations of temperature ... And the facts known to us from journeys by sea and land also confirm the conclusion that its length is much greater than its breadth. For if one reckons up these voyages and journeys, so far as they are capable of yielding any accurate information, the distance from the Pillars of Hercules to India exceeds that from Aethiopia to Lake Maeotis [Sea of Azov, adjoining the Black Sea] and the farthest parts of Scythia by a ratio greater than that of 5 to 3. Yet we know the whole breadth of the habitable world up to the uninhabitable regions which bound it, where habitation ceases on the one side because of the cold, on the other because of the heat; while beyond India and the Pillars of Heracles it is the ocean which severs the habitable land and prevents it forming a continuous belt around the globe.[28]

Aristotle's globe was divided into five climatic zones, or *klimata* (meaning 'slope' or 'incline'): two polar zones, two temperate, inhabitable zones either side of the equator, and a central zone, running around the equator, uninhabitable due to its fierce heat. It drew on the idea of *klimata* proposed by Parmenides, and made the first move towards establishing an ethnography of climate.[29] According to Aristotle, the 'climate', nor 'incline' of the sun's rays lessened the further north one travelled away from the equator. So, neither the unbearable heat of the equator, nor the freezing, 'frigid' northern polar zones could possibly sustain human life, which was only possible in the northern and southern 'temperate' zones. Aristotle's belief in the importance of experience and what he thought of as empirical facts in defining the width and breadth of the known world would have pleased Herodotus, but it also greatly expanded the extent of the known world in the light of the military conquests of Aristotle's most famous pupil, Alexander the Great, from the Balkans to India in 335–323 BC. Together with Ptolemy's later treatise, Aristotle's description of the earth would come to dominate geography for more than a thousand years.

Aristotle's *Meteorologica* represents the culmination of classical Greek theoretical speculation on the known world. His belief in trusting the senses and the importance of practical observation was a departure from the cosmologies of Anaximander and Plato, but Greek geography before him was not exclusively theoretical. There are scattered references (many retrospective) to the practical usage of maps as far back as the Ionian Revolt against the Persians. Herodotus tells the story of how Aristagoras of Miletus sought military aid against the Persians from Cleomenes, king of Sparta, and that he 'brought to the interview a map of the world engraved on bronze, showing all the seas and rivers', and 'the relative positions of the various nations'. The map's detailed geography of Lydia, Phrygia, Cappadocia, Cyprus, Armenia and 'the whole of Asia' seems to draw on far more than Anaximander's contemporary map, and included the Babylonian 'royal roads', the cleared tracks radiating outwards from Babylon, designed around 1900 BC to carry war chariots, and which also enabled trade and communication.[30] Aristagoras fails to enlist Cleomenes' military support when he admits the map reveals the prohibitive distance the Spartan army would have to travel from the sea: the story is thus one of the earliest examples of the political and military use of maps.

In a more light-hearted vein, Aristophanes' fifth-century comedy *The Clouds* depicts an Athenian citizen called Strepsiades quizzing a student and his academic paraphernalia. The student tells him, 'over there we have a map of the entire world. You see there? That's Athens.' Strepsiades' comical response is disbelieving: 'Don't be ridiculous,' he responds.

'Why, I can't see even a single lawcourt.' When the student points out the location of the enemy state of Sparta, Strepsiades tells him, 'That's much too close! You'd be well advised to move it further away.' All these examples imply that as early as the fifth century BC Greek world maps were physical, public objects, used in the arts of warfare and persuasion. They were extremely detailed, inscribed on brass, stone, wood, or even on the ground, and showed a certain level of geographical literacy. But they were also the preserve of the elite: Aristophanes satirizes common ignorance of the representational sophistication of maps, but his jokes only work on the assumption that the audience knows that the map is only a representation of territory, and that it was not possible to simply move countries across it if they seemed uncomfortably close.

This was the state of Greek geography in the fourth century BC. The military conquests of Alexander the Great propelled mapmaking into a more descriptive direction, based on direct experience and written records of faraway lands, that would ultimately culminate in the creation of Ptolemy's *Geography*. Alexander's conquests were not just significant for the ways in which they expanded Greek knowledge of the known world. Having learnt of the importance of empirical observation from his tutor Aristotle, Alexander appointed a team of scholars to gather data on the flora, fauna, culture, history and geography of the places they visited, and to provide written reports on the army's daily progress. Uniting the theoretical knowledge of Aristotle and his predecessors with the direct observation and discoveries of Alexander's campaigns would change how maps were made in the Hellenistic period that followed Alexander's death.

Where classical Greek mapping focused on cosmogony and geometry, Hellenistic mapmaking incorporated such approaches in what to us looks like a more scientific approach to mapping the earth. Alexander's contemporary, Pytheas of Massalia (Marseilles), explored the western and northern coastlines of Europe, travelling along the Iberian, French, English and possibly even the Baltic coastlines. His voyages established Thule (variously identified as Iceland, the Orkneys or even Greenland) as the northernmost limit of the inhabited world, and also correctly established the exact position of the celestial pole (the point at which the extension of the earth's axis intersects the celestial sphere). But perhaps most importantly for geography, he firmly established the

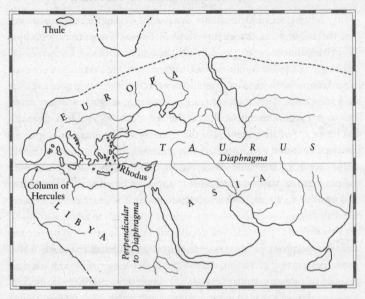

Fig. 2 Reconstruction of the world map of Dicaearchus, 3rd century BC.

connection between a location's latitude to the length of its longest day, and went on to project parallels of latitude running right round the globe.[31] At about the same time, Aristotle's pupil Dicaearchus of Messina (*fl. c.* 326-296 BC) developed a more sophisticated model of the size of the inhabited world, along with some of the earliest known calculations of latitude and longitude. In his lost *Circuit of the Earth*, Dicaearchus refined Aristotle by arguing that the ratio of the known world's length in relation to breadth was three to two, and made rudimentary latitudinal calculations by drawing a map with a parallel running from west to east through Gibraltar, Sicily, Rhodes and India, at approximately 36° N. Perpendicular to this parallel was a meridian running north to south through Rhodes.

Gradually, the inhabited world started to look like an incomplete rectangle, rather than a perfect circle. The Babylonian and early Greek philosophical and geometrical perceptions of the known world had supposed an ideal, abstract sphere, a finite space with a fixed circular boundary (the ocean), with a circumference defined by its centre, a location (Babylon, or Delphi) that defined their own cultures as shaping the world. The ideal early symmetry gives way to an irregular oblong inscribed within a rectangle. Gone is the exact centre of a circle based on geometry and faith, and instead calculations are made from a place like Rhodes simply because it stands at a point where the rudimentary lines of latitude and longitude bisect each other. Implicit in this shift is a changing mentality about the role of mapping. The titles of treatises describing the inhabited earth begin to change: works with titles such as *On the Ocean* and *On Harbours* supersede the more traditional *Circuit of the Earth*. Incremental geographical information slowly alters and expands the rectangular dimensions of the inhabited world, which are no longer perfectly delimited by the geometry of the circle. Conflating geometry with astronomical and terrestrial observation allowed Hellenistic thinkers to embark on a collective enterprise of adding new information on the calculation of latitude, the estimated length of the known world, or the location of a particular city or region. With this cooperative spirit came new ways of seeing maps as repositories of knowledge, encyclopedic compilations of information, or what one classical historian has called 'a great inventory of everything'.[32] A geographical treatise could encompass ideas of creation, astronomy, ethnography, history, botany or just about any other subject related to the natural world. 'The map', as Christian Jacob has argued, 'becomes a device for archiving knowledge about the inhabited world.'[33]

Whenever a culture begins to gather and archive its knowledge, it requires a physical location to safely accommodate such knowledge in whatever material form it comes in. For the Hellenistic world this was the Alexandria library, and it is no coincidence that one of its earliest librarians was the figure who, before Ptolemy, summarized Greek geography. Eratosthenes (*c.* 275–194 BC), a Greek born in Libya, studied in Athens before accepting an invitation from King Ptolemy III to work in Alexandria as tutor to the king's son and head of the royal library. During this time Eratosthenes wrote two particularly influential books (both lost): the *Measurement of the Earth*, and the *Geographica* – the first book to use the term geography as we understand it today, and the first text to plot a geographical projection across a map of the inhabited world.[34]

Eratosthenes' great achievement was to invent a method for calculating the circumference of the earth that united astronomical observation with practical knowledge. Using a gnomon, a part of a sundial that casts a shadow, Eratosthenes made a series of observations in Syene, modern-day Aswan, which he estimated as just over 5,000 stades south of Alexandria. He noticed that at midday on the summer solstice the sun's rays cast no shadow, and were therefore directly overhead. Taking the same calculation in Alexandria, Eratosthenes measured the angle cast by the gnomon at exactly the same time as one-fiftieth of a circle. Assuming that Alexandria and Syene lay on the same meridian, he calculated that the 5,000 stades between the two places represented one-fiftieth of the earth's circumference. Multiplying the two figures gave Eratosthenes a total figure for the circumference of the earth, which he estimated at 252,000 stades. Although the exact size of his *stadion* is unknown, Eratosthenes' final measurement probably corresponded to somewhere between 39,000 and 46,000 kilometres (most scholars believe it to be nearer the latter figure).[35] Considering that the actual circumference of the earth measured at the equator is 40,075 kilometres, Eratosthenes' calculation was extraordinarily accurate.

Although Eratosthenes' calculations were based on some erroneous assumptions – for instance, Alexandria and Syene were not on exactly the same meridian – they allowed him to calculate the circumference of any parallel circle around the earth, and to provide estimates of the length and breadth of the *oikoumenē*. Strabo tells us that, in his *Geographica*, Eratosthenes directly addressed the question of how to draw a map of the earth. Like the city from which he drew his knowledge of the world, Eratosthenes envisaged the world as shaped like a Greek chlamys, a rectangle with tapering ends. Drawing on Dicaearchus, he projected a parallel running from east to west from Gibraltar, through Sicily and Rhodes as far as India and the Taurus Mountains (which he placed too far east). Perpendicular to this parallel was a meridian running from Thule in the north to Meroë (Ethiopia) in the south, bisecting the parallel at Rhodes. Refining Dicaearchus' estimates, Eratosthenes calculated that from east to west the *oikoumenē* was 78,000 stades wide, and 38,000 stades from north to south. The width of the known world was, in other words, twice the length of its breadth. This led to some mistaken but tantalizing beliefs. If Eratosthenes' calculations were correct, the *oikoumenē* would have extended much too far eastwards, from the west coast of Iberia as far as modern-day Korea, at over 138° of longitude, rather than India, the limit of the Hellenistic world. In a remarkable moment of global imagining, Strabo quotes Eratosthenes maintaining that the earth 'forms a complete circle, itself meeting itself; so that, if the immensity of the Atlantic Sea did not prevent, we could sail from Iberia to India along one and the same parallel'.[36] Although such a claim was based on mistaken assumptions about the size of the earth and its eastward extent, such claims would have a significant impact on Renaissance explorers, including Columbus and Magellan.

Having made a calculation of the size of the earth and a rudimentary grid of parallels and meridians, Eratosthenes' final significant geographical innovation was to divide his *oikoumenē* into geometrical figures which he called *sphragides*, a term derived from the administrative term for a 'seal' or 'signet', designating a plot of land.[37] Eratosthenes attempted to match the size and shape of different regions to irregular quadrilateral shapes, drawing India as a rhomboid, and eastern Persia as a parallelogram. Although this method sounds like a retrograde step, it remained in line with the prevailing Greek tradition of projecting philosophy, astronomy and geometry onto the physical world. It also showed the unmistakable influence of Eratosthenes' predecessor as head of the Alexandria library, the Greek mathematician Euclid (*fl.* 300 BC).

In the thirteen books of his great mathematical treatise, the *Elements*, Euclid established the a priori principles, or 'elements', of geometry and mathematics. Explaining the basic rules of the theory of numbers and geometry, Euclid enabled thinkers like Eratosthenes to understand how anything (and everything) worked, based on the irreducible mathematical truths and reality of the universe. Beginning with the definitions of a point ('which has no part'), line ('breadthless length') and surface ('which has length and breadth only'), Euclid proceeded to the principles of plane and solid geometry. This posited a series of truths that still inform most secondary school geometry, such as that the sum of the angles in any triangle is 180°, or the Pythagorean theorem that in any right-angled triangle the area of the square whose side is the hypotenuse equals the total area of the squares of the two sides of the triangle that meet at right angles. Euclid's principles established a world shaped by the basic laws of nature as geometry. Although Euclid synthesized much earlier Greek thinking on the subject, taken together his *Elements* provided a perception of space that would endure for nearly two millennia, to Einstein's theory of relativity and the creation of a non-Euclidean geometry. For Euclid, space was empty, homogenous, flat, uniform in all directions, and reducible to a series of circles, triangles, parallels and perpendicular lines. The impact of such a perception of space on map-making was extremely important. It manifested itself initially in Eratosthenes' rather clumsy attempt to reduce all terrestrial space to a series of triangular calculations and quadrilateral shapes, but it then also allowed subsequent mapmakers to process empirical geographical data in completely new ways. All terrestrial space could, in theory, now be measured and defined according to enduring geometrical principles, and projected onto a frame made up of a mathematical grid of lines and points that represented the world. Euclidean geometry would thus form not only all subsequent Greek geography from Eratosthenes onwards, but also the Western geographical tradition until the twentieth century.

The Hellenistic response to Eratosthenes' astronomical and geo-graphical calculations was shaped by a shift in the political world throughout the third and second centuries BC. The rise of the Roman Republic, including its victories in the Punic and Macedonian wars, sig-nalled the decline of the Hellenistic empires and ultimately the destruction of the Ptolemaic dynasty in Alexandria. It is one of the great puzzles of cartographic history that hardly any world maps survive from either the Roman Republic or Empire. Although it is dangerous to extrapolate from the limited evidence of Roman cartography that does survive in the form of stone and bronze cadastral (or land-surveying) maps, floor mosaics, engineering plans, topographical drawings, written itineraries and road maps imply a relative indifference towards the more abstract preoccupations of Hellenistic geography. Instead, the Romans favoured the more practical use of maps in military campaigns, coloni-zation, land division, engineering and architecture.[38]

However, this apparent division between a more theoretical, abstract Hellenistic mapping tradition and a practical, organizational Roman geography is to some extent illusory, especially as the two traditions met and fused from the second century BC. Other centres of learning in the Hellenistic world were by then starting to challenge Alexandria's cultural pre-eminence. By 150 BC the Attalid dynasty, closely allied to the rise of Rome and with its capital in Pergamon, established a library second only to its Ptolemaic rival, run by the renowned philoso-pher and geographer Crates of Mallos. Strabo tells us that Crates constructed a terrestrial globe (since lost) with four symmetrical inhabited continents, separated by a vast cross of ocean running east to west across the equator and north to south through the Atlantic. The northern hemisphere featured the *oikoumenē* but also the *perioikoi* ('near dwellers') to the west, with the *antoikoi* ('opposite dwellers') and *antipodes* ('those with feet opposite') in the southern hemisphere.[39] Crates' globe was a fascinating combination of established traditions of Greek geometry with the developing ethnography of the Roman Republic, formalizing the geography of the *antipodes* and anticipating later Renaissance voyages to discover the 'fourth part' of the world.

But not everyone accepted Eratosthenes. The astronomer Hipparchus of Nicaea (*c.* 190–120 BC) wrote a series of treatises in Rhodes, including three books entitled *Against Eratosthenes*, in which he criticized his predecessor's use of astronomical observations in drawing maps. 'Hipparchus', Strabo tells us, 'shows that it is impossible for any man, whether layman or scholar, to attain to the requisite knowledge of geography without a determination of the heavenly bodies and of the eclipses which have been observed.'[40] Hipparchus' detailed astronomical observations of more than 850 stars allowed him to point out the inaccuracies of Eratosthenes' calculation of latitude, as well as acknowledging the problems of measuring distances from east to west – lines of longitude – other than through precise comparative observations of eclipses of the sun and moon. This was a problem that would only be satisfactorily resolved in the eighteenth century by means of the chronometer and the accurate measurement of seaborne time, but Hipparchus offered his own rudimentary calculations of both latitude and longitude in the first known astronomical tables.

Those who challenged Eratosthenes were not always right. One of the most influential revisionist geographers was the Syrian mathematician, philosopher and historian Posidonius (*c.* 135–50 BC). As well as running a school in Rhodes, he was befriended by distinguished Romans like Pompey and Cicero, and wrote several treatises (all lost) refining and revising various elements of Hellenistic geography. He proposed seven climatic zones running around the earth rather than Aristotle's five, based on astronomical and ethnographic observations, which included some of the most detailed information on the inhabitants of Spain, France and Germany drawn from the recent Roman conquests of these regions. More controversially, Posidonius questioned Eratosthenes' method of calculating the circumference of the earth. Starting from his adopted home of Rhodes, Posidonius argued that it was on the same meridian as Alexandria, and at a distance of just 3,750 stades (a serious underestimation, whatever his value of a *stadion*). He then observed the height of Canopus, in the Carina constellation, and claimed it was exactly on the horizon at Rhodes, but rose 7½° or one-forty-eighth of a circle at Alexandria. Multiplying the figure of 3,750 stades by forty-eight, Posidonius estimated the earth's circumference as 180,000 stades. Unfortunately, his estimate of the angle of inclination between the two places was wrong, as well as his calculation of the distance between Rhodes and Alexandria. His calculations provided a gross underestimation of the size of the earth, but they would prove to be remarkably enduring.

Historically, Posidonius represented the moment when Hellenistic and Roman mapping traditions came together. It was a development that reached its climax in Strabo's *Geography*, written between AD 7 and 18. The seventeen books of Strabo's *Geography*, most of which still survive, encapsulate the ambiguous state of geography and mapmaking prior to Ptolemy, as the Roman Empire came to dominate the Mediterranean, and the Hellenistic world went into its long decline. Strabo, a native of Pontus (in modern Turkey), was intellectually influenced by Hellenism, but politically shaped by Roman imperialism. Although generally following Eratosthenes' calculations, Strabo reduced the size of the *oikoumenē*, giving it a latitudinal range of less than 30,000 stades, and a longitudinal breadth of 70,000 stades. He sidestepped the problem of projecting the earth onto a plane surface by recommending the creation of 'a large globe', at least 3 metres in diameter. If this also proved impossible he accepted drawing a flat map with a rectangular grid of parallels and meridians, claiming rather breezily that 'it will make only a slight difference if we draw straight lines to represent the circles', because 'our imagination can easily transfer to the globular and spherical surface the figure or magnitude seen by the eye on a plane surface'.[41]

Strabo's *Geography* acknowledged the importance of philosophy, geometry and astronomy in the study of geography, while also praising 'the utility of geography' for 'the activities of statesmen and commanders'. For Strabo, 'there is need of encyclopedic learning for the study of geography', of everything from astronomy and philosophy to economics, ethnography and what he called 'terrestrial history'. In keeping with Roman attitudes, Strabo's version of the subject was a highly political version of human geography, and of how the earth is appropriated by mankind. This was practical knowledge concerned with political action, as it allowed rulers to govern more effectively, or, as Strabo put it, if 'political philosophy deals chiefly with the rulers, and if geography supplies the needs of those rulers, then geography would seem to have some advantage over political science'.[42] Strabo was no mapmaker, but his work marks an important change between Hellenistic and Roman geography. The Hellenistic world had established geography as the philosophical and geometrical study of the *oikoumenē*, the 'living space' of the known world; the Romans now perceived geography as a practical tool to comprehend their version of it: the *orbis terrarum*, or 'circle of lands', a space regarded from the period of Emperor Augustus onwards as coextensive with Rome as *imperium orbis terrarum*, or 'empire of the world'.[43] In one of the earliest and most daring syntheses of geography and imperialism, the *orbis terrarum* came to define the world and Rome as one and the same thing.

Virtually none of these changes in the intellectual and political world are immediately discernible when first reading Ptolemy's *Geography*. There is scant acknowledgement that the astronomer was writing at the culmination of a thousand-year tradition of Greek mapmaking, and little trace of the impact of Roman geography on his writing, despite the generations of Roman imperial administration of Alexandria since Augustus' conquest in 30 BC. Nor is there any mention in Ptolemy's work of the Alexandria library, which by the middle of the second century was a pale shadow of its glory under Eratosthenes, after a fire in 48 BC destroyed many of its books and buildings. Instead, Ptolemy's work reads like a timeless scientific treatise of high Hellenistic scholarship, serenely indifferent to the changes in the world around it. Ptolemy followed a well-established geographical tradition: establishing his astronomical credentials and then writing a treatise that, just like Strabo's *Geography* and Hipparchus' *Against Eratosthenes*, spends most of its time explaining itself in opposition to its immediate predecessors.

Ptolemy had already completed a monumental treatise on astronomy, a compilation of mathematical astronomy in thirteen books that became known as the *Almagest*. It provided the most comprehensive model of a geocentric universe, and would endure for more than 1,500 years before being challenged by Nicolaus Copernicus's heliocentric thesis, *On the Revolutions of the Celestial Spheres* (1543). Ptolemy's cosmology marked a decisive turn away from Plato's and the idea of divine heavenly bodies. The *Almagest* expanded the Aristotelian belief in a geocentric cosmology shaped by a mechanical physics of cause and effects. Ptolemy argued that the spherical, stationary earth lies at the centre of a spherical celestial universe, which makes one revolution around the earth every day, revolving from east to west. The sun, moon and planets follow this celestial procession, but they trace different motions from the fixed stars. Ptolemy also listed the planets according to their proximity to the earth, beginning with the moon, followed by Mercury, Venus, the sun, Mars, Jupiter and Saturn. Developing Hipparchus' astronomical observations and Euclid's geometrical principles, Ptolemy catalogued 1,022 stars arranged into 48 constellations; he explained how to make a celestial globe; and he used trigonometry (and in particular chords) to understand and accurately predict eclipses, solar declination, and what appeared to be the irregular or retrograde motion of the planets and stars from a geocentric perspective.[44]

Like Hipparchus and many of his Greek forebears, Ptolemy believed in the 'affinity of the stars with mankind and that our souls are a part of the heavens'.[45] From this spiritual statement emerged a more practical approach to the study of the cosmos: the more accurate the measurement of the movement of the stars the more precise the calculations of the size and shape of the earth. In the second book of the *Almagest*, while explaining how gathering astronomical data can produce more accurate measurement of terrestrial parallels, Ptolemy admitted:

> What is still missing in the preliminaries is to determine the positions of the noteworthy cities in each province in longitude and latitude for the sake of computing the phenomena in those cities. But since the setting out of this information is pertinent to a separate, cartographical project, we will present it by itself following the researches of those who have most fully worked out this subject, recording the number of degrees that each city is distant from the equator along the meridian described through it, and how many degrees this meridian is east or west of the meridian described through Alexandria along the equator, because it was for that meridian that we established the times corresponding to the positions [of the heavenly bodies].[46]

The *Almagest* was probably written shortly after AD 147. The need for a 'separate, cartographical project' based on the astronomical observations recorded in the *Almagest* was the spur for Ptolemy's subsequent text, the *Geography*: an exposition, in the form of tables supplemental to the larger astronomical work, that would provide coordinates of key cities. After completion of the *Almagest*, as well as writing treatises on astrology, optics and mechanics, Ptolemy completed the eight books of this second great work.

The finished text was substantially more than the promised table of key geographical coordinates. Ptolemy chose not to gather data himself or through agents, but instead collated and compared every text available to him in Alexandria. He stressed the importance of travellers' tales, but warned about their unreliability. The *Geography* acknowledged the need 'to follow in general the latest reports we possess' from pre-eminent geographers as well as historians. These included etymological and historical sources – Roman authors like Tacitus and his description of northern Europe in the *Annals* (*c.* AD 109) and *periploi* of uncertain origin, such as the anonymous *Periplus of the Erythraean Sea* (*c.* first century AD), a merchant's guide to places in the Red Sea and Indian Ocean. The most important author quoted in the *Geography* was Marinus of Tyre, whose work has since been lost, but who according to Ptolemy, 'seems to be the latest [author] in our time to have undertaken this subject'.[47] The first book defined the subject of geography and how to draw a map of the inhabited world. Books 2 to 7 presented the promised table of geographical coordinates, but now enlarged to include 8,000 cities and locations, all listed according to their latitude and longitude, beginning in the west with Ireland and Britain, then moving eastwards through Germany, Italy, Greece, North Africa, Asia Minor and Persia, and ending in India. Book 8 suggested how to divide the *oikoumenē* into twenty-six regional maps: ten of Europe, four of Africa (still called 'Libya') and twelve of Asia, a running order that would be reproduced in the earliest Byzantine copies of his book illustrated with maps, and most subsequent world atlases.

The wealth of geographical information contained in Ptolemy's tables included not only the scholarly tradition of geographical enquiry, but also astronomical calculations and the written testimony of travellers. From the beginning of the *Geography*, Ptolemy made it very clear that 'the first step in a proceeding of this kind is systematic research, assembling the maximum of knowledge from the reports of people with scientific training who have toured the individual countries; and that the inquiry and reporting is partly a matter of surveying, and partly of astronomical observation'. Such 'systematic research' was only possible thanks to the consultation of the Alexandria library's *Pinakes*, or 'Tables', the first known library catalogue indexed according to subject, author and title, created by Callimachus of Cyrene, *c.* 250 BC. The *Geography* was an immense data bank, compiled by the first acknowledged armchair geographer, a 'motionless mind'[48] operating from a fixed centre, processing diverse geographical data into a vast archive of the world.

For Ptolemy, there was no space for speculative cosmogonies on the origins of the universe, or attempts to establish the indeterminate and shifting geographical and political boundaries of the *oikoumenē*. The *Geography*'s opening statement set the tone, with its enduring definition of geography as 'an imitation through drawing of the entire known part of the world together with the things that are, broadly speaking, connected with it'. Ptolemy regarded geography as a comprehensive graphic representation of the known world (but not, we should note, the whole earth), in contrast to what he called, with a nod towards the Roman preoccupation with land surveying, 'chorography', or regional mapping. Whereas chorography requires skill in 'landscape drawing', Ptolemy said that global mapping 'does not require this at all, since it enables one to show the positions and general configurations [of features] purely by means of lines and labels', a geometrical process in which mathematical method 'takes absolute precedence'.[49] Using a telling corporeal metaphor to contrast the two geographical approaches, Ptolemy believed that chorography provides 'an impression of a part, as when one makes an image of just an ear or an eye; but the goal of world cartography is a general view, analogous to making a portrait of the whole head'.

Having established his methodology, Ptolemy then proceeds to discuss the size of the earth and its latitudinal and longitudinal dimensions through a detailed critique of Marinus of Tyre's methods, before providing his own geographical projections for drawing world maps. One of the most significant aspects of Ptolemy's calculations involved the size of the whole earth in relation to its inhabited realm, the *oikoumenē*. Revising the calculations of Eratosthenes and Hipparchus, Ptolemy divided the globe's circumference into 360° (based on the Babylonian sexagesimal system, in which everything was measured in units of sixty), and estimated the length of each degree as 500 stades. This gave him the same circumference of the earth as Posidonius: 180,000 stades. This was certainly too small, possibly by as much as 10,000 kilometres, or more than 18 per cent of the earth's actual circumference, depending on the length of *stadion* used. But if Ptolemy believed the earth was smaller than predecessors like Eratosthenes imagined, he went on to argue that its inhabited dimension was much larger than many believed: his *oikoumenē* stretched from west to east through an arc of just over 177°, starting from a prime meridian that ran through the Fortunate Isles (the Canary Islands), to Cattigara (believed to be somewhere near modern-day Hanoi in Vietnam), a distance he estimated as 72,000 stades. Its breadth was calculated at just over half the length, covering just under 40,000 stades, running from Thule, situated 63° N, to the region of 'Agisymba' (in sub-Saharan Africa), located by Ptolemy at 16° S, a latitudinal range, on his measurements, of just over 79°.[50]

Such measurements naturally lead to the question of how Ptolemy arrived at his calculations of latitude and longitude. He calculated parallels of latitude according to astronomical observations of the longest day of the year at any given location. Starting at 0° on the equator with a longest day of twelve hours, Ptolemy used quarter-hour increments for each parallel until he reached the parallel representing the longest day as fifteen and a half hours, at which point he switched to increments of half an hour, up to the limit of the *oikoumenē*, which he estimated as lying along the parallel of Thule, with a longest day of twenty hours. Drawing on this method of measurement, as well as Hipparchus' calculations based on astronomical observations of the sun's altitude on the solstice, Ptolemy drew up his tables of latitude, although the relative simplicity of his method of observation meant that many of them were inaccurate (including Alexandria).

The calculation of longitude proved even more difficult. Ptolemy believed that the only way of determining longitude was to measure the distance between meridians west to east according to time, not space, using the sun as a clock: all places on the same meridian will see the noon sun crossing the plane of the meridian at the same time. Ptolemy therefore began his calculation of longitude at his westernmost point, the Fortunate Isles, and drawing each meridian moving east at intervals of 5°, or a third of an equinoctial hour, and encompassing twelve hours, represented as 180°. His measurements may have been inaccurate, but his was the first systematic method to provide consistent data that allowed subsequent mapmakers to project a grid of latitude and longitude over the inhabited earth, a graticule composed of temporal rather than spatial calculations. We tend to think of mapmaking as a science of spatial representation, but Ptolemy was proposing a world measured not according to space, but by time.[51]

Towards the end of book 1 of the *Geography*, Ptolemy begins to move away from Marinus to explain his other great geographical innovation: a series of mathematical projections designed to represent the spherical earth upon a plane surface. Although acknowledging that a globe 'gets directly the likeness of the earth's shape', Ptolemy points out that such a globe would need to be extremely large to be of any use in seeing the earth and plotting movements across it with any precision, and it would not in any case allow a view 'that grasps the whole shape all at once'. Instead, Ptolemy suggests that 'drawing a map on a plane eliminates these difficulties completely', by creating the illusion of seeing the entirety of the earth's surface at a glance. Nevertheless, he admits that it brings its own problems, and 'does require some method to achieve a resemblance to a picture of a globe, so that on the flattened surface, too, the intervals established on it will be in as good proportion as possible to the true intervals'.[52] Ptolemy here encapsulates one of the major challenges which has faced mapmakers ever since.

Marinus had tried to resolve the problem by creating a rectangular or 'orthogonal' map projection which, according to Ptolemy, 'made the lines that represent the parallel and meridian circles all straight lines, and also made the lines for the meridians parallel to one another'. But when a geographer projects a geometrical network of imagined parallels and meridians onto a spherical earth they are in effect circles of varying length. Marinus neglected this fact in favour of prioritizing the measurements made along his prime parallel running through Rhodes at 36° N, and accepting increasing distortion north and south of this line. He accepted a centrifugal representation of terrestrial space, where the accuracy emanates outwards from a definable centre, dissipating the further one moves towards the margins, and finally leading to absolute distortion. Like a good Euclidean, Ptolemy wanted his terrestrial space to be homogenous and directionally uniform, and quickly dismissed

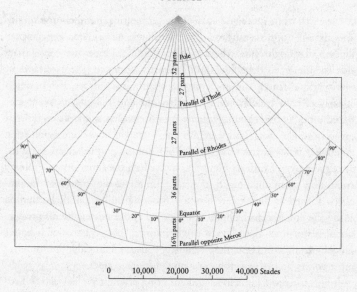

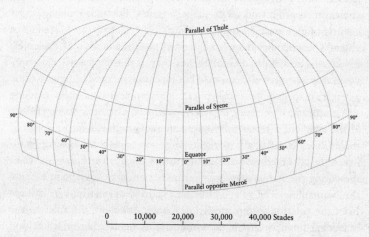

Fig. 3 Diagrams of Ptolemy's first and second projections.

Marinus' projection. But even Ptolemy was unable to square the circle of map projection, and acknowledged that a compromise was required.

With Euclid clearly still in his mind, he turned to geometry and astronomy for a solution. Imagine, wrote Ptolemy, looking at the centre

47

of the earth from space and envisaging geometrical parallels and meridians drawn upon its surface. The meridians, he argued, 'can give an illusion of straight lines when, by revolving [the globe or the eye] from side to side, each meridian stands directly opposite [the eye], and its plane falls through the apex of the sight'. In contrast, the parallels 'clearly give an appearance of circular segments bulging to the south'. Based on this observation, Ptolemy proposed what is known as his first projection. The meridians were drawn as straight lines converging at an imaginary point beyond the North Pole, but the parallels were depicted as curved arcs of different lengths, centred on the same point. Ptolemy could now maintain a more accurate estimate of the lengths of the parallels, as well as their relative ratios, focusing on the parallels running along the equator and Thule. The method could not excise all proportional distortions along every single parallel, but it provided a better model of conformation that retained consistent angular relations at most points on the map than any previous projection.

It was the most influential and enduring attempt yet devised to project the earth onto a plane surface. This was the first example of a simple conical map projection, as its shape suggests, although Ptolemy's cone also resembles another, more familiar shape: that of the Macedonian chlamys, the iconic image that shaped the foundations of Ptolemaic Alexandria and inspired Eratosthenes' map of the *oikoumenē*. Ptolemy's projection also provided a simple but ingenious method of how to draw a world map and then incorporate geographical data into it. Using straightforward geometry, he describes how to 'fashion a planar surface in the shape of a rectangular parallelogram', within which a series of points, lines and arcs are marked using a swinging ruler. Having established the basic geometrical outline, the mapmaker then takes the ruler measuring the radius of a circle centred at an imaginary point beyond the North Pole. The ruler is then marked with gradations in latitude from the equator to the parallel of Thule. Attaching the ruler to the imaginary point so that it can swing free along an equatorial line divided into 180° of hour-long intervals, it would be possible to locate and mark any location on a blank map by referring to Ptolemy's tables of latitudinal and longitudinal coordinates. The ruler was simply swung to the required longitude listed along the equatorial line, and, according to Ptolemy, 'using the divisions on the ruler, we arrive at the indicated position in latitude as required in each instance'.[53] The geographical outlines on such a map were relatively insignificant: what characterized it were not contours but a series of points, established by his coordinates of latitude and longitude. A point is of course the first defining principle of Euclidean geometry: it is 'that which has no part'; it is indivisible, with no length or breadth. To create an accurate map projection, Ptolemy went right back to the basics of Euclidean geometry.

This first projection still had its drawbacks: on a globe, parallel lines diminish south of the equator, but if drawn on Ptolemy's projection they actually increase in length. Ptolemy effectively went against the consistency of his own projection by solving this problem with meridians forming acute angles at the equator. This gave the projection the appearance of a chlamys, but it was hardly ideal. Ptolemy regarded this as only a minor drawback, as his *oikoumenē* only extended 16° S of the equator, but it would cause serious problems in later centuries when travellers began to circumnavigate Africa. Nevertheless, the first projection still projected straight meridians, which, as Ptolemy acknowledged from the outset, only corresponded to a partial perspective on the globe from space; like the parallels, meridians trace a circular arc around the globe, and their geometrical reality should retain such a curvature on a plane map. He therefore proposed a second projection. 'We could', he wrote, 'make a map of the *oikoumenē* on the planar surface still more similar and similarly proportioned [to the globe] if we took the meridian lines, too, in the likeness of the meridian lines on the globe.'[54] This projection, he said, was 'superior to the former', because parallels *and* meridians were represented as curved arcs, and because virtually all of its parallels retained their correct ratios (unlike the first projection, where this was only achieved for the parallels running through the equator and Thule). The trigonometry involved was more complicated than the first projection, and Ptolemy still had problems retaining uniform proportionality along his central meridian. He also acknowledged that it was far more difficult to construct a map based on the second projection, as the curved meridians could not be drawn with the aid of a swinging ruler.

Following the exhaustive descriptions of both map projections, Ptolemy concluded book 1 of the *Geography* with some remarkably sanguine observations. Although preferring the second projection, he appreciated that 'it might be inferior to the other with respect to the ease of making the map', and he counselled future geographers to 'hold on to descriptions of both methods, for the sake of those who will be attracted to the handier one of them because it is easy'. His advice would influence the response of scholars and mapmakers to the revival of the *Geography* from the thirteenth century onwards.

Ptolemy's predecessors used geography to try to understand cosmogony, the explanation of the creation of everything. In his *Geography*, Ptolemy turned away from this quest. There are no myths, and few political boundaries or ethnographies in his book. Instead, he recreates the origins of his subject in two enduring principles of Alexandrian learning: Euclid's principles of geometry and Callimachus' bibliographical method of classification. Ptolemy's innovation was to establish a repeatable methodology for mapping the known world according to recognized mathematical principles. His map projections allowed anyone with a basic understanding of Euclidean geometry to create a map of the world. His innovation of tables of latitude and longitude, drawn from the Alexandrian *Pinakes*, established the coordinates of locations throughout the *oikoumenē*. These tables enabled mapmakers to plot the positions of every known location upon a map with utter simplicity, and by refusing to place explicit boundaries upon his *oikoumenē*, Ptolemy encouraged future mapmakers to plot ever more locations upon the surface of their world maps.

Ptolemy's claim to objectivity and exactitude in the gathering of geographical and astronomical materials was of course an illusion. The measurement of any substantial distance in the second century was notoriously imprecise, astronomical observations were compromised by limited and unwieldy instruments, and much of Ptolemy's data on the location of places was based on what the Greeks call *akoē*, or 'hearsay' – the claims made by a certain merchant, the reported observations of an astronomer passed down through centuries or the anonymous records of *itineraria*. His projections were also limited to just half the earth, an inhabited surface that was only 180° wide, even though he and his contemporaries understood that there was a world elsewhere beyond the limits of the *oikoumenē*.[55] In many ways this was just an incitement to future speculation and projection. Having provided the methodological toolkit for making a map, Ptolemy invited others to revise his tables and relocate their places. Regional mapping, chorography, was an art, but the mapping of the world was now a science. The outline of a region or position of a place could be changed if new information appeared, but the methodology of marking a point upon the map's surface according to certain enduring mathematical principles was, he believed, immutable.

One puzzle in assessing Ptolemy's importance for mapmaking remains. Throughout the *Geography*, there are no explicit references to maps illustrating the text. As we have seen, the earliest surviving text only appears in Byzantium in the late thirteenth century, more than a thousand years after it was first written. These early texts included world maps (mainly based on the first projection), but it is unclear if these maps were copies of Ptolemy's original illustrations, or Byzantine additions based on Ptolemy's written instructions. The question of whether Ptolemy ever drew maps to illustrate the original *Geography* has divided cartographic historians for decades; scholarly opinion now leans towards the belief that, although he may have done so, any such maps were not incorporated into the original *Geography*.[56] There are very few examples of maps in Graeco-Roman treatises on geography, and it was more common for them to be erected in public spaces, as in the case of maps placed on the wall of a portico in Rome in the early first century AD by the emperor Augustus' friend Agrippa.[57]

It is possible that the initial form of the *Geography* was responsible for its lack of maps. It was probably written with lampblack ink made from soot, and inscribed on a roll of papyrus cut from plants growing along the Nile Delta. Most papyrus rolls from this period were composed of conjoined sheets measuring an average 340 centimetres in length. However, the height of any roll rarely measured more than 30 centimetres.[58] Such dimensions suited Roman itineraries, such as the so-called 'Peutinger Map', a twelfth- or thirteenth-century copy of a fourth-century AD Roman map showing the world from India, Sri Lanka and China to Iberia and the British Isles. These itineraries described the movement across terrestrial space in linear terms, a one-dimensional representation with little sense of depth, relief or scale, primarily due to the limitations of the medium. The 'Peutinger Map' is inscribed on a parchment roll with a length of over 6 metres, but a width of just 33 centimetres, creating obvious lateral distortion. These dimensions made it effectively impossible to reproduce either the world or regional maps described in such detail by Ptolemy without improbable reduction and distortion. Ptolemy's solution was either to draw maps separate from his book (but if so, none have survived), or, in the explanation offered by the most recent translators of the *Geography*, he decided to 'encode the map in words and numbers'.[59] If this was the case, then Ptolemy's approach was to provide the geographical data and the mathematical method, and leave the rest to future generations.

'Look on my works, ye mighty, and despair!' exclaimed Shelley's Egyptian pharaoh Ozymandias. In Shelley's poem on the hubris of imperial might, 'nothing beside remains' of the tyrant's kingdom and all its glittering monuments, except the ruins of his statue. Similarly, today, most traces of the Ptolemaic dynasty and their rule over Egypt have disappeared, submerged beneath the waters of Alexandria's harbour. The library is long gone, most of its books looted and destroyed. Its loss has haunted the Western imagination ever since and historians of different ideological persuasions throughout the ages have blamed everyone from Romans and Christians to Muslims for its destruction. It remains a romantic memory of endless possibilities, a source of speculation and myth, a 'might have been' in the development of learning and civilization, and a lesson on the creative as well as destructive impulses that lie at the heart of all empires.[60]

But some of the 'works' survived and migrated and they included Ptolemy's *Geography*. Although Ptolemy's writing seems remarkably untouched by the events that surrounded him, his text betrays a desire to transmit his ideas in a more enduring form than maps or monuments. The *Geography* was the first book that, either by accident or design, showed the potential of transmitting geographical data *digitally*. Rather than reproducing unreliable graphic, analogue elements to describe geographical information, the surviving copies of the *Geography* used the discrete, discontinuous signs of numbers and shapes – from the coordinates of places across the inhabited world to the geometry required to draw Ptolemy's projections – to transmit its methods. This first rudimentary digital geography created a world based on a series of interconnecting points, lines and arcs grounded in the Greek tradition of astronomical observation and mathematical speculation, that stretched through Eratosthenes and Euclid all the way back to Anaximander. Ptolemy threw a net across the known world, defined by the enduring abstract principles of geometry and astronomy and the measurement of latitude and longitude. One of his greatest triumphs was to make all subsequent generations 'see' a series of geometrical lines crisscrossing the globe – the poles, the equator and the tropics – as if they were real, rather than man-made geometrical projections upon the earth's surface.

Ptolemy's scientific methods sought to make the world comprehensible through the imposition of geometrical order onto the chaotic variety of the world 'out there', while also retaining a sense of wonder at its infinite variety. His vision, enshrined in one of the *Geography*'s earliest statements on the geometrical measurement of the earth, would inspire generations of geographers even beyond the Renaissance, all the way to the age of manned space flight:

> These things belong to the loftiest and loveliest of intellectual pursuits, namely to exhibit to human understanding through mathematics both the heavens themselves in their physical nature since they can be seen in their revolution about us, and the nature of the earth through a portrait since the real earth, being enormous and not surrounding us, cannot be inspected by any one person either as a whole or part by part.[61]

# 2

# Exchange

*Al-Idrīsī, AD 1154*

## Palermo, Sicily, February 1154

On 27 February 1154 Roger II, 'king of Sicily, of the duchy of Apulia and of the principality of Capua', died aged 58 in his palace, the Palazzo Reale, situated in the heart of his royal capital of Palermo. He was buried with due ceremony in the south aisle of Palermo Cathedral, where, twenty-four years earlier, he had been crowned king on Christmas Day 1130. His death brought to an end an extraordinary reign on the island that to modern eyes represents one of the great moments of medieval *convivencia*, the Spanish term for the peaceful coexistence of Catholics, Muslims and Jews under one rule.

Descended from the Hauteville dynasty that originated in Normandy's Cotentin peninsula, Roger and his forebears led a series of spectacular Norman conquests across Europe, Africa and the Middle East in the late eleventh century. As the Byzantine Empire declined in the face of first Persian then Arabic Muslim challenges to its authority, the Normans exploited the international disarray of medieval Christendom and soon established their rule over parts of southern Italy, Sicily, Malta and North Africa. They went on to conquer England and even created a principality in Antioch (straddling present-day Turkey and Syria) before the First Crusade of 1095.[1]

At each stage of their military conquests the Normans assimilated the cultures they conquered (with varying degrees of success). In 1072 Roger's father, Roger Guiscard, captured Palermo and appointed himself Count of Sicily, ending more than 100 years of Arab control of the island. Before Arab rule, Sicily had been governed by first the Greeks, then the Romans, and finally the Byzantines. It was a heritage that left the Normans in control of one of the most culturally diverse and strate-gically important islands in the whole of the Mediterranean. When Roger II was crowned king in 1130, he pursued a policy of political accommodation and religious toleration towards Muslims and Jews that quickly established Sicily as one of the most highly organized and culturally dynamic kingdoms in the medieval world. Roger's kingdom was primarily administered by a royal chancery that employed Greek, Latin and Arabic scribes. His court produced a trilingual psalter, and the liturgy was reportedly sung in Arabic.

Roger's death represented the end of an era. Of those mourners who gathered at his interment in 1154, none had more to lament Roger's passing than one of his closest confidants, Abu Abdallāh Muḥammad ibn Muḥammad ibn Abdallāh ibn Idrīs al-Sharif al-Idrīsī, more com-monly known as al-Sharīf al-Idrīsī. Just weeks before Roger's death, al-Idrīsī had finally completed a vast geographical compendium over which he had laboured for more than a decade since its commission by Roger in the early 1140s. The book provided a comprehensive summary of the known world, and was illustrated with seventy regional maps of the world – and one small but beautifully illuminated world map.

Written in Arabic, and completed (according to its introduction) in the month of Shawwāl in the Islamic calendar, or 14–15 January 1154, the book was entitled *Kitāb nuzhat al-mushtāq fī khtirāq al-āfāq*, the 'Entertainment for He Who Longs to Travel the World'. So close was the relationship between Roger and al-Idrīsī that the *Kitāb nuzhat al-mushtāq* (hereafter the *Entertainment*) became known simply as *The Book of Roger*. Few rulers had taken such a close personal interest in their patronage of maps or their makers. Originally commissioned as a statement of Roger's imperial and cultural ambitions, within weeks of its completion al-Idrīsī's book became a commemoration of the dead king's legacy and a powerful statement of his kingdom's syncretic tradi-tions, complementing the palaces and cathedrals he built throughout his reign. But with the death of his patron, al-Idrīsī and his newly com-pleted book faced an uncertain future.[2]

Its geographical range and painstaking detail made the *Entertain-ment* one of the great works of medieval geography, and one of the finest descriptions of the inhabited world compiled since Ptolemy's *Geography*. Al-Idrīsī's book and the maps that accompanied it drew on Greek, Christian and Islamic traditions of science, geography and travel to produce a hybrid perspective on the world based on the exchange of cultural ideas and beliefs between different faiths. There is an obvious appeal today in seeing al-Idrīsī's work as the product of a rapprochement between Christianity and Islam, when both learnt from each other in an apparently amicable exchange of ideas. But the twelfth-century world of Norman Sicily and the aspirations of individuals like Roger II and al-Idrīsī were more strategic and provisional than such thinking might hope. Muslims were granted only limited rights under Roger, and the Normans continued to provide support for the Crusades against the Muslims in the Holy Lands to the east. From the perspective of Islamic theology, the known world was divided into two: the *dār al-Islām*, or House of Islam, and the *dār al-ḥarb*, or House of War, inhabited by all non-Muslims. Until Muhammad's divine revelations were universally accepted, a state of perpetual war existed between the two houses.

But not all non-Muslims were the same. Both Christians and Jews were regarded as *ahl al-kitāb,* or 'people of the book', adhering to a revealed faith explained through a standard scriptural book of prayer (the Bible, the Torah and the Qur'ān). The belief in one common God led to a range of cultural encounters between the three religions, as each tried to assert its theological superiority over the others, more often than not producing exchanges and encounters characterized by conversion and conflict, rather than dialogue and diversity.[3] Nevertheless, discussion happened, debates took place, and in the midst of such competitive exchanges emerged al-Idrīsī's *Entertainment*.

The story of al-Idrīsī's relationship with Roger II and the creation of his maps is not one of the Muslim East encountering the Christian West on equal terms. Instead it reveals a world where these geopolitical distinctions were only just beginning to develop, and where dynastic conflicts and religious divisions ensured that the labels 'Muslim' and 'Christian' were fluid categories, characterized by schism, conversion and apostasy, rather than unconditional doctrinal belief. Its chapters unfolded against the backdrop of the wider Mediterranean world, where the Byzantine Empire fell in inverse proportion to the rise of the Muslim Caliphate, and a divided and relatively insignificant Latin Christianity was caught somewhere in the middle, trying (but often failing) to assert some vestige of political autonomy and control.

Only ten manuscript copies of al-Idrīsī's *Entertainment* remain, the earliest made in 1300, and the latest at the end of the sixteenth century.

As with Ptolemy's *Geography*, we are working with a book and its maps that were produced hundreds of years after their original creation. In one of the best preserved manuscript copies of the *Entertainment*, held in the Bodleian Library's Pococke Collection, and dated 1553, there is a circular world map, beautiful in its simplicity, which appears to show how al-Idrīsī represented the world in the middle of the twelfth century. The most startling aspect of the map is that it is oriented with south at the top.

Etymologically, 'orientation' stems from the original Latin root *oriens*, which refers to the east, or the direction of the rising sun. Virtually all ancient cultures record their ability to orient themselves according to an east–west axis based on observations of the rising (eastern) and setting (western) sun, and a north–south axis measured according to the position of the North Star or the midday sun.[4] Such orientation was as much symbolic and sacred as directional. In polytheistic sun-worshipping cultures, the east (*oriens*) was revered as the direction of renewal and life, closely followed by the south, while the west was understandably associated with decline and death, and north with darkness and evil. The Judaeo-Christian tradition developed these associations by orienting places of worship as well as maps towards the east, which was ultimately regarded as the location of the Earthly Paradise. In contrast, the west was associated with mortality, and the direction faced by Christ on the cross. The north became a sign of evil and satanic influence, and was often the direction in which the heads of excommunicants and the unbaptized faced when they were buried.[5] As the next chapter shows, virtually all Christian world maps (or *mappaemundi*) put east at the top of their maps until the fifteenth century.

Islam and mapmakers like al-Idrīsī inherited a similar reverence for the east, although it developed an even stronger interest in the cardinal directions with the Qur'ānic injunction to its believers to pray in the sacred direction of Mecca, regardless of their location on the globe; finding the direction (known as *qibla*, or 'sacred direction') and distance to both Mecca and the Kā'aba inspired some of the most complicated and elaborate maps and diagrammatic calculations of the medieval period.[6] Most of the communities who converted to Islam in its early phase of rapid international expansion in the seventh and eighth centuries lived directly north of Mecca, leading them to regard the *qibla* as due south. As a result, most Muslim world maps, including al-Idrīsī's, were oriented with south at the top. This also neatly established continuity with the tradition of the recently conquered Zoroastrian communities in Persia, which regarded south as sacred.

There are virtually no cultural traditions that place west at the top of the map, as it is almost universally associated with the sun's disappearance, a symbol of darkness and death, exemplified in the phrase 'to go west', meaning to die. The final cardinal direction, north, placed at the top of the Babylonian world map, has an even more complicated lineage. In China, north was accorded primacy as the sacred direction. Across the empire's wide plains, the south brought sunlight and warming winds, and so was the direction towards which the emperor looked down on his subjects. When everyone gazed up at the emperor from a position of subjection, they consequently faced north. Etymologically, the Chinese 'back' is synonymous with 'north', because the emperor's back faces that direction. Chinese world maps were oriented accordingly, one of the many reasons that their maps look at first glance to be remarkably modern. The Gnostic and Dualist beliefs of various ancient Mesopotamian communities also celebrated the north as the sacred direction, regarding the Pole Star as a source of light and revelation, and it is possibly for this reason that the Babylonian world map is oriented northwards.

On al-Idrīsī's world map the four cardinal directions are marked just outside the map's frame which, taking its inspiration from Qur'ānic verses, is composed of a fiery golden aureole. The map itself shows a world indebted to the Greek *oikoumenē*. The Mediterranean and North Africa are represented in detail, as are a fantastic jellyfish-shaped mountain range with its tributaries in central Africa. Named 'The Mountains of the Moon', the range was believed to be the source of the Nile. Egypt, India, Tibet and China are all labelled in Arabic, as are the Caspian Sea, Morocco, Spain, Italy and even England. The map retains a classically vague understanding of southern Africa and south-eastern Asia, although it departs from Ptolemy in showing a circumnavigable Africa, with the entire globe surrounded by an encircling sea.

Perhaps the most peculiar aspect of this world map is just how at odds it is with the book in which it sits. In contrast to the teeming human geography described in the other maps and the text of the *Entertainment*, the world map is a purely physical representation of geography. There are no cities, and virtually no discernible traces of the impact of humanity upon the earth's surface (with the exception of the fabled barrier erected by Alexander the Great in the Caucasus Mountains to keep the mythical monsters Gog and Magog at bay, represented in the map's bottom left-hand corner). This apparent contradiction between the *Entertainment*'s evocative description of the earth's regions and its geometrical world map can only be understood by turning to an explanation of what Roger wanted when he employed al-Idrīsī: the fruits of the preceding 300-year-old tradition of Islamic mapmaking.

The term 'Islamic maps' is something of a misnomer. The geographical traditions and cartographic practices that gradually coalesced following the rise of Islam in the Arabic peninsula in the late seventh century were too regionally, politically and ethnically diverse to merit being described as a unified body of mapmaking (although the same can be said to a degree of 'Greek' or 'Christian' maps). None of the early Islamic languages possessed a definitive noun to define 'map'. As in Greek and Latin, a variety of terms were used to describe what would today be called a map. These included *ṣūrah* (meaning 'form' or 'figure'), *rasm* or *tarsīm* ('drawing'), and *naqsh* or *naqshah* (painting).[7] Like the Bible, the Qur'ān offered little direct assistance to mapmakers. It does not have a definable cosmology with a clear account of the size and shape of the earth within a larger universe, despite offering a series of intriguing allusions. The sky is described as a canopy spread over the earth, which is held in place by mountains and illuminated by the sun and the moon. God 'created seven Firmaments and of the earth a similar number', although the specific dimensions of these worlds are unexplained.[8] References to an apparently disc-shaped earth encircled by water, and the description of the Mediterranean and Arabian seas separated by a barrier, appear to draw on early Babylonian cosmology, although allusions to the 'Sun setting in a spring of murky water' imply awareness of the Atlantic, a notion inherited from the Greeks.[9]

It was not until the Abbasid Caliphate established itself as the centre of the Islamic Empire in Baghdad by the end of the eighth century that a recognizably Islamic practice of mapmaking can be detected. The foundation of the imperial capital of Baghdad in AD 750 by the second Abbasid caliph al-Mansur represented the successful culmination of a bitter struggle with the Umayyad Caliphate, which had ruled from Damascus since 661. The shift in power eastwards had a significant effect on Islamic culture, diminishing the earlier tribal Arabic basis of Islamic authority, and bringing the caliphate into closer contact with the scientific and artistic traditions of Persia, India and even China, so complementing the initial assimilation by Islam of Christian, Greek and Hebraic cosmologies. At the same time, the empire's contact with Latin learning diminished, compounded by the subsequent rise of the rival Umayyad Caliphate established in al-Andalus. The move to Baghdad also centralized Islamic power and authority more effectively than that of any other empire of the period. The ruling caliph became all-powerful, and tribal alliances were absorbed into an absolutist monarchy that appointed a high-ranking minister, or *vizier*, to oversee *diwans*, or ministries, which controlled all aspects of public and political life. Almost inevitably, the Abbasid caliphs began to commission geographical descriptions of their dominion.[10]

The first recorded commission of a world map in Baghdad came in the reign of the seventh Abbasid caliph, al-Mā'mūn (813–33), who patronized an institute of scientific study which became known as the 'House of Wisdom' (*bayt al-ḥikma*). Referred to by its contemporaries as *al-ṣūrah al-ma'mūnīyah* after its patron, the map has not survived. But a few eyewitness descriptions of it have survived, and provide a startling insight into the level of intellectual exchange taking place at al-Mā'mūn's court, which included extensive knowledge of Ptolemy's *Geography*. The Arabic historian and traveller al-Masūdi (d. 956) recalled with admiration seeing the map 'that al-Mā'mūn ordered to be constructed by a group of contemporary scholars to represent the world with its spheres, stars, land and seas, the inhabited and uninhabited regions, settlements of peoples, cities, etc.' He concluded that 'this was better than anything that preceded it, either the *Geography* of Ptolemy, the *Geography* of Marinus, or any other'.[11] While the Latin West continued to remain in ignorance of Ptolemy's *Geography* for another 400 years, and completely lost Marinus' manuscript, al-Mā'mūn's court was busy incorporating Ptolemy (as well as many of his other works on astronomy and optics) into their world maps.

The Baghdad court did not limit its research to Greek texts. Al-Masūdi observed that al-Mā'mūn's world map adopted Ptolemy's concept of longitudinal climates (from the Greek *klimata*, translated into Arabic as *aqālīm*, or *iqlīm*) to divide the known world into seven regions, a tradition that would shape al-Idrīsī's geographical thinking. Ptolemy had drawn on Aristotle for his notion of *klimata*, but in the creation of their map, al-Māʾmūn's scholars had modified this model by drawing on the Persian concept of dividing the world into seven *kishvars*, or regions. This was in turn derived from archaic Babylonian and Indian cosmographical perceptions of the world as a lotus petal, with regions surrounding a primary zone, usually representing a sacred area or capital city.[12] The result was a system that placed Baghdad in its central region – the fourth – around which the other six regions were grouped from north to south. Although not explicitly located at the centre of the map, Baghdad and Iraq were regarded as lying at the heart of the earth, where 'moderation in all things', from climate and natural beauty to personal intelligence, could be found in a compelling mix of geography, astronomy and climate.[13]

What all this led to is regrettably unknown. Al-Māʾmūn's court created one of the many lost maps in world history, and probably the most important in the early Muslim world. Perhaps it was circular, reflecting prevailing Islamic cosmological belief that the universe and the earth were both spherical. But if it incorporated the thinking of Ptolemy and Marinus, it could also have been rectangular, and modelled on one of Ptolemy's two projections.

One clue to what the map might have looked like comes from a much later diagram in a manuscript entitled the 'Marvels of the Seven Climates to the End of Habitation', written by the little-known scholar calling himself Suhrāb who lived in Iraq in the first half of the tenth century. This treatise, one of the first comprehensive accounts in Arabic of how to draw a map of the world, is an invaluable source of early Islamic conceptions of the inhabited earth, as well as a tantalizing insight into what al-Māʾmūn's map might have looked like. Although the diagram in Suhrāb's treatise lacks any physical geographical features, it provides a rectangular frame within which to plot the known world. Suhrāb begins by advising aspirant makers how to construct a world map. 'Let its width', he wrote, 'be half its length.' He then describes adding 'four scales' in the map's borders to represent longitude and latitude. But his primary interest was in 'the latitudes of the seven climates, beginning the enumeration from the terrestrial equator towards the north'.[14] As in Ptolemy, Suhrāb's climates were determined by accompanying tables of maximum daylight. The result is a diagram that depicts the seven climates running from 20° S of the equator (shown on the left) to 80° N (on the right), with north facing the reader (at the bottom of the diagram). This assumes that Suhrāb plotted his world map with south at the top. Suhrāb's coordinates are distinctly Ptolemaic (although he actually expands Ptolemy's latitudinal range of the inhabited world), but his overall projection on a rectangle with intersecting lines at right angles is closer to that of Marinus. Suhrāb also substantially reproduced the coordinates in the *Kitāb ṣūrat al-arḍ* ('Picture of the earth'), written by al-Khwārazmi (d. 847), another member of al-Mā'mūn's 'House of Wisdom', a further indication that the caliph's world map may have been rectangular, as well as oriented with south at the top in line with prevailing Muslim beliefs.

Suhrāb's diagram provides an insight into the possible shape and orientation of al-Mā'mūn's map, although the improved calculations on the size of the earth undertaken subsequently by the caliph's scholars indicate that further progress was being made in mapping the earth. In response to the caliph's reported wish 'to know the size of the earth',[15] surveyors were dispatched into the Syrian desert to measure the sun's angle of elevation in relation to the cities of Palmyra and Raqqa – a replay of Eratosthenes' famous attempt to measure the earth's circumference. Most of the surveyors concluded that the length of a degree of longitude was 56⅔ Arab miles. Based on current calculations of the length of an Arabic mile as equal to 1¹⁄₁₅ of a modern mile, this estimate has been converted into a global circumference of just over 40,000 kilometres (25,000 miles). If the equivalence is correct, this meant that al-Ma'mūn's surveyors came within less than 100 kilometres of the correct circumference of the earth measured at the equator. The result was all the more astonishing when contrasted with Ptolemy's massive underestimation of the earth's circumference, at just under 29,000 kilometres, (18,000 miles).

All the surviving evidence from the 'House of Wisdom' suggests an evolving world picture heavily indebted to Greek scholarship, suffused by Indo-Persian traditions that produced a map based on climatic divisions oriented with south at the top. Although scholars like al-Khwārazmi appropriated Ptolemy to establish a genre of world maps using the generic term *Ṣūrat al-arḍ*, the *Geography* was only partially (and often erroneously) translated from Greek into Arabic. Al-Khwārazmi and his followers focused almost exclusively on Ptolemy's tables of latitudes and longitudes, improving on many of his mistakes and omissions. They provided a more accurate measurement of the Mediterranean and also depicted the Indian Ocean as flowing into what would now be seen as the Pacific Ocean, no longer landlocked. But they did not make an explicit connection with Ptolemy's method of projecting the earth onto a graticule of longitude and latitude, and Suhrāb's diagram offered no more than a revised version of Marinus' rectangular projection, which had been so heavily criticized by Ptolemy. Nor did the division of the earth into continents particularly appeal to the early Muslim scholars. The Islamic Caliphate took mapmaking in a different direction instead.

One of the earliest indications of this cartographic change is apparent in the works of Ibn Khurradādhbih (c. 820–911), the director of posts and intelligence in Baghdad and Samarra. Around 846, Ibn Khurradādhbih produced one of the first books known under the title *Kitāb al-masālik wa-al-mamālik* ('Book of Routes and Provinces'). Although his book openly acknowledged Ptolemy and contained no maps, it marked a change in Islamic geographical awareness about the appearance of the known world. In contrast to the *ṣūrat al-arḍ* tradition, the *Kitāb al-masālik* reflects Ibn Khurradādhbih's involvement in the movement of trade, pilgrims and postal correspondence throughout the provinces of the *dār al-Islām* and the growth of the empire under a centralized authority. The book shows little interest in the regions of non-Islamic sovereignty, known as *dār al-ḥarb*, and virtually no trace of the Greek *oikoumenē*. Instead, it concentrates on postal and pilgrimage routes, as well as measuring distances throughout the Islamic world. The sea route to China is described, but otherwise Ibn Khurradādhbih is primarily interested in places that have a direct bearing on the Islamic world.[16]

By the end of the ninth century Islam found itself being pulled in two different geopolitical directions. At the same time as centralizing itself under the Abbasid Caliphate in Baghdad, Islam's rapid expansion across the inhabited world inevitably led to division and secession. The most obvious conflict came with the rise of the Umayyad Caliphate in al-Andalus, but tenth-century dynasties like the Fatimids, the Seljuk Turks and the Berber Almoravids all created their own hereditary states that began to challenge Abbasid supremacy. By the time al-Idrīsī was compiling his *Entertainment*, the *dār al-Islām* was composed of at least fifteen separate states.[17] Although each was nominally Muslim, many were either openly hostile or indifferent to political or theological rule from

A HISTORY OF THE WORLD IN TWELVE MAPS

Baghdad. This dispersal of a centralized authority had obvious consequences for mapmaking, the most significant of which was the further
erosion of Greek traditions and the heightened interest in the portrayal
of routes and provinces recommended by Ibn Khurradādhbih, which
now became more important than ever in understanding an increasingly
diffuse Muslim world. The result was a noticeably different kind of
world mapping that no longer centred itself on the Abbasid Caliphate
in Baghdad, but placed the Arabian peninsula in the centre of the world,
with Mecca and the Kā'aba, the holiest site in the Islamic faith, at its
heart.

This tradition of mapmaking is usually referred to as the Balkhī
School of Geography, taking its name from a scholar born in north-
eastern Iran, Abū Zayd Aḥmad ibn Sahl al-Balkhī (d. 934). Little is
known of al-Balkhī's life and career except that he spent most of his life
in Baghdad and wrote a short commentary on a series of maps, entitled
Ṣuwar al-aqālīm (or 'Picture of the Climates'), none of which have survived. His work nevertheless influenced a later group of scholars who
all produced regional and world maps explicitly indebted to him.

The Balkhī tradition drew on Ibn Khurradādhbih's example of compiling detailed geographical itineraries, with the crucial difference that
they also added maps. One of al-Balkhī's disciples wrote that his master
'intended in his book chiefly the representation of the earth by maps',[18]
and the importance of these maps soon developed a format that looks
so close to a modern-day atlas that one critic described them as representing an 'Islam-atlas'.[19] Al-Balkhī's followers produced treatises that
contained a world map, preceded by maps of the Mediterranean, the
Indian Ocean and the Caspian Sea, and then up to seventeen regional
maps of the Islamic empire as it appeared in the tenth century. The
regional maps are rectangular with no projection or scale, although
they do offer distances between places, measured in terms of marḍalah,
or a day's journey. In contrast, the world maps are circular, although
they are similarly indifferent to longitude, latitude, scale or projection.
Geometry no longer informs their outlines, although the land and its
features are drawn using straight lines, circles, semicircular arcs, squares
and regular curves. The Greek klimata have been replaced by provinces
which are labelled iqlīm, a sign of just how far the Greek tradition had
been absorbed into Islamic conceptions of territory. The maps also
restricted themselves to depicting the Islamic world, with little or no

interest in the *dār al-ḥarb*. And virtually all of these maps, both regional and global, place south firmly at the top.

One of the Balkhī School's most sophisticated practitioners was Abū al-Qāsim Muḥammad ibn Ḥawqal (d. *c.* 367/977). Born in Iraq, Ibn Ḥawqal travelled extensively throughout Persia, Turkestan and North Africa. He is best known for his *Kitāb ṣūrat al-arḍ* ('Picture of the Earth'), which, in acknowledgement of its debt to the more recent Islamic geographical writing, is also known, like Ibn Khurradādhbih's book, as *Kitāb al-masālik wa-al-mamālik*.

As well as illustrating his text with regional maps, Ibn Ḥawqal also drew world maps, the first of which exemplifies the Balkhī School's perception of world geography, forgoing projections and climates and focusing almost exclusively on the Islamic world. The map is oriented with south at the top, although elements of Ptolemy are still recognizable. The world is surrounded by an encompassing sea, with the unseen, other side of the sphere understood as uninhabitable and composed purely of water. The inhabited world is divided roughly into three: the largest land mass, Africa, dominating the top half, Asia, occupying the bottom left-hand corner, and Europe, squeezed into the bottom right-hand side. In Africa, the most prominent feature is the Nile, curving up through East Africa to its apparent source in the Mountains of the Moon. Egypt, Ethiopia and the Muslim states of North Africa are all clearly labelled, in contrast to Europe, where only Spain, Italy and Constantinople are conspicuous. Unsurprisingly, Asia, including Arabia, the Red Sea and the Persian Gulf, is shown in considerable detail and broken down into separate administrative regions. Further east, as Islamic influence wanes, the geography becomes sketchier. Although China and India are shown, their outlines are utterly notional, and Taprobana (modern-day Sri Lanka), whose name originated with the Greeks, is not even depicted; indeed, the Indian Ocean is divested of any islands whatsoever. This is a new map of the world, dominated by Islam and shaped by its administrative and commercial interests.

As the world map of Ibn Ḥawqal shows, the geography of mapping provinces and focusing on religious sites and trade routes began to predominate. It became imperative to establish what the Balkhī mapmakers called *ḥadd* – defining internal boundaries between one Muslim state and the beginning of another. As Baghdad's political and theological power diminished, the Balkhī mapmakers shifted the centre of their maps away from the caliphate's capital, and, in a decisive moment of geographical Islamicization, placed Mecca at the centre of the known world. These mapmakers made the first sustained attempt to provide a detailed physical geography of the Islamic world, a region that since Ptolemy had been mapped with only limited degrees of success. It was this shift from Greek geometry to a definably Islamic physical geography that had such a noticeable effect upon the mapmaking of al-Idrīsī.

Of all the mapmakers described in this book, none boasts a more distinguished lineage than al-Sharīf al-Idrīsī. In Islam, the term 'sharīf' (meaning 'noble' or 'illustrious') denotes a descendant of the prophet Muhammad through his daughter Fatima. As his name suggests, al-Idrīsī was descended from the powerful Shi'ā Idrisid dynasty, founders of the first Islamic state in Iberia in AD 786, and rulers of much of Morocco throughout the ninth century, which traced its lineage back to the establishment of the Umayyad Caliphate in Damascus in the late seventh century. In 750, remnants of the Umayyad dynasty, including the Idrisids, fled Damascus after their defeat at the hands of the Abbasids, and settled in Iberia and North Africa, establishing a rival Caliphate in Córdoba. The new caliphate went on to conquer most of the Iberian peninsula, as well as absorbing the Idrisid dynasty in 985 as it declined under the pressure of internecine feuds. Al-Idrīsī's direct ancestors were the Ḥammūdids, rulers of the area around modern-day Malaga. By the time he was born, in 1100, probably in Ceuta, on the tip of North Africa (the Ḥammūdids's final stronghold), al-Idrīsī's family would have been very familiar with the violent dynastic and religious factionalism of their faith.

The records that remain of al-Idrīsī's life are sparse and often contradictory. Debate continues as to his place of birth, some suggesting Spain, others Morocco or even Sicily, but all the evidence suggests that he was educated in Córdoba. At its height during the eighth and ninth centuries as the capital of the Umayyad Caliphate, Córdoba was one of the largest cities in the world, with a population estimated at more than 300,000. It boasted the world's third largest mosque, founded in 786, and was home to what can claim to be the first university in Europe, which produced some of the greatest minds of the medieval world, including the Muslim philosopher Ibn Rushd (Averroes) and the Jewish rabbi, philosopher and physician Moses ben Maimon (Maimonides).[20] The city was another early example of *convivencia*, as Muslim,

Christian and Jewish scholars were given relative freedom to establish Córdoba as the intellectual (if no longer political) rival to Abbasid Baghdad.

According to one Islamic commentator of the time, Córdoba became 'the homeland of wisdom, its beginning and its end; the heart of the land, the fount of science, the dome of Islam, the seat of the imām; the home of right reasoning, the garden of the fruits of ideas'.[21] It was an understandable description: the Umayyads supported more than 400 mosques, 900 baths, 27 free schools, and a royal library of 400,000 volumes that rivalled the great collections in Baghdad and Cairo. As well as being a centre for the study and practice of Islamic jurisprudence, the city's schools and university taught science and a variety of other subjects, ranging from medicine and astronomy to geography, poetry and philology (including a thriving industry in translating classical Greek texts into Arabic).

Writing more than thirty years later in his *Entertainment* about the city where he was educated, al-Idrīsī called it 'the most beautiful jewel of al-Andalus'.[22] But by the time he arrived there the caliphate was a distant memory, having collapsed in 1031, giving way to a series of petty claimants, before being finally taken over in 1091 by the Almoravids, a Berber dynasty deeply distrusted by the Córdobans by the time al-Idrīsī began his studies, but who nevertheless represented the only hope of salvation in the face of the growing threat of the Christian *reconquista* moving southwards. While al-Idrīsī absorbed the multicultural learning the city had to offer, he also learnt how quickly the political geography of the Islamic world around him could change.

Al-Idrīsī's decision to leave Córdoba was a wise one. Caught between its Almoravid occupiers and the advancing Christian armies of Castile, the city's future must have looked bleak (and by 1236 it had fallen to Castile's armies). By the 1130s he was on the move. He travelled throughout Asia Minor, France, England, Morocco and the rest of al-Andalus. No contemporary records have survived to explain the reasons for his arrival in Sicily around 1138. It may have been that Roger's interest in al-Idrīsī was motivated more by political rather than by intellectual considerations: throughout his reign the Norman king annexed parts of the North African coastline (including Tripoli) and installed puppet rulers of Islamic descent; the possibility of using such a distinguished Muslim nobleman as al-Idrīsī in this way may have appealed to him.[23] Indeed, the Hautevilles already had a record of sheltering his Hammūdid kinsmen: when Muhammad ibn 'Abd Allah, the last of the Hammūdid rulers, fled Malaga in 1058, he was given refuge on Sicily by Roger II's father, Roger I, count of Sicily.[24] Writing in the fourteenth century, the Damascene scholar al-Safadī (1297–1362) gave one account of Roger's motives for sheltering al-Idrīsī:

> Roger, king of the Franks and lord of Sicily, loved learned men of philosophy, and it was he who had al-Sharif Al-Idrīsī brought to him from North Africa . . . When he arrived Roger welcomed his guest ceremoniously, making every effort to do him honour . . . Roger invited him to stay with him. To persuade him to accept, he told him: 'You are from the Caliphal house, and if you were under Muslim rule their lords would seek to kill you, but if you remain with me you will be safe'. After Al-Idrīsī had accepted the king's invitation, the latter granted him an income so large as to be princely. Al-Idrīsī was accustomed to ride to the king on a mule, and when he arrived Roger stood up and went to meet him, and then the two sat down together.[25]

This is the only surviving account of the first meeting between the two men, written nearly 200 years after the event. It is couched in the timeless language of the wise, beneficent patron and his silent, grateful subject. But it also grasps something of Roger's shrewd ability to conflate politics with learning, and his awareness that al-Idrīsī's lineage made him as much a target for his co-religionists as for the king. Both men had, for very different reasons, learnt to accommodate the customs and rituals of other cultures in an era which officially frowned on such behaviour. Both were strangers in a strange land, hundreds of kilometres away from their homelands. And both were far from orthodox in their approach to religion.

The ruler that al-Idrīsī met on his arrival in Palermo inherited an enduringly ambivalent relationship towards his religious faith, and a healthy scepticism towards the political claims made in its name. The Normans had claimed parts of southern Italy and Sicily from Byzantine rule since the mid-eleventh century, taking control of Calabria, Apulia, Reggio and Brindisi, despite continued opposition from virtually every power in Christendom, all with their own vested interests in these territories. The papacy was understandably suspicious of Norman domination of the states south of Rome, while the Hohenstaufen dynasty in

Germany, which also claimed areas of Italy, also objected to the Haute-villes encroaching on their territory. Even the Byzantine emperors of Constantinople reacted angrily to what they regarded as Hauteville usurpation of its traditional rights to Sicily, condemning Roger as a 'tyrant'.[26]

Despite the forces ranged against him, Roger had proved a wily opponent. In 1128, just before al-Idrīsī arrived in Palermo, Pope Hono-rius II refused to sanction Roger's claims to Apulia, and even went so far as to issue a bull of excommunication and encourage a crusade against him. When this failed, weakening Honorius's position, he reluc-tantly agreed to endorse Roger's Italian claims. Following Honorius's death in February 1130, Roger exploited the confusion that arose from the subsequent papal schism, supporting the Roman-based Anacletus II against his rival claimant Innocent II. In an attempt to ensure Roger's military support, the politically weakened Anacletus issued a papal bull later in 1130 conferring the title of king of Sicily upon him, but in 1138 Roger's kingdom was plunged into yet another crisis. Pope Anacletus died, leading to the accession of Innocent II with the support of the Ger-man rulers who were so hostile to Roger's Sicilian reign. Roger was confronted with yet another pope implacably opposed to him. The fol-lowing year Innocent excommunicated Roger once again, but in a subsequent military skirmish was captured by Roger's forces. He was faced with the humiliation of accepting the king's sovereignty and with-drawing support for any future challenges to his rule over Sicily.[27]

Opposition to Roger's rule rumbled on throughout the 1140s. Des-pite having neutralized papal opposition, Roger still faced attempts by the Byzantine and German rulers to oust him, but they all foundered. Then, just as the kingdom entered one of its few relatively stable periods of governance, the Norman ruler and his Muslim subject began work-ing together on the *Entertainment*.

As al-Idrīsī settled into his new life in Palermo, he found an island that enabled him as both a Muslim and a scholar to draw on a wide variety of intellectual traditions. Since Roman times Sicily had secured a reputation for its wealth and prosperity. Like Ptolemy's Alexandria, its position between different Mediterranean cultures and traditions ensured its commercial affluence and political importance. The island was a stopping-off point for political leaders travelling between Rome and Constantinople, and its ports welcomed traders of all faiths from across the Mediterranean. It also acted as a safe haven for both Christian and Muslim pilgrims. Spanish Muslims embarking on the *haj* to Mecca often broke their journey in Sicily's ports, as did European Christians heading for the Holy Land. Travelling through Sicily from Valencia to Mecca in 1183, the Spanish Muslim Ibn Jubayr wrote that 'the prosperity of the island surpasses description. It is enough to say that it is the daughter of Spain [al-Andalus] in the extent of its cultivation, in the luxuriance of its harvests, and in its well-being, having an abundance of varied produce, and fruits of every kind and species.' In describing the peaceful coexistence of the Muslim community with its Christian rulers, Ibn Jubayr even went so far as to approvingly quote a verse from the Qur'ān, observing that 'The Christians treat these Muslims well and "have taken them to themselves as friends" [Qur'ān, 20, 41], but impose a tax on them to be paid twice yearly'. He marvelled at the Norman court's 'splendid palaces and elegant gardens', and concluded that it exercised its legal, administrative and regal authority 'in a manner that resembles the Muslim kings'.[28]

Such a mixed heritage enabled Sicily to establish itself as a centre of learning by the time Roger was crowned king in 1130. Salerno was already a renowned centre for the diffusion of Greek and Arabic medical learning throughout the Latin-speaking world long before Roger annexed it as part of his Italian empire. Roger's chancery produced official proclamations in Latin, Greek and Arabic, which ensured that there was a steady stream of suitably qualified scholars able to continue a flourishing tradition of translating and disseminating such texts to and from all three languages. The Greek diplomat and archdeacon of Catania, Henry Aristippus, translated sections of Aristotle's *Meteorology* from Greek into Latin, and produced the first Latin translation of Plato's *Phaedo* during his time on the island. He was also responsible for bringing a Greek copy of Ptolemy's *Almagest* back to Sicily from Constantinople, where it was used as the basis for one of the earliest Latin translations of Ptolemy's astronomical treatise.[29] Roger also sheltered the Greek theologian Nilos Doxapatres, who fled from Constantinople to Palermo around 1140, and commissioned him to write a pro-Byzantine manuscript on 'The Orders and Ranks of the Patriarchal Thrones', described as an 'historical geography of the ecclesiastical world'.[30] In

Arabic, Roger patronized at least six poets to write in praise of his political and cultural achievements.[31]

The polyglot culture of Palermo, and the diverse range of intellectual traditions on which it could draw, made it the ideal place to complete the ambitious task that Roger was about to set al-Idrīsī. In the preface to the *Entertainment*, al-Idrīsī described the genesis of the king's commission. It was, unsurprisingly for Roger, initially conceived as an exploration of political geography. The king

> wished that he should accurately know the details of his land and master them with a definite knowledge, and that he should know the boundaries and the routes both by land and by sea and in what climate they were and what distinguished them as to seas and gulfs together with a knowledge of other lands and regions in all seven climates whenever the various learned sources agreed upon them and as was established in surviving notebooks or by various authors, showing what each climate contained of a specific country.

This was the most ambitious study of physical geography proposed since Ptolemy's tables identifying more than 8,000 places throughout the inhabited world, including the subsequent surveys undertaken by the Romans (and since lost). The Romans had at least been able to draw on their vast empire, and a relatively unlimited access to Greek geographical texts, to undertake such a project. Roger's tiny kingdom lacked the resources and manpower to complete such a survey, but it could draw on a diverse collection of texts written in Greek, Arabic and Latin. Al-Idrīsī concentrated on two main sources: Ptolemy's *Geography* (available in the original Greek and Arabic translations), and the writings of the early Christian theologian Paulus Orosius. Like al-Idrīsī, Orosius was an itinerant scholar, who had lived and worked throughout Iberia, North Africa and the Holy Land, and whose *History against the Pagans* (416–17) offered a geographical history of the rise of Christianity.

In a determined bid to unify past, present and evolving conceptions of geography, the king took what could be gleaned from Ptolemy and Orosius, put it together with the geographical knowledge of al-Idrīsī and his team of court scholars, and then supplemented it with newly commissioned travellers' reports drawn from across the inhabited world:

They studied together, but he did not find much extra knowledge from [other scholars] over what he found in the aforementioned works, and when he had convened with them on this subject he sent out into all his lands and ordered yet more scholars who may have been travelling around to come and asked them their opinions both singly and collectively. But there was no agreement among them. However, where they agreed he accepted the information, but where they differed, he rejected it.[32]

Over subsequent years, Roger's scholars painstakingly collated information. Where there was agreement on particular matters, the results were entered on a large drawing board, from which a vast map of the world slowly began to emerge:

He wished to make sure of the accuracy of what these people had agreed upon both of longitudes and latitudes [and in measurements between places]. So he had brought to him a drawing board [lauḥal-tarsīm] and had traced on it with iron instruments item by item what had been mentioned in the aforementioned books, together with the more authentic of the decisions of the scholars.[33]

The first result of these labours was not a gazetteer in the Ptolemaic tradition, but an enormous circular map of the world, made of silver. Al-Idrīsī tells us that Roger ordered that

a disk [dā' ira] should be produced in pure silver of a large extent and of 400 Roman raṭls in weight, each raṭl of 112 dirhams and when it was ready he had engraved on it a map of the seven climates and their lands and regions, their shorelines and hinterlands, gulfs and seas, watercourses and places of rivers, their habited and uninhabited parts, what [distances] were between each locality there, either along frequented roads or in determined miles or authenticated measurements and known harbours according to the version appearing on the drawing board.[34]

Neither this extraordinary silver world map nor the geographical drawing board have survived, but al-Idrīsī explains that, following the completion of the map, Roger commissioned 'a book explaining how the form was arrived at, adding whatever they had missed as to the conditions of the lands and countries'. This book would describe 'all the wonderful things relating to each [country] and where they were with regard to the seven climates and also a description of the peoples and their customs and habits, appearance, clothes and language. The book would be called the *Nuzhat al-mushtāq fī khtirāq al-āfāq*. This was all completed in the first third of January agreeing with the month of Shawwāl in the year A. H. 548.'[35]

The completed book is all that remains of Roger's geographical ambitions. Leafing through it today, it is obvious why the king wanted al-Idrīsī's help. As well as drawing on Greek and Latin geographical sources like Ptolemy and Orosius, the book incorporates the third crucial tradition that al-Idrīsī brought to the project: more than 300 years of Arabic geographical learning. The *Entertainment* represents the first serious attempt to integrate the three classical Mediterranean traditions of Greek, Latin and Arabic scholarship in one compendium of the known world.

Befitting someone not necessarily trained in astronomy and cosmography, al-Idrīsī spent little time describing the origins of the earth, beyond asserting that it was spherical, with a circumference estimated at a reasonably accurate 37,000 kilometres (23,000 miles), and that it remained 'stable in space like the yolk in an egg'. Little of what he said in his preface was particularly searching or innovative, and remained close to the standard Greek and Islamic authorities; it was his method of arranging the diverse information gathered by Roger's contributors that was so unprecedented. Drawing on Ptolemy, al-Idrīsī divided the rest of his book into seven longitudinal climates running east to west, but oriented his map with south at the top. The first clime ran through equatorial Africa all the way to Korea. 'This first climate', he writes, 'begins to the west of the Western Sea, called the Sea of Shadows. It is that beyond which no one knows what exists. There are in this sea two islands, called al-Khālidāt (the Fortunate Isles), from which Ptolemy begins to count longitudes and latitudes.'[36] The final, seventh clime covered modern-day Scandinavia and Siberia. His most daring innovation was to then subdivide each climate into ten sections, which if put together would make a grid of the world composed of seventy rectangular areas. Al-Idrīsī never envisaged unifying his maps in this way – the assembled map would simply be too large to be of any use, even in a ceremonial situation – but it was a new way of executing a geographical description of the whole world. In the *Entertainment*, each of the seventy regional maps followed written descriptions of the regions described, allowing the reader to visualize the territory after first reading about it.

In his preface, al-Idrīsī provided the motivation behind his decision to divide the world in this way, which offers one of the most detailed pre-modern accounts of how maps supplement and enhance written geographical description:

> And we have entered in each division what belonged to it of towns, districts, and regions so that he who looked at it could observe what would normally be hidden from his eyes or would not normally reach his understanding or would not be able to reach himself because of the impossible nature of the route and the differing nature of the peoples. Thus he can correct this information by looking at it. So the total number of these sectional maps is seventy, not counting the two extreme limits in two directions, one being the southern limit of human habitation caused by the excessive heat and lack of water and the other the northern limit of human habitation caused by excessive cold.

This account demonstrated the power of the map to visualize places that the viewer could never imagine ever visiting because of the distances, and dangers, involved. But al-Idrīsī also acknowledged that his regional maps could only go so far in the information they provided. After reiterating the importance of describing physical geography, he goes on to say:

> Now it is clear that when the observer looks at these maps and these countries explained, he sees a true description and pleasing form, but beyond that he needs to learn descriptions of the provinces and the appearance of their peoples, their dress and their adornments and the practicable roads and their mileages and *farsangs* [a Persian unit of measurement] and all the wonders of their lands as witnessed by travellers and mentioned by roaming writers and confirmed by narrators. Thus after each map we have entered everything we have thought necessary and suitable in its proper place in the book.

This eloquent statement on both the power and the limitations of map-making acknowledged the importance of giving a 'form', or geometrical order to the inhabited world, just as Ptolemy had described it, but it also implicitly conceded the problem of *akoē* ('hearsay') provided by 'roaming writers'. Travellers' reports were clearly necessary for the detailed human geography that Roger wanted, but how could these reports be verified and 'confirmed by narrators'? To al-Idrīsī, the basic

geometry of the map was unquestionable and could be reliably reproduced, unlike the partial accounts provided by even the most experienced
traveller.

Al-Idrīsī was stuck with the same problem voiced by Herodotus more
than 1,500 years earlier. His solution went against the grain of the cartographic traditions inherited from the classical world and early Muslim
mapmaking, and took a non-scientific route to depict the local reality of
the inhabited world. This would produce one of the most exhaustively
detailed geographical descriptions of the medieval world, but would
also leave his work neglected and dismissed, as political ideology
embraced increasingly moralized cartographic visions of the world.

How al-Idrīsī responded to the preceding history of mapmaking
from al-Mā'mūn's court to Ibn Ḥawqal is complicated because he says
relatively little about his sources, and also because of the problems of
the circulation and exchange of ideas within his culture of manuscripts.
We are reliant on the later scribal copies of the *Entertainment* (as well
as its maps) to assess his achievements. Similarly, his education and
early career at the westernmost limits of the Islamic world makes it hard
to interpret just which texts might have reached him, either in Córdoba
or Sicily. Is his apparent silence on the influence of someone like
al-Masūdī pure ignorance, or does it represent some more obscure intellectual or ideological conflict? We may never know. But by piecing
together the sources he does cite, along with his maps and written geographical descriptions, it is possible to offer some idea of what he was
trying to achieve.

In the preface to the *Entertainment* al-Idrīsī claims that, among
his other sources, he has drawn on Ptolemy, Paulus Orosius, Ibn
Khurradādhbih and Ibn Ḥawqal.[37] It is a revealing list: a Greek, a Christian and two Muslims, one an administrator, the other an inveterate
traveller. Reading al-Idrīsī and looking at the maps drawn from his text,
it appears that no one source predominates. He borrows from everyone,
while tacitly acknowledging their limitations by reaching his own conclusions. Having drawn on Ibn Khurradādhbih for his theoretical
understanding of the earth's shape, circumference and equatorial dimensions, he then turns back to Ptolemy in describing and drawing climes
and by extension the regional dimensions of his maps.

In the ensuing text and maps describing his seventy regions, al-Idrīsī
moves seamlessly between Ptolemy and his Muslim sources, often

describing places and estimating their locations at variance to their position on his maps. The written chapters describe routes and distances between places located on each map, for instance, 'Mecca to Medina, also called Yathrib, by the most convenient route, is 6 days' journey', or 415 kilometres. The conclusion of the route shows how closely al-Idrīsī turns from Ptolemy back to Ibn Khurradādhbih, this time drawing on his predecessor's administrative and practical interests:

> From Sabula to Mêlée, a halting place where there are springs of sweet water, 27 kilometres.
> From there to Chider, a meeting place for the inhabitants of Medina inhabited by a small number of Arabs, 19 kilometres.
> From Chider to Medina, 11 kilometres.[38]

The map showing Mecca betrays few signs of its sacred significance, and neither does the description that accompanies it. 'Mecca', writes al-Idrīsī, 'is a town so old that its origins are lost in the night of time; it is famous and flourishing, and people come there from every corner of the Muslim world.' The description of the Kā'aba is similarly prosaic. 'Tradition relates that the Kā'aba was the dwelling of Adam and that, being constructed of stone and clay, it was destroyed in the Flood and remained in ruins until God commanded Abraham and Ishmael to rebuild it.'[39] This is not the sacred geography of either contemporary Christian *mappaemundi* (discussed in the next chapter) with Jerusalem as the divine centre of the world, or the Mecca-centred mapmaking of the Balkhī School. Instead it offers a naturalistic description of the physical world, full of marvels and miracles, but with little apparent interest in a founding act of divine creation.

When al-Idrīsī turned his attention to the capital of the caliphate, Baghdad, his account was similarly muted. 'This great city', he writes, 'was established on the west bank of the Tigris by the caliph al-Mansur, who divided the surrounding territory into fiefs that he then distributed among his friends and followers.'[40] In direct contrast, the great cities of Christendom are celebrated in minute detail. Rome is described as 'one of the pillars of Christianity and first among the metropolitan sees', celebrated for its classical architecture, thriving markets, beautiful squares, and more than 1,200 churches, including St Peter's. Al-Idrīsī also writes of 'the palace of the prince called the pope. This prince is superior in power to all the kings; they respect him as if he were equal to the Divinity. He rules with justice, punishes oppressors, protects the poor and weak, and prevents abuses. His spiritual power exceeds that of all the kings of Christianity, and none of them may oppose his decrees.'[41] If al-Idrīsī was deliberately downplaying Islamic locations in favour of Christian ones to please Roger, this was hardly the version of papal authority that the king wanted to hear.

But it is in his account of Jerusalem that a subtly syncretic perspective on geography begins to emerge from al-Idrīsī's's book. He chronicles the city's entwined Jewish, Christian and Muslim theological histories, including repeated references to Christ as the 'Lord Messiah', describing his life geographically from the Nativity to the Crucifixion. In a remarkable passage on the Temple Mount, or the Noble Sanctuary in Islam, al-Idrīsī describes it as

> the holy dwelling that was built by Solomon, son of David, and which was a place of pilgrimage in the days of Jewish power. This temple was then snatched from them and they were expelled from it at the time of the arrival of the Muslims. Under Muslim overlordship it was enlarged, and it is today the mosque known to the Muslims by the name of Masjid al-Aqsa. There is not one in the world that exceeds it in greatness, with the exception of the great mosque of Córdoba in Andalusia; for, according to report, the roof of that mosque is bigger than that of Masjid al-Aqsa.[42]

Here is the holiest site in Judaism, the third holiest site in Islam after Mecca and Medina, the 'Farthest Mosque', named after the Prophet's visionary journey from Mecca to Jerusalem city on a flying horse, following which it was briefly adopted as the Muslim *qibla*. But in describing the edifice on which the mosque was erected, al-Idrīsī also reminds his readers that in 1104 'the Christians took possession of it by force and it has remained in their power until the time of composition of the present work'. As in al-Idrīsī's career, no one religion predominates; his identity as a Muslim is stated throughout the *Entertainment*, but he appears indifferent to the valorization of one intellectual or religious tradition over another.

The *Entertainment* clearly magnifies Roger's place on the world map. Sicily – described as 'a pearl of pearls' – looms larger than any other island in the Mediterranean, its ruler eulogized as 'adorning imperium and ennobling sovereignty'.[43] But this is the result of political exigency, and a typical example of egocentric mapping, whereby al-Idrīsī magnifies both his own location and that of his sovereign. At a more basic level, neither the geometry of Ptolemy nor the sacred geography of the Balkhī School of mapmaking takes precedence in the *Entertainment*. None of al-Idrīsī's maps contains a scale or consistent measurement of distances. In contrast to the maps drawn by Ibn Ḥawqal, al-Idrīsī's maps depict a world without *ḥadd*, the Islamic term for limits, boundaries or the end of a particular city, country or land mass.[44] Roger's continued patronage of the project over so many years indicates he was pleased with it as political geography, but for al-Idrīsī his *Entertainment* was clearly something else: *adab*, the refined and cultured pursuit of scholarly works of edification, recreation – or entertainment. An *adīb* – someone who possessed *adab* – sought to know something about everything, and the encyclopedic geography book represented one of the best vehicles for its expression.[45]

The much-vaunted spirit of *convivencia*, the multi-cultural exchange and transmission of objects, ideas and beliefs that gave rise to al-Idrīsī's *Entertainment,* was a transitory phenomenon. As it began to unravel towards the end of Roger's life, al-Idrīsī's geographical achievements were left stranded, a result of the growing ideological polarization between Christians and Muslims which left little room for a Muslim mapmaker at a polyglot Christian court. In 1147, as al-Idrīsī was compiling the *Entertainment*, Roger enthusiastically supported plans for the Second Crusade, with the ultimate aim of driving the Muslims out of Jerusalem. Cunning as ever, Roger was planning to exploit his involvement in the crusade to further his own political cause, but it was also a sign of the times that he found it increasingly difficult to sidestep the growing confrontation between the two faiths.

At his death in 1154, Roger was succeeded by his son, William I. Although William continued his father's enthusiastic patronage of learning, he lacked Roger's political acumen. According to one contemporary account of William's reign, 'after only a short time, all this tranquillity slipped away and disappeared', and the Sicilian kingdom soon collapsed into factionalism and internecine conflict.[46] Perhaps, just as al-Idrīsī had fled Córdoba as a young man, he understood that a moment had passed, and he left Sicily on one final journey, back to North Africa, probably Ceuta, where he died in 1165, aged 65. His departure coincided with growing Muslim rebellion against their Norman masters. Roger's nephew Frederick II, Holy Roman Emperor and king of Sicily (r. 1198–1250), took a very different approach to the island's Muslim community, deporting many of them. He also took up the mantle of Holy Crusade, leading the Sixth Crusade which culminated in his coronation as king of Jerusalem in 1229. By the time of his death the last of the island's Muslims were either in exile or had been sold into slavery. The Norman experiment of *convivencia* on the island had come to a bitter end and with it the eradication of the Muslim presence in Sicily for ever.[47]

The shifting cultural boundaries of the late twelfth-century Mediterranean world and the climate of amicable intellectual exchange they once created meant that al-Idrīsī's geographical legacy was limited. It is difficult to imagine how such a large and complex book like the *Entertainment* could have been easily transmitted from Sicily throughout the Islamic world, and in any case many Muslim scholars regarded al-Idrīsī as a renegade from his own faith. Some later Islamic writers drew on his writing and copied his maps, including the famous North African scholar Ibn Kaldūn (1332–1406), whose family had also fled the slow disintegration of al-Andalus. His monumental world history, the *Kitāb al-'ibar*, compares al-Idrīsī's maps with Ptolemy in describing 'the mountains, seas, and rivers to be found in the cultivated part of the world'.[48] Otherwise, circulation of al-Idrīsī's work was confined to the scholarly circles of North Africa. Although an abridged Latin version of the *Entertainment* was printed in Rome in 1592, it was by then regarded as an historical curiosity, and dismissed as an example of the backwardness of Islamic geography.

In the late twentieth century, as scholars began to reconsider the significance of Islamic cartography, al-Idrīsī's reputation was slowly rehabilitated. The importance of his mapmaking, and in particular the significance of his circular world map, might have continued to grow, were it not for an extraordinary recent discovery. In June 2002, the Department of Oriental Collections of the Bodleian Library in Oxford acquired an Arabic manuscript that shed new light on the development of Arabic geography, and challenged established assumptions about al-Idrīsī's world map. Based on its author's political and dynastic references, the original manuscript can be dated to the eleventh century, but it survived as an early thirteenth-century copy, probably made in Egypt.

Its author remains unknown, but the title, when translated into English, puts it tantalizingly in the same descriptive genre as al-Idrīsī's *Entertainment*.

Entitled *The Book of Curiosities of the Sciences and Marvels for the Eyes*, the book is composed of thirty-five chapters written in Arabic describing the celestial and terrestrial worlds. Of even greater significance is the fact that the treatise contains no fewer than sixteen maps, depicting the Indian Ocean, the Mediterranean, the Caspian Sea, the Nile, the Euphrates, the Tigris, the Oxus and the Indus. Other maps include Cyprus, North Africa and Sicily. The earliest chapters are also illustrated with two world maps, one rectangular and one circular, both remarkable in their own right. The rectangular world map is unlike any other known Islamic map. It is highly schematic, oriented with south at the top, showing the world effectively composed of two vast continents, Europe to the right and Asia conjoined with a limitless Africa to the left. The Arabian peninsula is particularly prominent, with Mecca depicted like a golden horseshoe. The map also contains a scale bar that bears a striking resemblance to Suhrāb's method for projecting a world map onto a plane surface. It runs from the map's top right to left, ending somewhere along the East African coast. Although the copyist clearly did not understand the graticule (it is incorrectly numbered), its presence suggests a hitherto unknown level of sophistication in the measurement of distances and the application of scale to Islamic world maps.[49]

The circular map is more familiar: it is virtually identical to the world map found inserted into at least six copies of al-Idrīsī's *Entertainment*. As the map in the *Book of Curiosities* predates the *Entertainment* by at least a century, it completely undermines the traditional attribution to al-Idrīsī. There are two possibilities to explain its appearance in the *Entertainment*. Either al-Idrīsī copied this map without acknowledging his source and included it in his treatise, or, even more intriguingly, later copyists took the liberty of adding the *Book of Curiosities* map, believing that it somehow complemented the rest of the *Entertainment*. Considering that al-Idrīsī's text never refers to a world map, and as its purely physical representation of the earth is at odds with the rest of the *Entertainment*'s interest in regional human geography, the second of these seems most likely. Whatever the truth, the appearance of the *Book of Curiosities* reveals that the circulation and exchange of maps and geographical ideas across the medieval Muslim world were much earlier and far more extensive than historians have previously believed. Our understanding of medieval mapmaking, of whatever religious denomination, continues to evolve.

The existence of the circular map in the *Book of Curiosities* changes how we see al-Idrīsī's geographical achievements. His method of regionally mapping the inhabited world is one of the great examples of non-mathematical mapping in the pre-modern world, the product of exchanges between not only Christians and Muslims but also Greeks and Jews. Its conventions may not look objective in the modern sense, but they pursued a kind of realism in the ways they mapped space as uniform and relatively free of the religious rhetoric that defined so many maps of his time. Although al-Idrīsī's regional maps and his descriptions of towns, cities, communities, commodities, trade routes and distances across the inhabited world reflect his attempt to unify elements of Christian and Islamic mapmaking, he appears reluctant to endorse either religion's cosmogony, or their claims to universal sovereignty.

Like Ptolemy, al-Idrīsī was drawn to creating a map of the world as an intellectual exercise, a task that ambitious patrons like Roger demanded. But what seems to have excited him was the potentially infinite possibility of regional mapping; he resisted unifying all seventy of his local maps into one global image, because such an image would inevitably beg the question of its creation based on the beliefs of one faith or another. Incrementally mapping the wonder of the earth's physical diversity was unacceptable to subsequent courts and rulers, Christian or Muslim, across the Mediterranean. By the thirteenth century, both sides had turned away from al-Idrīsī, instead demanding maps that provided unequivocal support for their particular theological beliefs. Despite his geographical innovation, neither Christians nor Muslims appreciated the value of his maps, and religious belief triumphed over geographical description.

# 3

# Faith

*Hereford* Mappamundi, c. *1300*

*Orvieto, Italy, 1282*

On 23 August 1282, the bishop of Hereford, Thomas Cantilupe, died at
Ferente, near Orvieto in Italy. A former Chancellor of England and of
Oxford University, canon of London and York, and personal adviser to
King Edward I, Cantilupe was one of the most influential figures in thirteenth-century English ecclesiastical life. In the last years of his life he
became embroiled in a bitter controversy with his superior, John Pecham,
archbishop of Canterbury. Born into the ruling baronial class, Cantilupe
was a firm believer in the established rights of senior clergy to hold multiple benefices – land and property attached to religious titles – a practice
commonly known as pluralism. Pecham was a vociferous critic of pluralism, along with what he regarded as indiscipline, absenteeism and
unorthodox theological teaching. Upon his appointment as archbishop
in 1279, Pecham made it clear to senior clergy, including Cantilupe, that
he intended to stamp out such practices. Pecham represented a new kind
of ecclesiastical authority. He was a firm supporter of the decrees laid
down at the Fourth Lateran Council held in Rome in 1215, which
wanted to formalize Christian doctrine by strengthening the power of its
ruling elite, who were given increased authority to disseminate the basic
points of doctrine to the laity.[1] Pecham enthusiastically endorsed such
reforms, expanding his jurisdiction over the dioceses, but in the process
eroding the authority and privileges enjoyed by many of his bishops.

Pecham was particularly concerned about bringing the Welsh clergy
into line on the issue of pluralism. This was as much a political as a religious matter. Throughout the 1270s and 1280s King Edward was
involved in a long and bitter conflict with independent Welsh rulers in

an attempt to incorporate the realm within England. Situated in the Marches (border regions) between England and Wales, the diocese of Hereford represented the furthest extent of English political and ecclesiastical authority, and Pecham was keen to ensure it abided by his reforms. While Cantilupe remained loyal to King Edward on political matters, he rejected Pecham's attempts to challenge pluralism and other practices deeply embedded in English religious life, and resisted the archbishop's attempts at reforming his diocese. Matters came to a head in February 1282, when the archbishop dramatically excommunicated Cantilupe at Lambeth Palace. The disgraced bishop went into exile in France, and by March 1282 was heading to Rome, to make a direct appeal to Pope Martin IV against his excommunication.[2]

Throughout the summer of 1282 Cantilupe met the pope and made his case. But before the matter could be resolved, Cantilupe's health began to deteriorate, and by August he had departed for England. Shortly after his death at Ferente, Cantilupe's heart was removed and his body boiled to separate the flesh from the bones. The flesh was interred in a church in Orvieto, the heart and bones carried back to England. On their return, Pecham refused permission for Cantilupe's bones to be interred in Hereford until early 1283. Thanks to the efforts of Richard Swinfield, Cantilupe's protégé and successor as bishop of Hereford, the former bishop's bones were finally laid to rest in the cathedral in 1287. The tomb was decorated with soldiers standing with their feet on monstrous beasts, an image of the Church Militant, fighting sin and protecting the virtuous Cantilupe, lying within the Garden of Paradise, and protected by Christ's battalions.[3]

The shrine was the beginning of a concerted effort by Swinfield to have his mentor canonized, and he cultivated Cantilupe's tomb as a site of pilgrimage for the faithful from across the country. Between 1287 and 1312, more than 500 'miracles' were associated with it, ranging from cure of the mad and the crippled, to the miraculous revival of children believed drowned, the recovery of a knight's favourite falcon trampled to death by his squire, and the restoration to a Doncaster man of the power of speech even though his tongue had been cut out by robbers. Finally, in 1320, after repeated petitions to the papal curia, Cantilupe was granted saintly status, the last Englishman before the Reformation to receive such an honour.

*

The story of Cantilupe's career and his conflict with Pecham over matters of ecclesiastical authority encapsulates the vicissitudes of faith in thirteenth-century Catholic England. But today, Cantilupe's life, and his final resting place, the base of which can still be seen in the north transept of Hereford Cathedral, is largely forgotten. Most tourists who make the secular pilgrimage to the cathedral walk straight past Cantilupe's tomb and head instead for the modern annex behind the church, designed to hold its most famous relic: the Hereford *mappamundi*.

The term *mappamundi* comes from the Latin *mappa* – a tablecloth or napkin – and *mundus* – the world. Its development in the Christian Latin-speaking West from the late eighth century did not always refer specifically to a map of the world; it could also designate a written geographical description. Similarly, not all world maps from this period were called *mappaemundi* (the plural of *mappamundi*). Other terms were also used, including *descriptio, pictura, tabula*, or, as in the case of the Hereford map, *estoire*, or history.[4] Just as geography was not recognized at this time as a distinct scholarly discipline, so there was no universally accepted noun in Latin or European vernacular languages to describe what we would now call a map. Of all the terms in circulation, however, *mappamundi* became the most common term to define a written and drawn account of the Christian earth for nearly 600 years. Of the 1,100 *mappaemundi* that survive today, the vast majority are to be found in manuscript books, some just a few centimetres in size, illustrating the writings of some of the most influential thinkers of the time: the Spanish cleric and scholar Isidore of Seville (*c*. 560–636), the late fourth-century writer Macrobius and the fifth-century Christian thinker Paulus Orosius. The Hereford *mappamundi* is unique; it is one of the most important maps in the history of cartography, and the largest of its kind to have survived intact for nearly 800 years. It is an encyclopedic vision of what the world looked like to a thirteenth-century Christian. It offers both a reflection and a representation of the medieval Christian world's theological, cosmological, philosophical, political, historical, zoological and ethnographic beliefs. But although it is the greatest medieval map in existence, it remains something of an enigma. We do not know exactly when it was made, nor its exact function within the cathedral; nor are we certain why it is to be found in a small cathedral town on the Anglo-Welsh border.

Going to Hereford today and walking into the cathedral's annex to examine the *mappamundi*, the visitor is first struck by just how alien it appears as an object, never mind as a map. Shaped like the gable end of a house, the map undulates and ripples like some mysterious animal – which, in effect is what it is. Measuring 1.59 metres (5 feet 2 inches) high, and 1.34 metres (4 feet 4 inches) wide, the map was made from one enormous animal skin. The shape of the animal is still discernible, from its neck, which forms the map's apex, to its spine, which runs down the middle of the map. At one glance, the map can look like a skull, or cross-section of a cadaver, with its veins and organs on display; with another look it could be a strange, curled animal. Gone are the grids of measurement found in Ptolemy and al-Idrīsī. Instead, this map emanates an almost organic aura, embodying a chaotic, teeming world, full of wonders, but also edged with horrors.

Most of the parchment contains a circular depiction of the world, portrayed within one vast sphere, encircled with water. Looking at the map's distribution of land masses and geographical orientation only leaves the modern viewer alienated and confused. The earth is divided into three parts, picked out in gold leaf on the map as 'Europa', 'Asia' and 'Affrica'.[5] The titles of Europe and Africa have been transposed, which says something either about the limitations of thirteenth-century geographical knowledge, or that the map's scribe experienced profound embarrassment when it was finally unveiled (unless there is a more obscure intention to show a deliberately confused image of the world in contrast to reality). The cardinal directions are represented on the map's outer ring from the top, moving clockwise, as *Oriens* (east, the rising sun), *Meridies* (south, the position of the sun at midday), *Occidens* (west, the setting sun), and *Septemtrio* (north, from the Latin for seven, referring to the seven stars of the Plough in the Great Bear, by which the direction of north was calculated). Where the world map in al-Idrīsī had placed south at the top, the Hereford *mappamundi* reorients the world with east at the top. But just like al-Idrīsī's map, Asia fills nearly two-thirds of the whole sphere on the Hereford *mappamundi*. To the south in the right-hand corner of the map is Africa, with its southern peninsula incorrectly shown as joined to Asia. Europe is to the west in the bottom left-hand corner, with present-day Scandinavia to the north. Asia takes up the rest of the map.

To reorient the *mappamundi* according to today's geography the viewer has to mentally turn it 90° clockwise, with the apex facing to the right, but even then its topography remains unfamiliar. Most people standing in front of the *mappamundi* try to get their bearings by look-ing for Hereford, but this hardly helps. The town is on the map, as is the River Wye (labelled 'wie'), alongside important thirteenth-century set-tlements like Conway and Carnarvon, but it lies on a barely recognizable, sausage-shaped island labelled 'Anglia', squashed into the bottom left-hand corner. Although the whole of the British Isles seems incom-prehensible to a modern eye, its toponymy reveals some strikingly modern conflicts over regional and national identity which are still with us today. Anglia is written in red to the north-east of Hereford, but fur-ther to the south the same island is also labelled 'Britannia insula', or the island of Britain. Wales, or 'Wallia' looks as if it is hanging to Eng-land (or Britain?) by a thread, while Ireland ('Hibernia') floats off on the very border of the map like a sinister crocodile, and appears to be almost split in two. To the north, Scotland ('Scotia') is shown as com-pletely separate from England.

Crossing the narrow arc of water to 'Europa', things do not get any clearer. The continent is also barely recognizable, a horn-shaped wedge riven by waterways snaking through the land, which is mainly distin-guished by the depiction of mountain ranges, trade routes, religious sites and major cities like Paris, curiously slashed and scratched (perhaps due to age-old anti-French sentiment), and Rome, emblazoned as 'head of the world'. The base of the map shows an island on which sit two clas-sical columns, with the legend, 'The Rock of Gibraltar and Monte Acho are believed to be the Columns of Hercules', established by the Greek hero as the westernmost point of the known classical world. Just to their left, on mainland Spain, just above Córdoba and Valencia, a legend reads 'Terminus europe'. From the Columns of Hercules, the Mediterra-nean runs back up the map's spine, littered with islands labelled with a mishmash of classical information. Minorca is described as the place where 'slings were first discovered', while Sardinia is, according to the map, 'called "Sandaliotes" in Greek from its similarity to the human foot'. The most prominent island is Sicily, the home of al-Idrīsī, floating off the African coast and directly adjacent to a castle portraying 'Mighty Carthage'. The island is depicted as an enormous triangle, with a legend offering precise distances between its three promontories. Just above Sicily lies Crete, dominated by what it describes as 'the labyrinth: that is, home of Daedalus'. In classical mythology the Athenian inventor Daedalus built the labyrinth to imprison the Minotaur, the monstrous offspring of Queen Pasiphae, wife of the island's king, Minos. Above Crete the Mediterranean divides itself: to the right, it flows out of the Nile; to the left into the Adriatic and the Aegean. Passing Rhodes and the remnants of its Colossus, one of the seven wonders of the ancient world, the map reaches the Hellespont, the modern-day Dardanelles, and directly above it the capital of the Byzantine Empire, Constantinople. The city is shown in oblique perspective, with its formidable walls and fortifications reproduced with impressive accuracy.

Moving further away from the centre, the map and modern geographical reality increasingly part company. The further up the map one looks, the more settlements become scattered, legends more elaborate, and strange monsters and effigies begin to rear their heads. A lynx stalks across Asia Minor, and we are told 'it sees through walls and urinates a black stone'. Noah's Ark sits further up in Armenia, above which two fearsome creatures march back and forth across India. On the left, a tiger, on the right, a 'manticore', sporting 'a triple set of teeth, the face of a human, yellow eyes, the colour of blood, a lion's body, a scorpion's tail, a hissing voice'. Moving deeper into Asia the map portrays the Golden Fleece, the mythical griffin, scenes of grotesque cannibalism, and an account of the fearsome Scythians, who are said to live in caves and 'make drinking cups out of the heads of their enemies'. Finally, at the left-hand shoulder of the map, at the very limits of the known world, a legend concludes that:

> Here are all kinds of horrors, more than can be imagined: intolerable cold, a constant blasting wind from the mountains, which the inhabitants call 'bizo'. Here are exceedingly savage people who eat human flesh and drink blood, the accursed sons of Cain. The Lord used Alexander the Great to close them off, for within sight of the king an earthquake occurred, and mountains tumbled upon mountains all around them. Where there were no mountains, Alexander hemmed them in with an indestructible wall.

The legend conflates well known biblical and classical versions of the origins of the 'savage people', the tribes of Gog and Magog. These were the monstrous descendants of Noah's son Japheth, scattered to the northernmost parts of the known world. The Book of Revelation predicts that in the Last Days Satan will gather the tribes of Gog and Magog from 'the four quarters of the earth', in a futile assault against Jerusalem (Revelation 20: 8–9). Early Christian and Qu'rānic versions of the exploits of Alexander the Great claim that when the king reached the Caucasus Mountains he forged gates of brass and iron to keep Gog and Magog at bay – a barrier reproduced on the circular world map attributed to al-Idrīsī. For all these traditions, Gog and Magog were the ultimate barbarians, on the literal and metaphorical margins of Christianity, a permanent threat to any civilization.

Moving across to the right-hand side of the map's portrayal of Asia, the map imagines a world no less marvellous and terrifying. Crocodiles, rhinoceroses, sphinxes, unicorns, mandrakes, fauns and a very unfortunate race of people 'with a prominent lip, with which they shade their face from the sun' inhabit the regions to the south-east. In the map's top right-hand corner the red claw-shaped ingress depicts the Red Sea and the Persian Gulf, with Sri Lanka (labelled 'Taphana', or Taprobana according to classical sources) floating at their mouth, rather than off the south-east coast of India. Moving back down the map, a tadpole-shaped river runs along the southern African coast, representing the Upper Nile (wrongly believed to flow underground before rejoining the Lower Nile, portrayed on the map further inland).

To the right of the Nile runs a fantastically elongated Africa, virtually devoid of settlements with the exception of Mount Hesperus on the north-west coast, to the monasteries of St Anthony in the top right-hand corner (in southern Egypt). The portrayal of Africa bears no relation to any geographical reality: its only function seems to be to explain the origin of the Nile and to depict a world of another 'monstrous' people; not Gog and Magog, but their diametrically opposed counterparts on the map's southernmost point. Moving south from Mount Hesperus, the map portrays a range of fantastic creatures, with bizarre features and behaviour, starting with the 'Gangines Ethiopians', who are shown naked, holding walking sticks and pushing each other away. The legend tells us that 'with them there is no friendship'. Hardly monstrous, more antisocial. But further south the map depicts 'Marmini Ethiopians' with four eyes; an unnamed people who 'have mouth and eyes in their shoulders'; the 'Blemmyes', with 'their mouth and eyes in their chest'; the Philli, who 'test the chastity of their wives by exposing their new-borns to serpents' (in other words, murdering illegitimately conceived offspring); and the Himantopods, who bear the misfortune of having to 'creep along more than walk'.

Moving south of where a modern map would locate the equator, the races take on even more monstrous and bizarre characteristics. A bearded figure wearing a turban with a woman's breast and male and female genitalia is labelled a people of 'either sex, unnatural in many ways', above an unnamed individual with 'a sealed mouth', who can only eat through a straw; below are 'Sciapodes, who though one-legged are extremely swift and are protected in shade by the soles of their feet; the same are also called Monoculi'. The map portrays the Sciapodes as not only possessing one leg (with an extra three toes), but also sporting just one eye. Finally, the catalogue of monstrous races ends off the east coast of Africa with 'a people without ears, called Ambari, the soles of whose feet are opposed'.

This is not a map as we understand it in any modern sense. Instead, it is an image of a world defined by theology, not geography, where place is understood through faith rather than location, and the passage of time according to biblical events is more important than the depiction of territorial space. At its centre stands the place that is so central to the Christian faith: Jerusalem, the site of Christ's crucifixion, graphically depicted above the city itself, which is represented with circular walls, rather like a giant theological cog. It takes its position at the heart of the map from God's pronouncement in the Old Testament's Book of Ezekiel: 'This is Jerusalem: I have set it in the midst of the nations and countries that are round about her' (Ezekiel 5: 5). The layered theological geography of al-Idrīsī's description of the city has gone, and is instead replaced by an exclusively Christian vision.

Tracing the map's topography outwards from Jerusalem in terms of theology rather than geography, we begin to see a clearer logic to its shape. Asia is covered with locations and scenes from the Old Testament. Surrounding Jerusalem are Mount Ephraim, the Mount of Olives and the Valley of Jehoshaphat; further north stand the Tower of Babel, and the cities of Babylon, Sodom and Gomorrah. To the right are Joseph's 'barns' – a medieval rendition of the Egyptian pyramids – and Mount Sinai, where Moses is shown receiving the Ten Commandments from God's hand. The map also weaves a mazy itinerary of the Exodus, wandering through the Dead Sea and the River Jordan before reaching Jericho, passing a series of fabled sites along the way, including Lot's wife, turned to a pillar of salt.

In the midst of all this wealth of geographical, biblical, mythical and classical detail, the viewer's eye is inexorably drawn upwards towards the map's apex, and its shaping theology. At the top, just below the circular border, lies the Garden of Eden, the Earthly Paradise, shown as a fortified circular island irrigated by four rivers, and inhabited by Adam and Eve, portrayed at the moment of the Fall. Just to the south, the couple is shown being expelled from Eden, cursed to roam the terrestrial world that lies beneath them. Directly above this scene, beyond the worldly frame of human time and space, sits the resurrected Christ, presiding over the Day of Judgement. Around him a legend reads 'Behold my witness', a reference to the marks of the Crucifixion (the stigmata and spear wound in his right breast) that testify to his status as the promised Messiah. To Christ's right (the viewer's left), an angel resurrects the saved souls from their graves, proclaiming 'Arise! You shall come to joy everlasting.' To Christ's left, the damned are led away to the gates of Hell by an angel brandishing a flaming sword, declaring 'Arise! You are going to the fire established in Hell.'

Between these contrasting scenes, a bare-breasted Mary gazes up at her son. 'See, dear son, my bosom, in which you took on flesh,' she tells him, 'and the breasts at which you sought the Virgin's milk.' 'Have mercy,' she implores him, 'as you yourself have pledged – on all those who have served me, since you made me the way of salvation.' Mary's appeal is probably designed as a mnemonic. It evokes the exchange in the Gospel of Luke where 'a certain woman of the company' called out to Jesus, 'Blessed is the womb that bare thee, and the paps which thou hast sucked.' Viewers of the map would be versed in Jesus's response: 'Yea rather, blessed are they that hear the word of God, and keep it' (Luke 11: 27–8). They would understand that the Final Judgement is based on strict adherence to the word of God.

The whole biblical scene of resurrection and judgement stands at the top of the *mappamundi*, where a modern reader might look for a gloss or explanation of a world map or atlas. But instead of a written title, the Hereford *mappamundi* provides its audience with a visual image of the drama of Christian creation and redemption. It shows how the world was created by God, and how it will come to an end with the Day of Judgement and the creation of 'a new heaven and a new earth' (Revelation 21: 1). This is a map of religious faith, with a symbolic centre and monstrous margins, which bears little resemblance to either Ptolemy's geometrical project of the terrestrial sphere created in Alexandria nearly a millennium earlier, or al-Idrīsī's world maps made in Palermo just 100 years previously. In the period between Ptolemy and the Hereford *mappamundi*, Christianity emerged as a global religion that also manufactured a new and compelling idea of the world made in its own theological image. The Hereford *mappamundi* is an enduring example of this ambitious new world picture shaped not by science but primarily by faith. Within the map's unfamiliar geography, and what seems to a modern eye its bizarre ethnography and eccentric topography, it is possible to trace a development from classical Graeco-Roman civilization, and the rise of Christianity, a religion which only reluctantly embraced geography, but nevertheless adopted *mappaemundi* from the eighth century as its defining image of the world for the next 600 years.

The Hereford map is a classic example of a *mappamundi* that emanated from centuries of conflict and gradual accommodation between Graeco-Roman attitudes concerning the earth and its origins, and the new monotheistic Christian faith and belief in a divinity that created the world and promised everlasting salvation to mankind. Although Greece and Rome were regarded as 'pagan' societies, inimical to Christianity's story of Creation, they provided the only geographical accounts available through which to understand the Bible's varied (often vague, even contradictory) pronouncements on the shape and scope of the earth. As a result, the early Fathers of the Church, responsible after the death of the Apostles for defining the tenets of the Christian faith, had to step carefully, celebrating the classical world for its intellectual achievements, but castigating it for its paganism.

Nevertheless, it was Rome that provided Christianity with its earliest geographical knowledge. One of the great enigmas of early *mappaemundi* is their repeated inference of the existence of a standard Roman map of the world, a lost original which provided the basis for all subsequent Roman and early Christian mapmaking. On the Hereford *mappamundi* the outer pentagonal frame in the top left-hand corner contains a legend that reads, 'The terrestrial landmass began to be measured under Julius Caesar.' This is a reference to Julius Caesar's decision in 44 BC to survey the entire earth by dispatching consuls to map each cardinal direction – Nicodoxus (the east), Teodocus (the north), Policlitus (the south) and Didymus (the west), and to return with a world map to be publicly displayed in Rome. The first three men are all given their own legends in the eastern, northern and southern corners of the map, and reappear again in the map's bottom left-hand corner illustration. Above them sits Augustus Caesar, Julius' adopted son, enthroned and wearing a Christian papal triple tiara, who presents the three men with a scroll on which is written, 'Go into all the world and make a report to the Senate on all its continents: and to confirm this [order] I have affixed my seal to this document.' Above this scene another legend reads 'Luke in his Gospel: "There went out a decree from Caesar Augustus that all the world should be described".' In the King James Version the phrase is translated as 'all the world should be taxed', but this interpretation was not followed by subsequent translations, and the reference on the *mappamundi* is clearly to topography, not population.[6]

Whatever the scientific achievements of Roman surveying and map-making may have been, many of the Latin Fathers – including Tertullian, St Cyprian, St Hilary and St Ambrose – had little interest in such innovations. The third-century Christian martyr St Damian certainly dismissed such pursuits. 'What can Christians', he asked, 'gain from science?'[7] The more intellectually adventurous Fathers, like St Augustine (354–430) and his near contemporary St Jerome (c. 360–420), had a rather different attitude. Augustine acknowledged that the classical study of *physica*, the created world, was necessary to understand *sapientia*, what Augustine defined as 'the knowledge of divine things'.[8] According to Augustine, without a knowledge of 'the earth, the heavens, and the other elements of this world', we cannot understand the Bible, nor, by implication, can we be good Christians. He argued that biblical time and history should be studied alongside space and geography for a better understanding of divine creation. In his book *On Christian Doctrine*, Augustine skilfully argued for the study of both geography and history, without suggesting that this in any way showed man challenging God. 'Thus,' he suggested, 'he who narrates the order of time does not compose it himself,' and similarly: 'he who shows the location of places or the natures of animals, plants or minerals does not show things instituted by men; and he who demonstrates the stars and their motion does not demonstrate anything instituted by himself.' Such observations only reflected on the glory of God's creations, and allowed those who undertook such study to then 'learn or teach it'.[9]

St Jerome took up Augustine's suggestion of listing biblical locations. Jerome is better known today for translating and standardizing the Vulgate, a Latin version of the Bible, from its various early Hebrew and Greek versions. But around 390 he also produced a book *On the Location and Names of Hebrew Places*, often referred to simply as the *Liber locorum*, which provided an alphabetical description of place names from the Bible. Jerome's book was based on the writing of an earlier Church Father, Eusebius (*c.* 260–340), bishop of Caesarea, who wrote one of the earliest histories of the Christian Church; he also acted as an adviser to Constantine I (272–337), the founder of the capital of what became the Byzantine Empire, Constantinople, and the first Roman emperor to convert to Christianity. Around 330 Eusebius completed his Greek text *Onomasticon*, 'a list of proper nouns naming people or places', a topographical dictionary listing nearly a thousand biblical locations. Jerome corrected and updated Eusebius' text to provide a comprehensive Latin gazetteer of biblical place names, so that someone 'who knows the sites of ancient cities and places and their names, whether the same or changed, will gaze more clearly upon Holy Scripture'.[10]

Eusebius, Augustine and Jerome, like all the other early Church Fathers, were living in the shadow of the decline of the classical Roman Empire and its gradual Christianization. The emperor Constantine's conversion around 312 gave ultimate sanction to the faith, but the adoption of Christianity took place against a backdrop of the erosion of Rome's military and political dominance, and Constantine's decision to split the empire into eastern and western spheres, with Constantinople as its eastern imperial capital. The Sack of Rome by the Visigoths in 410 made some realize what had for centuries seemed unthinkable: that Rome might not be eternal after all. This caused further problems for the Church Fathers. Until Constantine's conversion, Rome had represented the pagan, repressive past, but by the end of the fourth century Rome had adopted Christianity as its official religion. Many now worried that the empire's political decline was somehow connected to its newly adopted faith. Augustine provided a theologically and intellectually profound answer in *The City of God*, written as a direct response to the Sack of Rome. Augustine used the metaphor of the city of Rome to propose that there were two cities: the earthly city of men, represented by Rome, its pagan gods and pursuit of glory; and the eternal city of God, a religious community of earthly pilgrims temporarily inhabiting this world, dedicated to the divine capital of Heaven. For Augustine, Rome, and earlier earthly cities and empires (such as

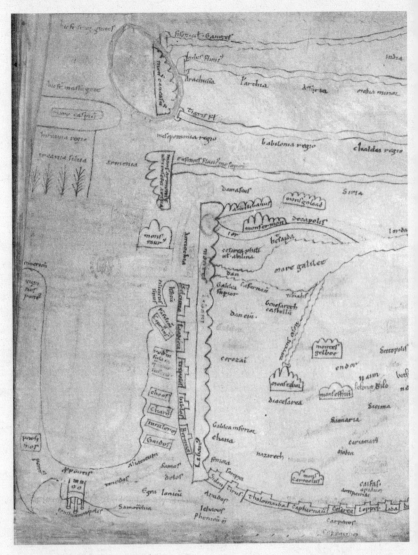

Fig. 4 Map of Palestine, St Jerome, *Liber locorum*, twelfth century.

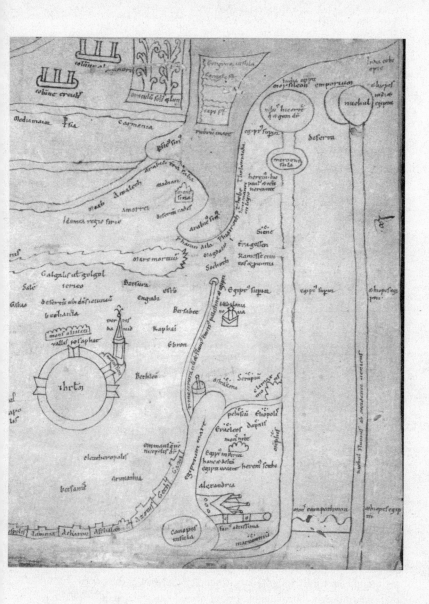

Babylon and Persia), were necessary historical prefigurations of the ultimate creation of the City of God. This account of faith and salvation would become central to subsequent Christian theology.

For Christians, the City of God was a spiritual community, rather than a physical location, so how did thinkers like Jerome and Augustine visualize the terrestrial world so as to be consistent with Scripture? How did they represent the Christian world on a flat map? Jerome offered one response in his *Liber locorum*. Later twelfth-century copies of the book made in Tournai contain regional maps of Palestine and Asia, designed to illustrate Jerome's catalogue of places. Jerome's text, and the maps that accompany it, influenced *mappaemundi* like Hereford's in their use of biblical place names and their geographical location. In Jerome's map of Palestine Jerusalem stands at the centre, a fortified circle distinguished by the tower of David. To the right is Egypt, with the two versions of the Nile which reappear on the Hereford *mappamundi*. Above Jerusalem, the Ganges, Indus, Tigris and Euphrates are shown flowing down from the Caucasus and Armenia, where a legend notes that Noah's Ark came to rest, which is again reproduced on the Hereford map. Although this is an explicitly biblical map, with most of its 195 locations drawn from Scripture, it also shows the rather garbled influence of Graeco-Roman mythology. At the top of the map in India stand Alexander's altars, next to the prophetic or 'oracular' trees he consulted during his time in the East.

The Jerome maps focused primarily on one part of the known world. But there were other mapping traditions available to the Church Fathers that claimed to represent the whole of the earth's surface and which would have a decisive influence upon the shape of the Hereford *mappamundi*. The first is now known as the T-O map, which is composed of a 'T' within a circle containing three continents, Asia, Europe and Africa, surrounded by water. The land masses are divided by three waterways which make up the 'T': the Don (usually labelled as the Tanais) dividing Europe and Asia, the Nile separating Africa and Asia, and the Mediterranean dividing Europe and Africa. Most *mappaemundi*, including Hereford's, inherited the orientation of east at their apex from the T-O tradition. The classical origin of these maps remains obscure. One possible source is the Judaic belief in the peopling of the three continents by Noah's sons – Japheth (Europe), Shem (Asia) and Ham (Africa),

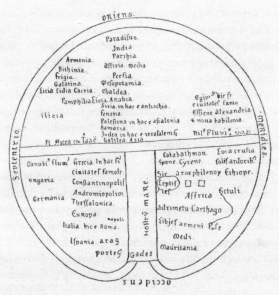

Fig. 5 T-O Map, from Sallust, *The Jugurthine War*, thirteenth-century manuscript.

but there are no known surviving examples of this specifically Jewish tradition.

The earliest surviving examples of T-O design date from the ninth century and are used to illustrate manuscripts of classical Roman history. Historians like Sallust (86–34 BC) and Lucan (AD 39–65) used geographical descriptions to situate their written histories of the battles and struggles for power that defined the period surrounding the death of the Roman Republic and the rise of the empire. In *The Jugurthine War* (40 BC), Sallust describes the unsuccessful rebellion of the Libyan King Jugurtha against the Republic in 118–105 BC. In chapter 17 he pauses to reflect that 'my subject seems to require of me, in this place, a brief account of the situation of Africa, and of those nations in it with whom we have had war or alliances'. Discussing debates over the division of the earth, Sallust continues, 'most authorities recognise Africa as a third continent', although he goes on to concede that 'a few admit only Asia and Europe as continents, including Africa in Europe'.

Sallust then provides two chapters describing what he calls 'the aborigines of Africa, the immigrant races, and the cross-breeding that took place', before returning to his commentary on Jugurtha's rebellion.[11] Sallust's geographical references were limited, but they offered one of the few classical accounts of what we would today call human geography: how humans interact with and shape their physical environment. The book and its geographical content were popular: between the ninth and twelfth centuries 106 manuscript copies survived, with more than half illustrated with a T-O map.[12]

The second cartographic tradition known to the Church Fathers, and which had a more intangible influence upon the Hereford *mappamundi*, was the zonal map. This method of world mapping has a clearer lineage, which we have seen stretching back even further than the T-O map, through Arabic astronomy to Ptolemy, Aristotle, Plato and the early Greek cosmographers. Its most influential exponent in the early Christian period was the fifth-century writer Macrobius and his *Commentary on Scipio's Dream*.[13] Little is known about Macrobius' life. He may have been Greek, or more likely an African-born Roman administrator in North Africa. His book provided a commentary on the closing section of Cicero's *Republic*, which was itself a response to Plato's *Republic*, but, instead of exploring the idea of utopia, Cicero used Rome's Republic as a model for the ideal commonwealth. Much of Cicero's text was subsequently lost, but Macrobius inherited its later section, known as 'Scipio's Dream', which he interpreted as an astronomical and geographical text.

In the *Commentary*, Macrobius describes a classical, geocentric world picture. 'The earth', he argues, is 'fixed in the middle of the universe', around which seven planetary spheres rotate from west to east. The terrestrial sphere is 'divided into regions of excessive cold or heat, with two temperate zones between the hot and cold regions. The northern and southern extremities are frozen with perpetual cold', and, Macrobius believes, cannot support life, 'for their icy torpor withholds life from animals and vegetation; animal life thrives upon the same climate that sustains plant life'. The central zone, 'scorched by an incessant blast of heat, occupies an area more extensive in breadth and circumference, and is uninhabited because of the raging heat'. Between the frozen extremities and the middle, torrid zone lie the temperate zones, 'tempered by the extremes of the adjoining belts; in these alone has nature permitted the human race to exist'. In anticipating the later discovery of

Australia (whose name derives from the Latin *auster*, or south wind), Macrobius argues that the southern temperate zone is inhabited because 'it has the same climate as our zone, but by whom it is occupied we have never been permitted to learn and never shall be, since the torrid zone lying between denies the people of either zone the opportunity of communicating with each other'.[14]

Where T-O maps proposed a simplified diagram of human geography, with mankind shaping the bare outlines of the division of the world into distinct continents, zonal maps of the kind described by Macrobius tried to provide some understanding of physical geography, or how the natural world dictated where humans dwelt upon its surface. For the Christian Fathers, both models required a certain amount of appropriation and manipulation to fit into their theological vision of the world. Zonal maps were particularly tricky, as they drew on a Greek tradition which claimed that mankind's place on the earth was primarily shaped by the physical environment. These maps also posited an unknown, inaccessible race in the southern half of the terrestrial globe. Was this race created by God? If so, why was it not mentioned in the Bible? Such questions remained unanswered, but continued to preoccupy theologians throughout this period.

However, zonal maps did allow the Church Fathers to claim a strand of Neoplatonic philosophy for the new Christian theology. Writers like Macrobius provided the Church Fathers with a crucial concept, which can be detected in the Hereford *mappamundi*. This was the belief in transcendence, in rising up above the earth in a moment of physical separation and spiritual insight. Interpreting Cicero's description of Scipio's dream, Macrobius argues that 'his reason for emphasizing the earth's minuteness was that worthy men might realize that the quest for fame should be considered unimportant since it could not be great in so small a sphere'.[15] For the Church Fathers, this insight appeared consistent with the redemptive belief in the Resurrection – of Christ rising up to Heaven, transcending the petty, local conflicts of the earth on which he looks down from his omniscient perspective above, offering the holistic scene of salvation we can see at the top of the Hereford *mappamundi*.

This Neoplatonic vision was developed by early Christian writers, including Paulus Orosius, one of the sources for both al-Idrīsī's *Entertainment* and the maker of the Hereford *mappamundi*. Orosius' *History against the Pagans* was commissioned by and dedicated to

St Augustine. Like Augustine's *City of God*, Orosius' book refuted the belief that Rome's collapse was due to the rise of Christianity. Orosius starts his history of what he calls 'the founding of the world to the founding of the City [of Rome]' with a moralized geography. 'I think it necessary', he tells his reader, 'to disclose the conflicts of the human race and the world, as it were, through its various parts, burning with evils, set afire with the torch of greed, viewing them as it were from a watch-tower, so that first I shall describe the world itself which the human race inhabits, as it was divided by our ancestors into three parts.' Orosius claims that such an approach is necessary so that 'when the locale of wars and the ravages of diseases are described, all interested may more easily obtain knowledge, not only of the events of their time, but also of their location'.[16]

T-O maps were easier for Christians to accommodate than zonal maps, and provided fewer philosophical difficulties for the Church Fathers, partly because of the simplicity of their appearance. Gradually, the T was appropriated as an image of the Crucifixion, and its location, Jerusalem, was placed at the centre of maps using this design, as well as *mappaemundi* like the Hereford example. The one figure most closely associated with Christianizing the T-O map, and another key source used in the making of the Hereford *mappamundi*, was Isidore of Seville. During his tenure as bishop of Seville (600–636), Isidore was instru-mental in a series of Church councils aimed at formalizing the principles of Christian belief and teaching. Today, he is better known for writing two of the most important encyclopedic texts of the early Middle Ages, both of which had a decisive impact on all subsequent Christian geog-raphy. Their titles emphasized Isidore's intellectual ambition: *De natura rerum* – *The Nature of Things* – was written *c.* 612–15, and, as its title suggests, attempted to explain everything, from the Creation, time and the cosmos, to meteorology and other divinely inspired natural phe-nomena. Isidore stressed that he was presenting his ideas 'as ancient writers have done and, even better, adding whatever one finds in the work of Catholic men'.[17]

Similarly, his *Etymologiarum sive originum libri XX* (622–33) – the *Etymologies*, also known simply as *Origins* – fused classical and biblical knowledge to argue that the key to all knowledge was language. 'When you see where a name has come from,' argued Isidore, 'you understand its meaning more quickly. For everything is known more plainly by the study of etymology.' Developing this method into the sphere of geography, book 14 of the *Etymologies* contains a detailed summary of the Christian world. In a move that would influence most subsequent *mappaemundi*, including the Hereford example, Isidore began his description of the world in Asia, with the location of Paradise, before moving westwards through Europe, Africa and, in an acknowledgement of the influence of classical zonal maps, describing a projected fourth continent, 'which is unknown to us due to the heat of the sun'.[18] Throughout his description, Isidore uses classical and biblical etymology to explain geography: Libya must be older than Europe, he observes, because Europa was the daughter of a king of Libya; Africa is named after Afer, a descendant of Abraham; and Assyria takes its name from Assur, son of Shem.[19] For Isidore, all natural phenomena reflect the divine creation of God. The seasons follow the vicissitudes of the Christian faith: winter represents tribulation, spring the renewal of faith. The sun represents Christ, and the moon the Church. Isidore even argues that the constellation of the Great Bear represents the seven Christian virtues.

Early manuscript copies of Isidore's books contain T-O maps, often little more than basic diagrams showing the tripartite division of the world. But from the tenth century more elaborate maps began to illustrate Isidore's works, until more than 600 were created, many showing Jerusalem at their centre. The written geographical accounts of writers like Orosius and Isidore were soon incorporated into the early medieval curriculum under the rubric of the seven liberal arts. The *trivium* involved the study of grammar, rhetoric and logic. But it was the introduction between the ninth and twelfth centuries of the other four arts, known as the *quadrivium* – arithmetic, geometry, music and astronomy – which allowed for the dissemination of the new Christian approach to geography. Although geography was not itself regarded as an academic discipline, the fifth-century pagan scholar Martianus Capella introduced the figure of Geometry as one of the seven personified liberal arts who speaks the language of geography. In Martianus' *The Marriage of Philology and Mercury*, Geometry explains, 'I am called Geometry because I have often traversed and measured out the earth, and I could offer calculations and proofs for its shape, size, position, regions and dimensions', before going on to offer a classical zonal account of the world.[20] Martianus' innovation provided a new outlet for the academic study of geography under the umbrella of geometry and the *quadrivium*. It also allowed Christian scholars to produce written accounts of the known world, which described the places and events depicted in *mappaemundi*. These were written versions of *mappae-mundi*, and they scoured classical geographical sources as a way of understanding the references to particular locations in the Bible.[21]

This new tradition of describing *mappaemundi* in written form introduced a *story* of the Christian creation into geography. The classical Graeco-Roman religions did not conceive themselves according to a chain of events of creation, salvation and redemption, nor did they have an account of the world with a beginning, middle and an end. The Christian Fathers from Jerome to Isidore understood the physical world according to a finite biblical story that begins with Genesis and ends in Revelation and Apocalypse. According to this belief, all earthly relations between time, space and individuals were connected along a vertical chain of narrative events which inevitably ended just as they began, with God's Divine Providence. In this approach, every human, terrestrial event anticipated, or prefigured, the fulfilment of God's Divine Plan. The Church Fathers' approach to biblical exegesis involved a clear distinction between a historical figure or event in time, and its wider fulfilment within God's plan. For example, the Old Testament story of the sacrifice of Isaac 'prefigures' the New Testament sacrifice of Christ. The former is a figure that anticipates the latter event, which fulfils (or justifies) the former. Their connection is through the logic of Divine Providence, as set out in the Scriptures.[22]

The impact of this new Christian philosophy of time on maps was acute. From the ninth century both visual and written *mappaemundi* began to appear not just in texts illustrating authors like Macrobius and Isidore, but also in school manuals, geographical treatises used in universities and monasteries, literary compositions in epic and romance poems, and in public spaces like monasteries and churches for more political and didactic purposes.[23] World maps emerged that conflated aspects of both zonal and T-O maps, as well as more detailed accounts of particular geographical locations. All this was done in the name of Christianity. Hardly any of these maps provided new geographical material on the world based on travel or exploration. Instead, they fused classical and biblical places to project a history of Christian creation, salvation and judgement onto the surface of a map. On most of these *mappaemundi*, viewers could trace the passage of biblical time vertically, from its beginning at the top of the map in the Garden of Eden in the east, to its conclusion in the west, with the end of time taking place outside its frame in an eternal present of the Final Judgement.

One early *mappamundi* that reflects these various traditions, and which also bears a close resemblance to the Hereford *mappamundi*, is the so-called Munich 'Isidore' world map, dated *c.* 1130. Made in Paris in the early twelfth century to illustrate a manuscript copy of Isidore's *Etymologies*, the map has a diameter of just 26 centimetres. This was a book, and a map, to be read in private by scholars, rather than seen in public by the laity. Nevertheless, the similarity to the Hereford *mappamundi* is striking. The general conformation of land masses is extremely close, and both maps are framed by the twelve winds, with islands floating round their circumference. The monstrous races of southern Africa are in the same positions, either side of an almost identical portrayal of the Upper Nile. Both maps agree on the location of the Red Sea, as well as the prominent islands of the Mediterranean, including a triangular Sicily. Although the Munich map is far smaller than the Hereford, and therefore lacks the elaborate depiction of the Earthly Paradise and extensive quotations from classical authors, it still fuses classical and biblical sources, tracing the travels of Alexander, the location of Gog and Magog, the whereabouts of Noah's Ark and the crossing of the Red Sea. The Munich 'Isidore' *mappamundi* shows how Christian scholars were gradually departing from classical and early Christian sources. Although it illustrates a copy of the *Etymologies*, the shape and detail of the Munich *mappamundi* bears little resemblance to Isidore's text. Instead, it represents the summation of the shape and outline of an evolving Christian world picture.

The Munich *mappamundi* is also based on the thinking of Hugh of Saint-Victor (1096–1141),[24] which exemplified the new approach to the use of *mappaemundi* in Christian teaching. Hugh was one of the twelfth century's most influential theologians, a follower of Augustine who used his position as head of the school at the abbey of St Victor in Paris to disseminate his scholastic writings like the *Didascalicon* (1130s), a textbook on the basic teachings of Christianity, where he argued that 'the whole sensible world is like a kind of book written by the finger of God'.[25] In his *Descriptio mappe mundi* (*c.* 1130–35), probably written as a lecture for St Victor's students, Hugh provided a detailed description of the earth and its regions along the lines of the Munich *mappamundi*.

Hugh's interest in geography was part of a larger understanding of God's creation, expounded in his mystical text *De Arca Noe mystica* (1128–9). In his treatise Hugh compares the earth to Noah's Ark, describing a cosmic plan which seems to have been painted on the wall of the cloisters at St Victor and used in his teaching. Although it no longer survives, it is possible to recreate this *mappamundi* in some detail, thanks to Hugh's detailed instructions. The painting depicted the body of Christ flanked by angels. He becomes an embodiment of the universe as he embraces it in an explicit reference to Isaiah's vision of God surrounded by the seraphims announcing 'the whole earth is full of his glory' (Isaiah 6: 3). Six circles emanate from his mouth, representing the six days of Creation. Moving towards its centre, Hugh's model portrays the signs of the zodiac and the months of the year, the four cardinal winds, the four seasons, and finally at its very centre a *mappamundi*, drawn according to the dimensions of Noah's Ark:

> the perfect Ark is circumscribed with an oblong circle, which touches each of its corners, and the space the circumference includes represents the earth. In this space, a world map is depicted in this fashion: the front of the Ark faces the east, and the rear faces the west . . . In the apex to the east formed between the circle and the head of the Ark is paradise . . . . In the other apex, which juts out to the west, is the Last Judgment, with the chosen to the right and the reprobates to the left. In the northern corner of this apex is hell, where the damned are thrown with the apostate spirits.[26]

Like the Hereford *mappamundi*, Hugh's world as an ark can be read as a story where the passage of time moves from top to bottom. At its apex is the literal godhead, overseeing the top (east) of the map and the Creation and Paradise. Moving downwards, from east to west, hell is to the north, Africa lies to the south with its monstrous races, while the westernmost point contains the Last Judgement, and the end of the world. For Hugh, the world as the Ark represents a prefiguration of the creation of the Church: just as the Ark saved Noah's family from the destruction of the Flood, so the Ark of the Church, built by Christ, will protect its members from death and eternal damnation. The Ark is a repository of all religious knowledge, part-book, part-building, in which 'are bountifully contained the universal works of our salvation from the beginning of the world until the end, and here is contained the condition of the universal Church. Here the narrative of historical events is woven together, here the mysteries of the sacrament are found.'[27]

Within this mystical theology is a unification of Christian time and space. The world as Ark both shows and tells a complete story of the Christian history of creation and salvation, stretching from the beginning to the end of time. Like Orosius and Augustine, Hugh proposed a version of Christian history based on a progression of time, starting in the east, and ending in the west. He claimed that 'in the succession of historical events the order of space and the order of time seem to be in almost complete correspondence'. He went on: 'what was brought about at the beginning of time would also have been brought about in the east – at the beginning, so to speak, of the world as space.' According to this belief, Creation took place in the east, as shown on the Hereford *mappamundi*. But following the Flood, 'the earliest kingdoms and the centre of the world were in the eastern regions, amongst the Assyrians, the Chaldeans and the Medes. Afterwards, dominion passed to the Greeks; then, as the end of the world approached, supreme power descended in the Occident to the Romans.' This movement can be seen on Hugh's *mappamundi*, which moves vertically, from the beginning of the world and time in the east at the top, to its anticipated end in the west, at the bottom.

This transfer of imperial power from east to west was also a summation of the prefiguration of both individual salvation and the end of the world. Or as Hugh put it, 'as time proceeded towards its end, the centre of events would have shifted to the west, so that we may recognize out of this that the world nears its end in time as the course of events has already reached the extremity of the world in space.'[28] For Hugh, who repeatedly turned to geography to define his theology, the vehicle for this unification of Christian time and space was the *mappamundi*, a space within which biblical time and the end of the world could be projected, and mankind could chart its final salvation – or damnation. His views may sound extreme, even eccentric, but the fifty-three surviving manuscripts of his book and extensive references to his work on medieval *mappaemundi* (including the Hereford map's debt to his descriptions of the 'splendid column' at Rhodes, and people riding crocodiles down the Nile) show that he was widely read and believed.[29]

At the zenith of this long, historical tradition stands the Hereford *mappamundi*. There are other *mappaemundi* contemporary with the Hereford example, but none survives that rival its scale and detail. Although earlier examples of *mappamundi* existed in England, there is no consistent explanation or contemporary accounts of how these texts were transmitted and influenced each other; nevertheless, they share striking topographical and theological similarities. The so-called 'Sawley Map', dated around 1190 and generally regarded as the earliest known English *mappamundi*, was discovered in the library of Sawley Abbey, a Cistercian monastery in Yorkshire. Like the Munich 'Isidore' *mappamundi*, this was a tiny map, illustrating a popular twelfth-century book on geography. Size may have limited its ability to portray Paradise and the Last Judgement, but the four angels in each corner of the map appear to derive from Hugh of Saint-Victor's cosmology, and represent the angels holding back the winds in Revelation.[30] The topography of the map is extremely similar to the Hereford map, from its biblical references and monstrous races in the far north, to the almost identical placement of rivers, gulfs and seas. Among surviving maps of the time, the Hereford *mappamundi*, however, is unique in assimilating so many diverse strands of classical and contemporary geographical and theological belief and in the process providing a comprehensive written and visual statement on the past, present and projected future of Christianity and its believers. The Bible, St Jerome, Orosius, Martianus Capella, Isidore and a range of other sources, from the 'marvels of the east' described in Pliny the Elder's *Natural History* (AD 74–9) to Caius Julius Solinus' book of wonders and monsters, *A Collection of Memorable Facts* (third century AD), are evoked (directly or implicitly) in the map's 1,100 inscriptions. These range from direct biblical quotations to reproducing Pliny's length and breadth of Africa and quoting Isidore's belief in unicorns ('monoceros').

It also registers a new and particularly Christian version of physical and spiritual travel: the pilgrimage. Pilgrimage routes to the Holy Land were well established in northern Europe by the twelfth century, and the pursuit of such a route was regarded as a statement of personal piety. The Hereford *mappamundi* shows three of the most important pilgrimage sites in Christianity – Jerusalem, Rome and Santiago de Compostela, identified on the map as 'The Shrine of St James'.[31] Each place is illuminated in bright red, and towns associated with the routes to each shrine are all carefully recorded. The *mappamundi* also retraces St Paul's journeys throughout Asia Minor, as well as reflecting contemporary experiences of pilgrimage to the Holy Land by reproducing fifty-eight place names in the region, twelve of which are not on any other maps of its time.[32]

Although the *mappamundi* was too large to act as a medieval pilgrimage routefinder, it appears to have been intended to inspire the faithful to contemplate pilgrimage, to admire the piety of those who undertook such a journey, and to reflect on the widely held medieval belief that the Christian life was itself an ongoing metaphorical pilgrimage. Homilies and sermons repeatedly reminded the faithful that their earthly life was a temporary exile from their ultimate destination and true, eternal home of heaven.[33] In St Paul's Epistle to the Hebrews the faithful are regarded as 'strangers and pilgrims on the earth' (Hebrews 11: 13), who 'seek a country' from which they came, and to which they seek to return. Earthly life is simply a stage in man's spiritual pilgrimage, replaying on an individual level the vast historical gulf between the exile from Eden and the quest for ultimate salvation and the return to the heavenly Jerusalem.

The essence of the Hereford *mappamundi* is contiguity, the proximity of one place to one another, each place charged by a specific Christian event. It is a map shaped by its religious history connected to specific places, rather than geographical space. The map offers the faithful a depiction of scenes from the Creation, the Fall, the life of Christ and the Apocalypse in an image of the vertical progression of Christian history from top to bottom in which they could grasp the possibility of their own salvation. The Hereford congregation or visiting pilgrims would read the *mappamundi* vertically according to the passage of preordained time, beginning with the Garden of Eden and Adam's expulsion, moving down through the growth of the great Asian empires, the birth of Christ and the rise of Rome, and ending with the prefiguration of the Last Judgement in the representation of the most westward point on the map, the Columns of Hercules. All these key historical moments, identified through their geographical locations, are placed equidistant from each other on the Hereford *mappamundi*. Each location is one further step in a religious story that anticipates divine revelation, which is represented at the apex of the map's pentagonal frame, outside of earthly time and space. The wonder of *mappaemundi*, both in general and in their particular manifestation at Hereford, is the ability to embody all of human history in one image, and simultaneously to provide a sequential account of divine judgement and personal salvation.

So this is a map that promises salvation; but it also prefigures its own destruction. Man is a pilgrim on earth, questing and anticipating the Last Judgement: the earth itself is a husk, a divinely created but ultimately expendable shell to be superseded at the end of time when 'the first heaven and the first earth were passed away' in preparation for 'a new heaven and a new earth' (Revelation 21: 1). *Mappaemundi* are created with a prefiguration of their own end; Christian salvation is predicated on an obliteration of the worldly individual and the world he or she inhabits. The theme of *contemptus mundi* (literally, 'contempt for the world'), the active renunciation of the terrestrial world in preparation for death and the world to come, pervaded medieval Christian belief. Pope Innocent III's *contemptus mundi* tract, *On the Misery of the Human Condition* (c. 1196) survives in more than 400 medieval manuscripts.[34] Its message, that the conclusion of the earthly pilgrimage was inevitably death and divine judgement, shaped religious observance, and suffused *mappaemundi*. Nowhere is this more graphically portrayed than in Hereford. It is there in the prefiguration of the arrival in heaven (or hell) at its top, to the rider at its bottom, waving farewell to the world before embarking on one final journey, to 'go ahead' as the legend says, into the eternal present of the afterlife. The *mappamundi* prefigures the end of its representation of the world with the Last Judgement, the terminus of the *contemptus mundi* tradition and the beginning of a new world of heaven and earth. This genre reached its zenith in the thirteenth century with the Hereford *mappamundi*. From the late fourteenth century the tradition began to decline, as a result not of the discovery of the new world of heaven, but of a whole host of new worlds discovered by more prosaic earth-bound travellers.

The Hereford *mappamundi* was therefore designed to work at various levels: to display to the faithful the wonders of God's created world; to explain the nature of creation, salvation and, ultimately, God's final judgement; to project the history of the world through locations, moving gradually from east to west, from the beginning of time to its end; and to describe the physical and spiritual world of pilgrimage, and the ultimate end of the world. All of this is built from the long historical, philosophical and spiritual tradition it inherits, stretching back through the early Christian Fathers to Roman times.

There is one final, more pragmatic dimension to the map's creation, one which leads all the way back to the life and death of St Thomas Cantilupe. In the bottom left-hand corner of the pentagonal frame, below Augustus Caesar's feet, is the legend: 'Let all who have this history – or who shall hear, read, or see it – pray to Jesus in his divinity to have pity on Richard of Haldingham, or of Lafford, who made it and laid it out, that joy in heaven may be granted to him.' The legend provides clues as to the authorship of the *mappamundi*, and its use once installed in Hereford Cathedral. There were in fact two closely related Richards who are relevant to the map's history. Richard of Haldingham and Lafford, also known as de Bello, held the prebendery of Lafford (today known as Sleaford in Lincolnshire) and was treasurer of Lincoln Cathedral until his death in 1278. The Latinized surname de Bello indicates his family name, and 'Haldingham' his birthplace – such alternative surnames were common in the thirteenth century.

There was also a second, younger Richard de Bello (or 'de la Bataille'). As his surname suggests, his family hailed from Battle, in Sussex, with another branch living in Lincolnshire, making the younger Richard a possible cousin of his older namesake, Richard of Haldingham. Richard de Bello took holy orders in Lincoln in 1294, but was subsequently appointed prebend in Norton in Herefordshire, going on to hold clerical positions in Salisbury, Lichfield, Lincoln and Hereford. He was in other words a pluralist, enjoying a series of non-residentiary benefices, just like both his patron, Richard Swinfield, who administered Lincoln Cathedral chancery in the late 1270s, and Swinfield's mentor, Thomas Cantilupe. It seems that Richard de Bellos, Richard Swinfield and Bishop Cantilupe were all clerical pluralists, connected by a web of ecclesiastical patronage, and all with good reason to oppose the reforming, anti-pluralist campaign of Archbishop John Pecham. In 1279 Pecham launched a fierce attack on Richard Gravesend, the bishop of Lincoln, insisting on a reformation of what he saw as a range of abuses, including the seizure of benefices. Swinfield appears to have been sent from Hereford to Lincoln by Cantilupe, who held a benefice in the diocese, to defend the pluralist case and oppose what he and his supporters saw as Canterbury's interference.[35]

These conflicts over ecclesiastical rights all point to a very specific earthly context for the *mappamundi*'s creation. It may even have been conceived not in Hereford but in Lincoln, by some combination of Richard Haldingham/de Bello senior, Richard Swinfield, and the younger Richard de Bello who briefed the craftsmen involved in the map's composition. These men, with unrivalled access to the great ecclesiastical libraries of thirteenth-century England, were able to assimilate the diverse strands of classical and biblical learning that are so evident throughout the map, as well as consulting contemporary *mappaemundi* held in other religious institutions across the country. Their combined wealth would have enabled them to appoint those responsible for making the map. These included the artist who first drew the map's illustrations and coloured them, the scribe who copied out the long and complicated written texts that cover its surface, and the expert limner who provided the finishing touches to the map's display script and vivid illumination.

Although the *mappamundi* does not provide specific theological support for Cantilupe's quarrel with Pecham and his defence of pluralism, the final scene in its frame seems to support the bishop in a different dispute that took place just a few years before Cantilupe's death. In 1277 Cantilupe protested against Earl Gilbert of Gloucester, who was accused of usurping the bishop's rights to hunt in the Malvern Hills. The royal justices who were asked to adjudicate found in the bishop's favour, and the earl's foresters were instructed to step aside and allow Cantilupe and his retinue to hunt as they wished. The *contemptus mundi* scene in the bottom right-hand corner of the *mappamundi* shows an elegantly attired rider on a richly caparisoned horse, followed by a huntsman leading a pair of greyhounds. The huntsman addresses the words 'Go ahead' to the rider, who turns and raises his hand as if to acknowledge the offer, as he trots forward, glancing upwards at the world above him. The scene is an invitation to the map's reader to 'go ahead' beyond the earthly realm into the heavenly world outside of time, space and the map's frame. But it perhaps also, more prosaically, evokes Cantilupe's local dispute with Gloucester. The huntsman represents Gloucester's men, allowing the rider, possibly Cantilupe himself, to 'go ahead' and hunt in their place.[36]

There is one final intriguing scenario that may connect Cantilupe with the creation of the Hereford *mappamundi*: that it represents an attempt to support the canonization of the highly controversial bishop. In the early 1280s Cantilupe's feud with Archbishop Pecham came to a head, leading to his excommunication, his travel to Italy, and finally his death in August 1282. In life, any prospective plan to create a *mappamundi* that celebrated Cantilupe was hardly original. But in death, it could represent a unique opportunity to memorialize him and put Hereford on the map of international Christianity. None of this would have been possible without Cantilupe's protégé, Richard Swinfield. It was Swinfield who succeeded Cantilupe as bishop of Hereford, and who as we have seen launched a campaign, in spite of Pecham's opposition, to have his mentor canonized and the cathedral established as an international centre of pilgrimage.

All pilgrimage sites required some kind of 'marvel', usually a tangible and recurring miracle. Where this was not possible, other wonders were needed to attract pilgrims and sanctify the object of their veneration. Swinfield soon began work on an elaborate shrine in the cathedral's north transept. It was here that the former bishop's remains were translated in a ceremony held over Holy Week, 1287. The latest archaeological evidence may even indicate that the *mappamundi* was initially installed on the wall next to Cantilupe's tomb, a novel and striking 'marvel' to what one critic calls the 'Cantilupe pilgrimage complex', a carefully orchestrated series of routes, sites and objects situated throughout the cathedral, designed to attract pilgrims and confirm Cantilupe's saintliness.[37]

An eighteenth-century drawing of the the Hereford *mappamundi* by the antiquary John Carter showed that it originally formed the centrepiece of a magnificently adorned triptych, presumably also commissioned by Swinfield, complete with folding side panels.[38] This was a particularly striking innovation, and one of the earliest known examples of a painted panelled triptych in western Europe – roughly contemporary with the paintings of the great early Italian Renaissance masters Cimabue and Giotto. Carter's drawing shows that the side panels of the Hereford triptych depicted the Annunciation, with the Archangel Gabriel on the inside left panel and the Virgin Mary on the right, intensifying the message of the central *mappamundi* panel. Experienced as an ensemble, the triptych invited pilgrims to meditate on the Annunciation's anticipation of Christ's First Coming, in contrast to the Second Coming represented at the apex of the *mappamundi*.[39] Where the side

Fig. 6 Drawing by John Carter, *c.* 1780, of the triptych containing the Hereford map.

panels celebrate life, the central panel spells out death – MORS – around its edge, confirming the *mappamundi*'s prefiguration, for those pilgrims who gazed on it of death and the end of the world, of the 'new heaven' and 'new earth' to come.

Many pilgrims who saw the Hereford *mappamundi* probably shared the approach towards spiritual pilgrimage as that voiced by the anonymous twelfth-century Benedictine monk living in Bèze Abbey, who prayed, 'May your soul leave this world, traverse the heavens themselves, and pass beyond the stars until you reach God.' Who, he asked, 'will give us wings like the dove, and we shall fly across all the kingdoms of this world, and we shall penetrate the depths of the eastern sky? Who then will conduct us to the city of the great king in order that what we now read in these pages and see only as in a glass darkly, we may then look upon the face of God present before us, and so rejoice?'[40] Such imagined journeys to the heavenly Jerusalem involve a rejection of the earthly world, and echo Macrobius' *Dream of Scipio* – transformed into a Christian vision of ascending the earth and looking down upon it from the heavens, grasping the insignificance of the earth and mankind's futile, mortal struggles upon its surface, when faced with divinity.

At some point during the late eighteenth century the Hereford *mappamundi* lost its side panels and its identity as part of a triptych. It now hangs in its own purpose-built extension, the subject of the scrutiny of a more secular pilgrim: the modern tourist. One consequence of this almost inevitable relocation of the *mappamundi* (whatever its original position might have been) is the distortion of our modern understanding of its original function. This is a map that celebrates religious faith, but it does so on a range of different levels, some abstract and universal, and some, as with the map's possible connection to Cantilupe, pragmatic and local. It is also a genre of map unique in the history of cartography that eagerly anticipates and welcomes its own annihilation. It looks forward to the moment of Christian Judgement when the terrestrial world as we know it will come to an end, all our travelling and peregrinations will cease, and salvation will be at hand. The Hereford *mappamundi* hopes and prays for the end of space and time – an eternal present in which there will be no need for either geographers or maps.

# 4

# Empire

*Kangnido World Map, 1402*

## The Liaodong peninsula, northeastern China, 1389

In 1389, the Korean military commander Yi Sŏnggye (1335–1408) stood poised to march his army into the Liaodong peninsula, on the border between China and Korea. Yi was part of a military expedition dispatched by the ruling Koryŏ dynasty to attack the forces of the recently founded Ming dynasty (1368–1644). The Koryŏ were indignant at the Ming threat to annexe a huge swathe of their northern kingdom, and ordered Yi to attack. As part of Manchuria, the Liaodong peninsula would see more than its fair share of bloody conflict over the next six centuries, but in 1389 Yi Sŏnggye stepped back from war. Yi was a pro-Ming critic of Koryŏ policy towards its powerful new neighbour, and opposed the decision to mobilize against them. On Wihwa Island at the mouth of the Yalu River, bordering Ming China, Yi called his army to a halt and made a fateful decision. He announced that instead of attacking the Chinese, the army would now be marching against the Koryŏ king U.

In the political coup that followed, Yi overthrew King U and his ruling elite, bringing to an end nearly 500 years of Koryŏ dynastic rule over the Korean peninsula. Pronouncing himself King T'aejong, Yi founded a new dynasty, the Chosŏn, which governed Korea for the next 500 years, the longest period of continuous rule by a single dynasty in any East Asian kingdom. The dominant Buddhist values of the Koryŏ had overcome archaic, tribal shamanistic practices, but in time Buddhist monasteries and their leaders, richly endowed with lands and exempt from taxation, generated a level of corruption and nepotism that many of the ruling elite could no longer support. From the ninth century the Chinese ruling dynasties became increasingly critical of

Buddhism, championing instead a revival of Confucianism, or 'Neo-Confucianism', that stressed the importance of practical rule and bureaucratic organization over the Buddhist retreat into spiritualism. As Koreans like Yi Sŏnggye adopted Neo-Confucianism, the impetus for change in Korea became irresistible.

Neo-Confucianism supported a programme of social and political renovation that drew on the classical texts of the sage-kings of Chinese antiquity. Opposed to the shamanistic and Buddhist principles that shaped Koryŏ society, Korean Neo-Confucianism taught that an active, public life was necessary to understand human nature and maintain social order. Pragmatic learning was preferred to esoteric study: where Buddhism cultivated the self, Neo-Confucianism embedded the individual within the management of the state. For the new Chosŏn elite, the contrast between the worldly outlook of Neo-Confucianism and the Buddhist message of spiritual liberation and the abandonment of worldly troubles provided a compelling justification for the sweeping programme of social reform and political renovation (or *yusin*) that took place from the 1390s.[1]

The transition from the Koryŏ to the Chosŏn dynasty is regarded as a key moment in Korean history that transformed its culture and society by reforming its political, legal, civic and bureaucratic structures. Power was concentrated within the hands of the king, and the kingdom's territory was consolidated with the creation of a new military infrastructure. Bureaucratic power was centralized and civil service examinations introduced in line with Neo-Confucian beliefs; land was nationalized; a new, fairer system of taxation was proposed; and Buddhism was all but abolished.[2] The rise of the Chosŏn was also part of a broader realignment of imperial and cultural geography. The foundation of the Ming dynasty in 1368 signalled the gradual demise of Mongol influence in the region. To the east, the region's other great power, Japan, was beginning to unify its northern and southern kingdoms, establishing a period of relatively peaceful and commercially prosperous relations with both the Ming and Chosŏn dynasties.[3]

In seeking to legitimize their usurpation of the Koryŏ dynasty, King T'aejong and his Neo-Confucian advisers drew on the classical Chinese concept of the 'Mandate of Heaven', which explained the rise and fall of dynasties. Only heaven could dispense the moral right to rule. As far as King T'aejong was concerned, part of this new mandate included not

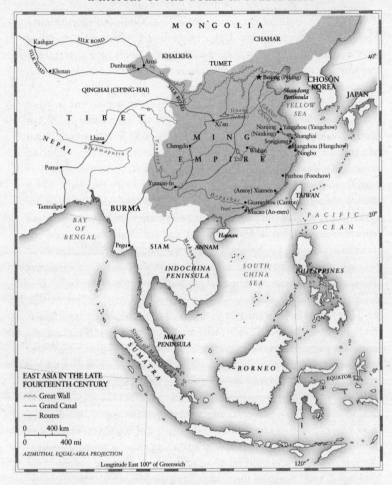

Fig. 7 Modern map of East Asia showing the regional situation in the late fourteenth century.

just a new ruler, but a new capital as well. The Chosŏn moved the capital from Songdo (today's Kaesŏng, in North Korea), to Hanyang (modern-day Seoul in South Korea), where T'aejong built his new residence, the Kyŏngbok Palace. The new administration also commissioned two new maps, one of the earth and the other of the heavens. The map of the heavens, entitled 'Positions of the Heavenly Bodies in their Natural

Order and their Allocated Celestial Fields', was engraved onto an enormous block of black marble over 2 metres high (a stela), and displayed in the Kyŏngbok Palace. It was based on Chinese star charts, and is unusual for reproducing the Chinese names for the Greek zodiacal signs, which reached China through its contacts with the Muslim world from the ninth century. Although it has many inaccuracies (many stars are misaligned), it showed the position of the heavens as they looked to King T'aejong and his astronomers in the early 1390s. This was a map that represented a new vision of the heavens for a new dynasty, a way of conferring cosmic legitimacy on the Chosŏn kingdom.[4]

By 1395, the star map had been completed by the king's team of astronomers led by Kwŏn Kŭn (1352–1409), a Neo-Confucian reformer and Assistant Councillor in the State Council, the highest position in the new Chosŏn regime. Kwŏn Kŭn was already at work on another map, this time of the entire world, and by 1402 it was finished. The original has not survived, but three copies are still in existence, all of them currently in Japan. The copy held in the Ryūkoku University Library in Kyoto and recently dated to the late 1470s or 1480s, is generally believed to be the earliest and best preserved, and includes the original preface written by Kwŏn Kŭn. Entitled *Honil kangni yŏktae kukto chi to*, or 'Map of Integrated Regions and Terrains and of Historical Countries and Capitals', it is better known simply as the Kangnido map (an abbreviation of its full title). It is the earliest surviving dated example of an East Asian map of the world, predating all Chinese and Japanese examples, the first cartographic representation of Chosŏn Korea, and the earliest Asian map to show Europe.[5]

The Kangnido map, exquisitely painted in ink on silk with gorgeous illuminated colours, is a beautiful and imposing object. The seas are olive-green and the rivers blue. Mountain ranges are marked by jagged black lines, with smaller islands shown as circles. All these features are offset against the rich yellow ochre of the earth. The map is criss-crossed with Chinese characters in black ink identifying cities, mountains, rivers and key administrative centres. Measuring 164 × 171 centimetres, and originally attached to a baton, allowing it to be unrolled from top to bottom, it was probably designed, like the star map, to hang on a screen or a wall in a prominent location such as the Kyŏngbok Palace. Just as the star map situated the Chosŏn dynasty under a new heaven, the Kangnido map located it on a new representation of the earth.[6]

Where, as we saw in Chapter 3, Christian maps placed east at the top, and many Islamic maps chose south, the Kangnido map is oriented with north at the top. The world is one continuous land mass, with no separate continents or encircling sea. Its rectangular dimensions, together with the land dominating the top of the map, seem to show a flat earth. At its centre is not Korea, but China, a large, pendulous land mass stretching from the west coast of India to the East China Sea. Indeed, China is so prominent that it seems to absorb the Indian subcontinent, which loses its west coast, while the Indonesian archipelago and the Philippines are reduced to a series of tiny circular islands bumping along the map's bottom. China's pervasive political and intellectual influence can also be seen in the inscription at the top of the map, directly below which is a list of historical Chinese capitals, followed by descriptions of contemporary Chinese provinces, prefectures and routes between them.

To the east of China lies the map's next largest land mass, Korea, surrounded by what appears to be a flotilla of small islands; these are in fact naval bases. At first glance, the mapmaker's depiction of his native country seems remarkably close to the modern outline of Korea, especially when contrasted with al-Idrīsī's portrayal of Sicily, or even Richard of Haldingham's illustration of England. Despite the flattened northern border, Korea is shown in astonishing detail. Its 425 identified locations include 297 counties, 38 naval bases, 24 mountains, 6 provincial capitals, and the new Chosŏn capital of Hanyang, prominently marked by a red, crenellated circle.[7]

In the bottom right-hand corner of the map floats the region's other major power, Japan, far south-west of its actual position. Its forked tip points menacingly up towards China and Korea. To compensate for this apparent threat, Japan's size is diminished relative to Korea, which appears three times its size, when in fact Japan is half as big again. Its westernmost island, Kyushu, is shown pointing northwards, and the actual position of the archipelago has been been rotated clockwise by ninety degrees.

Even more surprising to modern eyes is the map's portrayal of the world to the west of China. Sri Lanka looms large off its west coast (rather than south-east of India), but the wedge of the Arabian peninsula is quite recognizable, as is the Red Sea and the west coast of Africa. More than eighty years before the first Portuguese voyages discovered that the continent was circumnavigable, the Kangnido map shows Africa with its now familiar southern tip, although its overall size is massively underestimated (Africa's land mass is more than three times larger than that of modern-day China). A further peculiarity is that the continent is shown with what looks like an enormous lake at its centre, although this could also represent the Sahara desert. Many of the locations shown in Africa, Europe and the Middle East are Chinese transcriptions of Arabic place names, indicative of the pervasive reach of Islamic map-making even at this relatively early stage (Korea represented the limits of al-Idrīsī's geographical knowledge).[8]

Above Africa is an equally intriguing depiction of Europe. The Mediterranean is shown (although confusingly not shaded green like the rest of the map's seas) in a rudimentary but recognizable shape, as is the Iberian peninsula. Alexandria is represented by a pagoda-like object. A capital city, possibly Constantinople, is marked in red, and the outline of Europe contains an estimated 100 place names, most of which still await a convincing translation. Even Germany is shown, spelled phonetically as 'A-lei-man-i-a'.[9] At the very edge of the map is a tiny rectangle which appears to represent the British Isles, but is more likely to show the Azores, the westernmost point in the *Geography*, which are probably reproduced because of the partial transmission of Ptolemy's ideas.

The map's knowledge of the names and shapes of Africa and Europe might be inherited from Ptolemy, but that is where his influence ends. The Kangnido map contains no apparent graticule, scale or explicit orientation; not surprisingly, it offers a more detailed perspective on the South Asian region at the point where Ptolemy's coordinates dwindle into increasingly speculative geography, and his place names disappear. In contrast to medieval Christian and Islamic maps such as those produced in Hereford or Sicily, with their shared Greek heritage, the Kangnido map draws on very different cartographic conventions rooted in Korean and ultimately Chinese perceptions of the earth's place in the wider cosmos.

Unlike the disparate social and cultural inheritance of the Graeco-Roman world, which spawned a variety of competing religious beliefs and political worlds, pre-modern East Asia was broadly shaped by one universal empire: the Chinese. For centuries China saw itself as the unquestioned centre of legitimate imperial authority, ruled by an emperor who regarded himself as leader of the civilized world (or *tianxia*, 'all under heaven'). Satellite kingdoms like Korea were bit-part players in the grand Chinese scheme of things; peoples beyond the Chinese sphere were dismissed as largely irrelevant barbarians. Governing a vast and relatively well-defined empire required the creation and maintenance of one of the most sophisticated pre-modern bureaucracies ever created. The costly upkeep of its large (and continually shifting) imperial borders, alongside an intellectual conviction of innate political supremacy and geographical centrality, meant that, unlike late medieval Europe, China had little interest in the world beyond itself. The Buddhist and Confucian heritage that shaped Chinese beliefs was also profoundly different from that of the Christian and Muslim religions of the book that developed in the West after the collapse of the Graeco-Roman world. As universal religions, both Christianity and Islam believed they had a divine responsibility to spread their religion across the entire earth, a concept that was completely alien to both Buddhism and Confucianism.[10]

The result was a tradition of mapmaking which focused on the establishment of boundaries and the practical maintenance of empire, concerns pursued by bureaucratic elites much earlier than in the religious societies of the West. It did not try to project an imaginative geography beyond its borders that could be claimed on behalf of a particular religion or ideology, nor did it aim to encourage or enable long-distance travel and maritime expansion beyond the Indian Ocean (by the 1430s the Ming dynasty had permanently recalled its fleets from wider exploration). Where China led, Korea followed. Working as a client state of imperial China for much of its early history stretching back to *c.* 100 BC, Korea's mapmakers were similarly concerned to provide the kingdom's elite with practical maps for the administration of political rule. The Kangnido map did this from a very particular perspective. It was made first and foremost according to the Korean peninsula's distinctive physical geography, and to relations with its larger and infinitely more powerful neighbour.

Most maps offer some interplay between image and text, and the Kangnido map is no exception. Across its bottom is an extensive legend transcribed in forty-eight columns, written by Kwŏn Kŭn:

> The world is very wide. We do not know how many tens of millions of *li* there are from China in the centre to the four seas at the outer limits, but

in compressing and mapping it on a folio sheet several feet in size, it is indeed difficult to achieve precision; that is why [the results of] the mapmakers have generally been either too diffuse or too abbreviated. But the *Shengjiao guangbei tu* [Map of the vast reach of (civilization's) resounding teaching] of Li Zemin of Wumen is both detailed and comprehensive, while for the succession of emperors and kings and of countries and capitals across time, the *Hunyi jiangli tu* [Map of integrated regions and terrains] by the Tiantai monk Qingjun is thorough and complete. In the fourth year of the Jianwen era [1402], Left Minister Kim [Sahyong] of Sangju and Right Minister Yi [Mu] of Tanyang, during moments of rest from their governing duties, made a comparative study of these maps and ordered Yi Hoe, an orderly, to carefully collate them and then combine them into a single map. Insofar as the area east of the Liao River and our own country's territory were concerned, Zemin's maps had many gaps and omissions, so Yi Hoe supplemented and expanded the map of our country and added a map of Japan, making it a new map entirely, nicely organised and well worth admiration. One can indeed know the world without going out of his door! By looking at maps one can know terrestrial distances and get help in the work of government. The care and concern expended on this map by our two gentlemen can be grasped just by the greatness of its scale and dimension.[11]

Kwŏn Kŭn's preface seems to share similarities with al-Idrīsī's approach to the *Entertainment*: there is general uncertainty about the size and shape of the known world; to make a more comprehensive map, it is necessary to borrow from an established geographical tradition (in al-Idrīsī's case both Greek and Islamic, for Kwŏn Kŭn the Chinese); political and administrative patronage involving a team of experts is crucial to the endeavour; and the result inspires wonder and pleasure.

The preface raises two elements, both related, that provide a way to understand the map. The first is the political context of the map's creation, and the second the influence of Chinese mapmaking. Kim Sahyong (1341–1407) and Yi Mu (d. 1409) were part of the Chosŏn dynasty's cadre of Neo-Confucian advisers. Both men were involved in land surveys carried out on Korea's northern frontier in 1402, just months before the Kangnido map was made, and both travelled to China on diplomatic business; Kim's trip in 1399 possibly enabled him to obtain the Chinese maps mentioned by Kwŏn Kŭn. The Kangnido map is dated

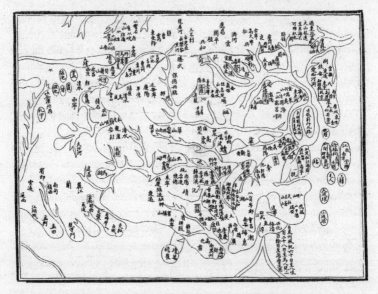

Fig. 8 Copy of Qingjun's map of China, from a mid-fifteenth-century commonplace book by Ye Sheng.

1402 by Kwŏn Kŭn not in relation to the foundation of the Chosŏn dynasty, but to the neighbouring Chinese Jianwen period of rule. The Jianwen emperor, Zhu Yunwen (r. 1398–1402) was the second ruler of the Ming dynasty and the grandson of its founder, the Hongwu emperor Zhu Yuanzhang (r. 1368–98). The Buddhist monk and mapmaker Qingjun was a close adviser to the Hongwu emperor and oversaw the rituals performed in Nanjing in 1372 to legitimize the new regime. A fifteenth-century reproduction of Qingjun's *Hunyi jiangli tu* shows that it provided both a geographical and historical description of the early Chinese dynasties, to which, as Kwŏn Kŭn notes, Yi Hoe 'supplemented and expanded' Korea to the east, and added the Arabian peninsula, Africa and Europe to the west.[12] Hoe (1354–1409) was a senior civil servant in the Koryŏ regime. He survived temporary exile at the orders of King T'aejong and by 1402 was back in the capital, making a map of the Chosŏn dynasty, and by the time he began work on the Kangnido map he was Legal Secretary (*kŏmsang*) in the new government (possibly as a result of his cartographic expertise).[13]

The Hongwu emperor's successor Zhu Yunwen was overthrown by

his uncle Zhu Di, the prince of Yan, who installed himself as the Yongle emperor after two years of bloody civil war.[14] By the time the Kangnido map was completed, Zhu Yunwen was already dead. Despite his explicit references to the Ming rather than the Chosŏn dynasty, Kwŏn Kŭn refers to the most militarily sensitive area in recent disputes between the two kingdoms when he points out the need to rectify the limitation of the Chinese mapmaker Li Zemin's mapping of Korea 'east of the Liao River'. His only other geographical observation is that the Kangnido adds a new map of Korea's other powerful and historically troublesome neighbour, Japan. The map clearly seeks to position the new Korean kingdom in the changing political world of early fifteenth-century East Asia.

Whatever the vicissitudes of regional dynastic politics between China and Korea evoked by Kwŏn Kŭn's preface, his admiration for the Chinese mapmaking which he cites as underpinning the map's creation is unquestionable. Both Li Zemin and Qingjun were making maps in the first half of the fourteenth century, but China's influence on Korean politics and geography goes back much further. Ever since Korea's emergence as an independent kingdom by the beginning of the fourth century BC, both its rulers and its scholars had looked to its larger and more powerful neighbour's civilization for inspiration in matters of statecraft, science and culture. This was never a purely passive relationship. Korea continued to assert its political independence from China while appropriating its cultural achievements wherever they seemed expedient.

Objects which can be described as maps can be found in China as early as the fourth century BC. But as in the case of any pre-modern society that produces manuscript maps across a large period of time and space, to speak of a Chinese cartographic 'tradition' over several millennia is problematic, even anachronistic. The first problem is one of surviving sources. Prior to the tenth century relatively few maps survive, making claims about the 'development' of Chinese mapmaking almost meaningless. Where written records have survived but maps have disappeared, it is difficult to speculate on what particular maps *might* have looked like. Too much interpretation rests on too few maps. Even those that have survived are plagued with the usual problems associated with the circulation and transmission of handmade maps, from unreliable copying and scholarly distribution to political injunctions against their wider dissemination.

Even more problematic is establishing just what is meant by a 'map'. As in Greek, Christian and Islamic societies, the Chinese term for 'map' is just as imprecise, and covers a range of different meanings and artefacts. In pre-modern Chinese, *tu* generally designates what in the West would be regarded as a map or a plan, although it can also refer to a wide variety of pictures, diagrams, charts and tables created in a range of different media (wood, stone, brass, silk and paper). *Tu* could be both word and image, and often combined graphic visual representations with written, textual descriptions (including poetry) which were seen as complementing each other. As one twelfth-century scholar put it, 'images (*tu*) are the warp threads and the written words (*shu*) are the weft ... To see the writing without the image is like hearing a voice without seeing the form; to see the image without the writing is like seeing a person but not hearing his words.'[15] The emotive resonances of the interplay here between *tu* and *shu* are largely absent from Western definitions of a map. As a verb *tu* describes planning, anticipating or thinking. It was sometimes even directly translated as 'difficulty in planning', which rather succinctly captures the practice of many early mapping activities, in this case both within and beyond China.[16]

Unlike the *pinax* of the early Greeks, the Chinese *tu* is a dynamic act rather than a physical medium, as recent Sinologists have argued by defining it as a 'template for action'.[17] And in contrast to the Greek *periodos gēs*, or 'circuit of the earth', it is not intimately related to prevailing cosmographical beliefs. Here again the Chinese developed a different approach from that of the Greeks. In early Chinese mythology there is no divine will authorizing the act of creation. With little religiously or politically authorized cosmogony (unlike the Judaeo-Christian and Islamic traditions), the Chinese developed an extraordinarily diverse range of beliefs about the origins of the earth and its inhabitants. Within this diversity, three schools of cosmological thought were particularly influential.

The most archaic was the *Kai t'ien* theory. This believed that the circular dome of the heavens sat like a bamboo hat on top of the earth. The earth was square like a chessboard, sloping down towards its four corners to form the rim of an encircling ocean. A more popular system was the *huntian*, or enveloping heaven theory, which emerged in the fourth century BC. It argued that the heavens encircle the earth, which lies at their centre (and it is intriguing that this idea developed at the same time as Greek theories of concentric celestial cosmography). One proponent of the *huntian* theory, Zhang Heng (AD 78–139) claimed that the 'heavens are like a hen's egg and as round as a crossbow bullet; the earth is like the yolk of the egg, and lies alone in the centre'.[18] The most radical belief was the more allusive *xuan ye shuo*, or infinite empty space theory: 'The heavens were empty and devoid of substance,' according to one later Han dynasty writer, and the 'sun, the moon and the company of stars float freely in the empty space, moving or standing still'.[19]

From the sixth century onwards official histories regarded the *huntian* theory as dominant, although strains of all three recur throughout Chinese astronomy, cosmology and cosmography, and the theory was not without its own ambiguities. Although its metaphor of the

Fig. 9 Depiction of the round heavens and square earth, from Zhang Huang, *Tushu bian*, 1613.

earth as a 'yolk' at the centre of heavens suggests a spherical world, the theory was often starkly illustrated showing a square, flat earth encircled by the heavens; and even this assumption was not absolute. Chinese astronomy was already using armillary spheres (depicting the heavens as a sphere), whose surviving calculations, based on detailed observations, assume a circular earth to represent the cosmos. Nevertheless, a foundational belief that runs through these theories is the conviction, the first surviving record of which is first found in a mathematical work from the third century BC, that 'Heaven is round, Earth is square'.[20]

This belief was based on an even more basic principle that pervaded early Chinese culture, which organized terrestrial space according to the 'nonary square', 'one of the great world-ordering discoveries or inventions'[21] of ancient China. A nonary square is divided into nine equal squares, creating a three-by-three grid. Its origins remain obscure, ranging from the archaic observation of the shape of a turtle shell (with its round carapace covering the square plastron), to the more convincing explanation that the vast plains of northern China inspired a rectilinear way of understanding and dividing space.[22] The Chinese celebration of the square was in direct contrast to the Greek philosophical (and geographical) ideal of the perfect circle. The nonary square also established the number nine as central to the classification of virtually every sphere of classical China: there were nine fields of heaven; nine avenues of the capital; nine divisions of the human body; nine orifices; nine viscera; nine wells in the realm of the dead, even nine branches of the Yellow River.

These divisions originated in one of the most important foundational texts of classical Chinese culture, the 'Yu Gong', or 'Tribute of Yu' chapter of the *Book of Documents*, compiled some time between the fifth and third centuries BC, regarded as the oldest surviving book of Chinese geography. The book describes the legendary ruler Yu the Great, founder of the Xia dynasty in remote antiquity (*c.* 2000 BC). Yu is said to have brought order to the world following the great flood by organizing fields and channelling the rivers.[23] Starting with the Yellow and Yangtze river basins, 'Yu disposed the lands in order. Going along the mountains, he put the forests to use, felling the trees. He determined the high mountains and the great rivers.'[24] The territory was demarcated into nine provinces (or 'palaces'), described in nine land and nine river itineraries. The nine provinces are described as a three times three grid, with the length of each side of the nine squares measuring 1,000 *li* (one *li* was equal to approximately 400 metres).[25]

As well as ordering the space of the known world according to the figure of nine, the 'Yu Gong' also offered a schematic division of the whole world into five concentric rectilinear zones, oriented according to the four cardinal directions based on the winds. This was a classic example of egocentric geography. Civilization resides at the very centre of the image, representing the royal domain. The degree of barbarism increases with each square outwards, from tributary rulers, the marches, the 'allied' barbarians, and finally the zone of cultureless savagery, which included Europe. Once again, the contrast between this scheme and its Graeco-Roman counterpart is striking. Although Western zonal maps are also rectilinear, they are based on latitudinal zones, and are not defined by a symbolic imperial centre as in the case of the 'Yu Gong'.[26]

The nonary square and its figure of nine enabled Chinese mapmakers to draw on a cosmological world view, and apply it to political administration and practical policy. At a symbolic level, the relationship between the circle and the square enabled scholars to recommend a particular way of running an empire. According to one Qin dynasty writer, '[w]hen the ruler grasps the round and his ministers keep to the square, so that round and square are not interchanged, his state prospers.'[27] At a more practical administrative level, the nonary square also drew on the so-called *jing tian*, or 'well-field' system of agricultural cultivation. The Chinese character for 'well' (*jing*) closely resembles the three times three grid, and was used as the basis for the allocation of agricultural land. A group of eight families would be allocated equal allotments of land, leaving the ninth (central) portion to be collectively cultivated. This orderly division of space was regarded as a basic element of social cohesion and effective government. 'Benevolent government must begin with demarcating boundaries,' argued the Confucian scholar Mencius (fourth century BC). 'Violent rulers are always sloppy in the demarcation of boundaries. Once boundaries are correctly demarcated, then the division of fields and the regulation of salaries can be fixed without exertion.'[28]

In the surviving records, the earliest descriptions of maps (or *tu*) are similarly associated with issues of dynastic rule and its administration. One of the first written references comes from the Zhanguo, or 'Period of the Warring States', *c*. 403–221 BC, when regional states battled for dynastic supremacy. In the *Shu jing* (*Book of Documents*, dating from the early years of this period), the duke of Zhou is recorded as turning to a map to choose the kingdom's capital city of Luoyi, today's Luoyang in the Henan Province, 800 kilometres south-west of Beijing:

> I prognosticated about the region of the Li River north of the He; I then prognosticated about the region east of the Jian River, and west of the Chan River; but it was the region of Luo that was ordered [by the oracle]. Again I prognosticated about the region east of the Chan River; but again it was the region of Luo that was ordered. I have sent a messenger to come [to the king] and to bring a *tu* [chart or map] and to present the oracles.[29]

The duke's prognostications about the location of the dynasty's capital were informed by political geography as well as providence. Following the pronouncements of the 'Yu Gong', the duke concentrates his attention on the agriculturally and politically pivotal areas of the Yellow and Yangtze river basins. Whatever the duke's 'map' actually showed, it was clearly being used as a supplement to these pronouncements for finding locations for the Zhou kingdom's new capital, in an attempt to unify newly conquered political space with the legendary geography of the ancient sages.

The iconography of maps played its part in subsequent key moments in Chinese dynastic politics. The Period of the Warring States came to an end in 221 BC with the rise to power of the Qin dynasty, unifying the Chinese kingdom under one rule. But it was not without a struggle: in 227 BC, prior to his accession, the first Qin emperor was attacked by an assassin using a dagger wrapped in a silk map of territories coveted by the Qin.[30] Nor was the dynasty necessarily secure: one third-century scholar advised the anti-Qin states that he had 'examined a map [*tu*] of the empire, according to which, the territory of the princes is five times larger than that of the Qin . . . If the six states were to join forces, head west and attack Qin, Qin would be smashed.'[31]

Alongside such explicitly political and symbolic functions, maps were also regarded as part of the administration of dynastic rule. 'The laws are codified in maps and books,' wrote the philosopher Han Feizi (d. 233 BC). They were, he claimed, 'kept in government offices, and promulgated among the people'. Despite such claims, other scholars were more sceptical. The Confucian philosopher Xun Qing (d. 230 BC) claimed that state officials 'preserve the laws and regulations, the weights and measures, the maps and books'. But unfortunately 'they do not know their significance, but take care to preserve them, not daring to decrease or increase them'.[32]

One of the earliest surviving maps of this period is an engraved bronze plate from the end of the fourth century BC found in the tomb of the King Cuo, ruler of the Zhongsan dynasty during the Period of the Warring States. It depicts a series of golden and silver inlaid rectangles and squares, interspersed with text, and is hardly recognizable as a map. It is in fact a mausoleum plan, or *zhaoyu tu*, offering a topography according to nonary principles of the carefully planned funerary rites of the Zhongsan ruler. The plate's outer rectangles represent two walls, between which stand four square buildings. Within the third rectangle is a raised mound, on which sit five square sacrificial halls, designed to cover the tombs of the ruler and members of his family. The *zhaoyu tu*, which absorbs the classical measurements of the nine provinces and the well-field system, is the earliest surviving Chinese example of a map-like object from a bird's-eye perspective. It is also drawn to scale: the plate's notes provide dimensions and distances measured in terms of *chi* (a foot, approximately 25 centimetres), and *bu* (paces, equal to around 6 *chi*).[33]

Why the map was in the tomb remains unclear. Traditionally, tombs contained precious objects imbued with arcane power, designed to convey ritual respects to ancestors.[34] The inclusion of the map could be a memorialization by a relatively sophisticated political administration of the Zhongsan ruler's control over his terrestrial space, just at the point when he enters the spiritual world of the afterlife.

Both the Qin dynasty and its successor, the Han dynasty (206 BC–AD 220) utilized maps in the drive towards political, administrative and military centralization. *Yudi tu*, or 'maps of the empire', were still used as ritualistic and commemorative devices, including the diplomatic exchange of maps with neighbouring kingdoms (such as Korea), and in the confirmation of military victories or the subjugation of subject states. But they were also beginning to permeate the administration of imperial governance. The *Zhou li* ('Ritual forms of Zhou') offered an ideal of Han bureaucracy in which maps were central to policy making. They were vital to water conservancy projects; in taxation, mining and the demarcation of roads; and the settlement of boundary disputes, the delineation of fields and the assessment of livestock; in auditing the distribution of the population; in maintaining the accounts of government officials, and in sustaining the loyalty of feudal states and their fiefdoms. In a sign of growing civic awareness of the importance of maps, two officials were appointed to keep their ruler informed about geography. Both men travelled with the emperor wherever he went. The *tuxun*, or royal scout, explained maps, while the *songxum*, or travel guide, deciphered local records where disputes arose.[35]

The best example of the Chinese approach to maps from this period is the work of Pei Xiu (AD 223–71). Pei Xiu is often referred to as the Chinese Ptolemy, primarily thanks to his establishment of what he called the six principles of mapmaking. Appointed as Minister of Works under the first emperor of the Jin dynasty (AD 265–420), Pei Xiu composed a study of ancient geography which drew on the 'Tribute of Yu' text, leading to the creation of his *Yu gong diyu tu* ('Regional Maps of the "Yu Gong"'), since lost. Pei's approach is recorded in the surviving *Jin shu* ('History of the Jin'), which describes how he 'made a critical study of ancient texts, rejected what was dubious, and classified, whenever he could, the ancient names which had disappeared'. The result was the eighteen-sheet *Yu gong diyu tu*, which he presented to the emperor, 'who kept it in the secret archives'. In making his map, Pei followed his six principles. The first was *fenlü*, the 'graduated divisions which are the means of determining the scale'. The second was *zhunwang*, a 'rectangular grid (of parallel lines in two dimensions)'. The third was *daoli*, or 'pacing out the sides of right-angled triangles, which is the way of fixing the lengths of derived distances'. The fourth was *gaoxia*, measuring 'the high and the low', the fifth was *fangxie*, measuring 'right angles and acute angles', and the sixth principle was *yuzhi*, measuring 'curves and straight lines'.[36]

To a Western reader, Pei Xiu's six principles appear to offer the foundations for modern scientific cartography, with their emphasis on the need for a graticule, the use of a standard scale, and the calculation of distance, elevation and curvature using basic geometrical and mathematical calculations. This was as good as anything the Greeks or the Romans had to offer at the time – but in China it did not translate into the development of a recognizably modern science of mapmaking, partly because Pei was not exclusively interested in this kind of mapping. His work was an early example of what Sinologists call *kaozheng*, or evidential research, which involved textual scholarship that recovered the past, paying particular attention to ancient texts as a guide to the present. This kind of scholarship also describes Pei's cartographic method. He acknowledged that his work involved 'a critical study of ancient texts', and his maps did not rely on any direct topographical measurements, but were based instead on reading textual sources. For Pei and the new Jin dynasty, the task was to superimpose a new, updated geography onto the authority of the classical text of the 'Yu Gong'. The reverence for and continuity with the past meant that Pei tried to combine the new with the old, to validate the past and legitimate the present in a graphic picture (and written description) of dynastic continuity.[37]

So powerful was this textual tradition that Pei's later followers even pointed to the limitations of visual descriptions of physical geography. 'On a map', wrote the Tang scholar Jia Dan (AD 730–805), 'one cannot completely draw these things; for reliability, one must depend on notes.'[38] For Pei, the textual, classical tradition is even present when he appears to be writing about space. 'When the principle of the rectangular grid is properly applied,' he writes, 'then the straight and the curved, the near and the far, can conceal nothing from us.'[39] This is simultaneously both a justification of his new quantitative principles of mapmaking, *and* a celebration of the classical textual tradition of dynastic administration based on the nonary square.

Like so many maps made before the accession of the Ming dynasty in 1368, Pei's have not survived. One that has is the famous 'Yu ji tu', or 'Map of the Tracks of Yu', made under the Song dynasty (907–1276) and dated 1136, which draws on the legendary exploits of Yu the Great. Joseph Needham has called it 'the most remarkable cartographic work of its age in any culture', arguing that anyone looking at contemporary European *mappaemundi* 'cannot but be amazed at the extent to which Chinese geography was at that time ahead of the West'.[40] The map was engraved on a stone stela 80 centimetres square and stood in the courtyard of a prefectural school in Xi'an, the capital of today's Shaanxi province. Like Pei's six principles of mapmaking, the 'Yu ji tu' initially appears strikingly modern. Its depiction of China's outline is in many cases remarkably accurate. It is also the first known Chinese map to use a cartographic grid to represent scale, as recommended by Pei. There are more than 5,000 squares, and the sides of each one represents 100 *li* (more than 50 kilometres). This gives the map an estimated scale of

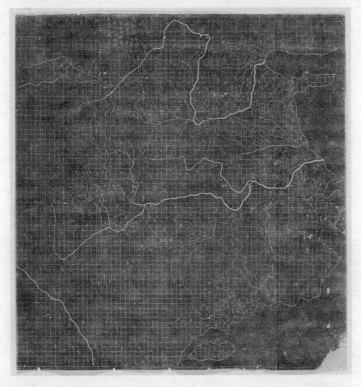

Fig. 10 'Yu ji tu' ('Map of the Tracks of Yu'), 1136.

1:4,500,000. But this grid is not the same as a Western graticule. The graticule plots a location through latitude and longitude relative to the rest of the globe's surface; the Chinese grid has no such interest in projecting a spherical globe onto a flat surface but simply helps to calculate distance and local area.

On the opposite side of the stela is another map, entitled 'Hua yi tu', or 'Map of the Chinese and Foreign Lands'. This map obviously complements the 'Yu ji tu' in some way, but how? Its scope is much broader, marking more than 500 places, including the rivers, lakes and mountains of the nine provinces, as well as the Great Wall in the north-east. It also depicts 'foreign lands' on the empire's borders (including Korea), listing more than 100 others on the copious notes running around its

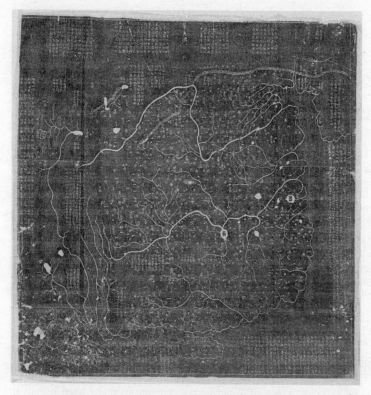

Fig. 11 'Hua yi tu' ('Map of the Chinese and Foreign Lands'), 1136.

edges. But this map is also very different from the 'Yu ji tu'. It lacks a grid, the coastline is extremely vague and often erroneous (especially in the crucial Liaodong peninsula), and its river-systems are inaccurate. To understand what is happening here, the viewer needs to walk back round the stela to examine the 'Yu ji tu' again.

Just as striking as the grid of the 'Yu ji tu' is its network of rivers criss-crossing its surface, with the Yellow River to the north, the Yangtze to the south, and the Huai midway between the two. Central to the map's toponymy are mountain names, but it also includes cities and provinces. The legend at the upper left suggests that, once again, textualism is as important to this map as quantitative measurement. It reads: 'Names of mountains and rivers from the *Yugong*, names of provinces and prefectures from past and present, and mountain and river names and toponyms from past and present.'[41] The 'Yu ji tu' represents contemporary geography by describing legendary times and places. It is marked by references to the foundational text of the 'Yu Gong' and its description of a mythical, unified China defined by rivers and mountains. For example, Yu is said to have guided the course of the Yellow River from a place called 'Jishi', which is reproduced on the map, even though thirteenth-century scholars knew that the river originated in the Kunlun mountain range in north-western China. The map retains 'Yu Gong' geography even where more recent Chinese mapmakers had shown this to be incorrect.

Rather than celebrating its use of scale and incorporating new geographical data, the 'Yu ji tu' conflates mythical geography with contemporary place – and for a very specific reason. For more than 100 years the Song dynasty had attempted to centralize military and administrative authority across the borders of classical China. Despite its political difficulties, or perhaps as a result of them, the Song encouraged a period of extraordinary cultural and economic reform, issuing one of the first examples of printed money, massively expanding its scholar-official class (the *shidafu*), and instituting one of the most innovative periods of woodblock and movable type printing since its invention in China at the end of the seventh century.[42] But by the early twelfth century the dynasty's northern territories were under threat by the Jurchen Jin (a confederacy of Tungusic tribes from northern Manchuria). In 1127 the Song capital of Kaifeng on the southern banks of the Yellow River fell to the Jin, and the Song retreated south of the Yangtze to their new capital of Hangzhou. In 1141 they signed a peace treaty with the Jin ceding nearly half their territory, and drawing a boundary line between the Yellow and Yangtze rivers. For the rest of the twelfth century until the dynasty's collapse in 1279, Song rulers and their *shidafu* dreamt of reunification with the lost northern territories and the recreation of classical imperial China.[43]

It never happened, but the 'Yu ji tu' offers just such a unity, as much through what it does not show (or say) as through what it does. It is a map without boundaries, on which there is no mention of the Jurchen Jin territories. Instead, the mythical geography of the 'Yu Gong' is conflated with the ideal geography of the Song dynasty, before the incursions of the Jurchen Jin. The Song try to represent themselves as not only unified, but the natural heirs of the original idea of a unified China, the nine provinces created by Yu the Great to which foreign rulers paid tribute. That political reality was so far removed from the idealized and nostalgic space depicted on the map only magnifies its apparent power to convince the Song audience of the possibility of such unification.

The 'Yu ji tu' and the 'Hua yi tu' represented two strands of Chinese imperial mapmaking that told the same story. The 'Yu ji tu' projected an enduring world free of contemporary political divisions, defined by the mythical unity of the nine provinces described by Yu. The 'Hua yi tu' drew on the same ideal, defining the empire as the 'Middle Kingdom', or *Zhongguo*, a reference to the northern Chinese provinces which lie at its centre, a reiteration of centralized power and authority in relation to foreign lands that was desperately needed during the turbulent period of the Southern Song. To Western eyes, both maps exhibit glaring topographical 'inaccuracies', but these were irrelevant to the projection of an ideal, imperial landscape based on classical texts like 'Yu Gong'.[44]

Poetry describing maps either side of the traumatic division of the Song also captures their power at first to acknowledge, and then lament the loss of territory. Writing more than 100 years earlier, the ninth-century Tang poet Cao Song describes 'Examining "The Map of Chinese and Non-Chinese Territories"':

> With a touch of the brush the earth can be shrunk;
> Unrolling the map I encounter peace.
>
> The Chinese occupy a prominent position;
> Under what constellation do we find the border areas![45]

On this occasion, the almost meditative act of unrolling the map and seeing a unified Chinese dynasty at its centre evokes emotions of security and assurance. Later Southern Song poets used a similar conceit, but with very different emotions. Writing in the late twelfth century, the celebrated Lu You (1125–1210) lamented:

> I have been around for seventy years, but my heart has
> remained as it was in the beginning,
> Unintentionally I spread the map, and tears come gushing forth.[46]

The map is now an emotive sign of loss and grief, and perhaps a 'template for action', a call to unite what has been lost.

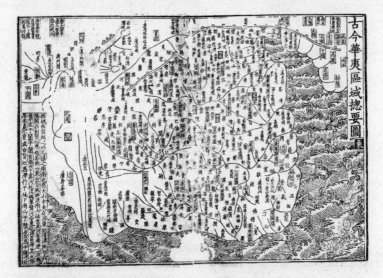

Fig. 12 The General Survey Map of China and Non-Chinese Territories from the Past to the Present, *c.* 1130.

The maps referred to by the Song poets include not just the stone 'Hua yi tu', but also other contemporary woodblock printed examples, such as 'The General Survey Map of China and Non-Chinese Territories from the Past to the Present' (*c.* 1130), one of the earliest surviving Chinese printed maps. The growth of the Song civil service in the twelfth and thirteenth centuries saw candidates for its examinations rise to as many as 400,000, and part of their preparation involved an understanding of the practical and administrative uses of maps. Commercial printers quickly capitalized on this new market, producing maps like the 'General Survey'. It is a sign of the map's popularity, and how far it was disseminated among the elite, that it went through up to six different editions, all of which were subject to updates and revisions by a variety of printers. The map's political function can be gleaned from some of its written legends, which describe 'administrative subdivisions past and present', 'the northern barbarians', and even 'the Great Wall', shown running across the top of the map. But as these descriptions indicate, this was a view of the empire rooted as much in the past as the present. Like the stone stelae, these printed maps created a vision of empire evoked through the rivers and mountains of an immutable

China. Although used by scholars and officials in the everyday administration of empire, they also contained a set of deeply held beliefs about
its enduring space.

To read the Kangnido map through this diffuse and discontinuous history
of Chinese mapmaking is fraught with difficulties. Nevertheless, the map's
textual references make it possible to trace certain continuities with Chinese methods: its reproduction of places, and reliance on a mythic, textual
geography. But it is also understandably full of specifically Korean preoccupations. Korea was unique in the pre-modern world for using a unit of
monetary exchange in the shape of the country's peninsula. In 1101 a
proclamation announced the circulation of a silver vase (ŭnbyŏng) that
'resembled the territorial outline of this country'.[47] In such a geographically distinct region, shaped by mountainous terrain and an almost
obsessive concern with its larger and more militarily powerful neighbours
China and Japan, a distinctive tradition of mapmaking emerged that
combined mythical spirituality with political security. In Korean, the
word for 'map' is chido, meaning 'earth chart' or 'land picture', and the
first written references to them date back to the early seventh century AD.
Although none of these early maps survives, virtually all remaining references to them indicate that, like many of the Chinese maps already
discussed, they were developed for administrative and imperial uses. In
628 the Korean kingdom presented a 'Map of the Infeudated Region'
(since lost) to the Chinese Tang court, a classic example of a subject
state using maps to pay tribute to its imperial superior.[48]

Equally important to Korean mapmaking was the age-old belief in
geomancy, or p'ungsu ('wind and rain', better known in Chinese as feng
shui), also referred to as 'shapes-and-forces'. Geomancy involved siting
graves, dwellings, monasteries and even cities in auspicious places where
they could harmonize with the natural flow of the earth's energy (or
ch'i), channelled through features such as mountains and rivers. As with
the Chinese use of the nonary square, geomancy involved a radically
different perception of physical space from that in Judaeo-Christian
tradition. Predating Buddhist beliefs, geomancy regarded the landscape
rather like the human body, with its practitioners acting as 'earth physi-
cians', taking the land's pulse and tracing its arteries through particularly
important mountains and rivers. Describing geomancy in Chinese landscape painting, the art critic Roger Goepper writes that 'every section of

nature in a particular countryside is, so to speak, a closed world in and of itself, a largely isolated microcosm in a greater fabric with which it is linked not so much spatially as by the common universal force of *ch'i*.[49]

In the distinctively shaped peninsula of Korea, where mountain ridges make up 70 per cent of the land's surface, geomantic mapping (*hyŏngse*, or 'shapes-and-forces') was even more prevalent than in China.[50] Geomancers regarded propitious areas for habitation as lying between the cosmically charged mountains of Paektu in the north and Chiri in the south, with cosmic power decreasing the further one travelled away from its mountainous origins. In mythology, Mount Paektu, a volcanic mountain in the peninsula's north-eastern region, represented both the origins of the Korean people, and the country's natural energy. Its importance was stressed by a typical geomantic description of the peninsula written in 1402 by the Korean official Yi Ch'ŏm. 'The central highland stretches down [from Mount Paetku] from which point neither the earth features nor the map scroll go any farther south, into the sea; rather, the pure and pristine matter here mingles and accumulates, which is why the mountains are so high and steep.' For Yi Ch'ŏm, the description of physical geography is a representation of spiritual shapes and forces. 'Primal matter here flows and there solidifies,' he continues, 'and the mountains and rivers form their separate zones.'[51] The founder of the Koryŏ dynasty, T'aejo (Wang Kŏn, r. 935–43), used similar geomantic principles as a basis of political rule, advising his son that 'the geographic harmony to the south is rugged and disharmonious, and it is easy for the people of that region also to lack a harmonious spirit', warning that if such people 'participate in the management of national affairs, they might cause disturbances and imperil the throne. So, beware.'[52] Although the Neo-Confucian approach of the Chosŏn dynasty was circumspect about what they regarded as the Buddhist (and specifically Zen) conventions of geomantic siting associated with the previous Koryŏ dynasty, such beliefs still persisted (particularly at a local level), albeit in diminished form. The Chosŏn Office of Astronomy and Geomancy used such beliefs when siting and building the new capital of Hanyang.[53]

None of these early geomantic maps has survived, but a copy of a 1463 official map of Korea (known as the *Tongguk chido*), made by Chŏng Ch'ŏk (1390–1475), a known shapes-and-forces specialist, reflects prevailing geomantic concerns. The entire map is characterized by the arterial network of rivers (in blue) and mountains (in green), all of which can be traced directly back to Mount Paektu, the ultimate source of cosmic energy. Each province is given its own colour, and significant towns are marked with circles, enabling viewers to assess their propitious geomantic siting in relation to the surrounding rivers and mountains. But as well as its geomantic influence, the map also shows the Korean preoccupation with national security. Despite extensive geographical knowledge of Korea's borders, this map grossly compresses the country's northern frontier, despite the region's geomantic importance as the location of Mount Paektu. In bringing together Korean mapmakers' distinctive preoccupations with geomancy and political security, the map's northern borders appear to have been deliberately distorted in case it fell into the hands of northern invaders like the Chinese or the Jurchen (which, given the diplomatic circulation of maps during this period, was a distinct possibility).[54]

The Kangnido map shows a remarkable fusion of these disparate cartographic elements; some diminished, others heightened. Its Chinese sources, the mid-fourteenth-century maps of Qingjun and Li Zemin, were the products of a textual and historical tradition of mapping that unified the two Song conventions represented by the 'Yu ji tu' and 'Hua yi tu'. But the Kangnido is intriguingly selective about which elements it borrows from this kind of map. It forgoes using a scaled grid, but it does depict 'foreign lands', rather than just writing about them in textual legends. With no investment in the Chinese tradition of reclaiming empire through its mythical foundations in texts like the 'Yu Gong', the Kangnido map is free to represent the world beyond China's borders as an act of curiosity, rather than anxiety. Nevertheless, its composition clearly accepts the cultural and political importance of China, placed right at its centre; and despite its lack of a grid, the map is rectangular, in an oblique acknowledgement of the nonary principles of Chinese cosmography.

Of all its Chinese influences, the map's northern orientation is perhaps the most striking. From ancient times, burial sites in Korea were oriented towards the east, a principle also adopted by the Mongolian and Turkic peoples to the north. But in archaic Chinese scriptural tradition, as we saw in Chapter 2, the king or emperor faced south from an elevated position above his subjects, who faced north and looked 'up' at the emperor, who always gazed 'down' on his subjects. As we have seen, the Chinese word for 'back' (in the anatomical sense) is synonymous with the word for north, both phonetically and graphically, because the emperor's back is always turned to the north. 'Recite' is also associated phonetically and graphically with the north, as students reciting a classical text must turn their 'backs' to the teacher, so they cannot see texts displayed in the classroom. In phraseology involving orientation, 'left' will indicate east and 'right' will indicate west, according to the perspective of the emperor. Even the Chinese compass was oriented southwards. It is referred to as a 'south pointer' (zhinan), because in conventional orientation the user will face south – unless the emperor is present – since that is the direction of warm winds and sun enabling crops to ripen, a factor which also influenced the geomantic siting of Chinese homes and graves.[55]

Despite the Korean obsession with geomancy, its influence on the map's depiction of Korea is surprisingly limited. Nowhere is more important to Koreans than Mount Paektu, but on the Kangnido map Paektu is hardly even highlighted, and, when compared to a modern map, is placed too far to the south-east. The peninsula's main mountain ranges are only faintly marked with jagged lines, as is the 'Baekdudaegan', the main range running down the spine of the kingdom's east coast, with its main arteries running westwards towards the major cities of Songdo and Hanyang. The rivers are accurately portrayed, running like veins across the country's surface. But compared to geomantic descriptions like Yi Ch'ŏm's, the shapes-and-forces tradition appears to be dramatically diminished within the map's broader international horizons.

Kwŏn Kŭn clearly understood the sensitivity of maps embracing a wider political perspective, and his involvement in a diplomatic mission in 1396–7 sheds new light on his motivation for creating the Kangnido map in a new era of Sino-Korean relations. Following the coup of 1389, the Chosŏn regime was anxious to retain its long-standing diplomatic relationship of sadae (or 'serving the great') with its Ming neighbour. Before he assumed the throne in 1392, Yi Sŏnggye dispatched letters to the Hongwu emperor Zhu Yuanzhang, justifying his actions and even consulting the Ming court on his kingdom's new name (the Chinese preferred Chosŏn because of its associations with the ancient Korean kingdom of Old Chosŏn). But in 1396, in an attempt to ensure Korea's subjection, the Ming court condemned Chosŏn correspondence as 'flippant and disrespectful', and detained its visiting envoys.[56] This inspired a diplomatic crisis known as the *p'yojŏn* dispute, which revolved around dynastic and textual definitions of empire and territory.

The political geography of Zhu Yuanzhang's official account of the perceived affront can almost be seen as a justification for the subsequent creation of the Kangnido map:

> Now Chosŏn is a country with a king [and] by his disposition he has sought to have close relations with us and rules accordingly, but the foolish and treacherous [envoys] do as they please and the document they brought requested seals and imperial mandates, which cannot be given lightly. Chosŏn is hemmed in by mountains and blocked by the sea, it has been fashioned by heaven and earth to be the land of the eastern Yi [barbarian] people where customs are different. If I bestow the official seals and mandates and order these here envoys be vassals, then in the eyes of the ghosts and spirits would I not be being exceedingly avaricious? Compared to sages of antiquity, I would certainly not have shown a measure of restraint.[57]

The rhetoric of withholding favour is classic diplomacy, but the Ming justification is based on Neo-Confucian principles of empire. Chosŏn is regarded as a 'barbarian' kingdom across the mountains and sea. Their 'customs are different', and arguably lay beyond the writ of the classical Chinese provinces. Should they be included in the realm of Chinese imperial influence, asks the emperor, or would such a claim offend the pronouncements of the classical sages?

The *p'yojŏn* dispute was only resolved with the intervention of Kwŏn Kŭn. During an eight-month stay in Nanjing, he developed an amicable personal relationship with the Hongwu emperor, negotiating the release of the detained envoys and re-establishing Ming–Chosŏn diplomatic relations. The two men even exchanged poems. Hongwu's became known as *ŏjesi* ('Poems of the Emperor'), and Kwŏn's *ŭngjesi* ('Poems written at Royal Command'). The stylized and metaphorical language used in the poems records the intricate manoeuvrings between the two states as they came to accommodate each other's political and territorial differences.

The first of Hongwu's poems focuses on the contested boundary of the Yalu River, the site of Koryŏ–Ming tensions in the 1380s, and the location of Yi Sŏnggye's pivotal military rebellion in 1389.

*Yalu River*

The clear waters of the Yalu mark the boundary of ancient fiefdoms,
[each of us is] strong now tyranny is no more and deception has
   ceased, we enjoy these times of harmony.
Refusing to accept fugitives gave a thousand years of dynastic stability,
cultivating rites and propriety gave a hundred generations of merit.
The Han expeditions can be clearly examined in the historical records,
evidence of Liao campaigns merely await checking the traces left behind.
Your King's kind thoughts have reached to Heaven's mind,
the river's strength is bereft of waves, yet it defends us and nobody
   is attacked.[58]

Like the earlier Song maps, Hongwu's poem applies the antique past to the present in asserting Ming dominance in the region. Classical Chinese texts defined the Yalu as the limit of China's sphere of influence, but also took credit for bringing civilization to the peninsula, and by implication, to Korea. The more recent expulsion of the anti-Ming Koryŏ, and the refusal to harbour imperial 'fugitives', has brought harmony and stability to the region. But Hongwu also reminds Kwŏn of the 'historical records' of Chinese claims to the Liaodong peninsula, stretching back to the Han conquest of the region in 109 BC, and including the more recent conflict in the late 1380s. Ultimately, the Yalu is regarded as a permeable natural boundary between the two kingdoms, currently free of political 'waves'.

In a subsequent poem entitled the 'Envoy Travels past Liaodong', Hongwu moves west of the Yalu and imagines a Korean diplomat crossing the peninsula into Ming territory. Filled with images of a peaceful, timeless society, it concludes, 'The boundary of *Zhonghua* [China] extends to heaven and the ends of the earth, / grains fill the fields and are reaped year after year.'[59] Kwŏn's poems responded in a more subservient tone, also describing the politically sensitive regions of the Yalu Liaodong. In 'Crossing the Amnok [Yalu] River', he avoids Hongwu's aggressive historical assertion of Chinese influence, instead posing a clever rhetorical question:

The Emperor's virtue knows no boundary between the realm of
   Ming and we Yi people,
[so] how can the land be divided into borders of this and that?[60]

Similarly, when he describes 'Passing through Liaodong', Kwŏn elides the region's fraught history of military occupation. His focus is exclusively on a Confucian 'journey'. 'The road stretches endlessly through plains', but 'with a fervour I am resolved to establish works of merit in the central plain'.[61]

At a geographical level, Kwŏn's poetic diplomacy describes what appears on the Kangnido map that was completed on his return to Korea. Both his poetry and the map reflect the shift in early Chosŏn Korea from Buddhism to Neo-Confucianism. Although China is placed at the heart of the map, 'in the central plain', it is a world free of political borders, stressing the close regional and cultural ties between the neighbouring Confucian kingdoms; and the political importance of the Yalu (in Korean Amnok-gang) is obvious, because it is one of only three rivers named on the map. Even in poems not directly related to resolving the p'yojŏn dispute, Kwŏn described a moral geography that resurfaces on the Ksangnido map. In 'Looking towards Japan', he describes the 'wickedness and treachery' of the Japanese, who 'plunder and raid their neighbour's border'.[62] Kwŏn's preface to the Kangnido map reminded its readers of the importance of the addition of a new map of Japan, although the correct orientation and size of its islands was clearly not the issue. What mattered was proximity, based on the relative threat or diplomatic opportunity represented by Japan. Kwŏn's consistent response to Japan, in both maps and poetry, enabled him to establish a common cause between China and Korea in their mutual fear of Japanese pirates and the diplomatic difficulties of dealing with the shoguns.

In its relations with Japan, the Chosŏn pursued a policy of kyorin (or 'neighbourly relations'), which involved educating the 'innately stubborn' Japanese through the principle of ritual, or ye.[63] By the time Kwŏn returned from his successful diplomatic mission and wrote rather modestly in his collected works, Yangch'on chip, that he 'enjoyably watched the making of the [Kangnido] map',[64] the Chosŏn dynasty's diplomatic and geographical position in the known world was established, as were relations with China and Japan. Anyone able to look at the Kangnido map could see them.

When we try to reconstruct the 1402 Kangnido map, the best surviving copy is the later fifteenth-century copy held in Ryūkoku University. The Ryūkoku Kangnido, which has recently been dated to between

1479 and 1485, seems to reflect the later fifteenth-century anxieties of the Chosŏn dynasty. Its toponymy incorporates several civil and administrative initiatives carried out by the Chosŏn during this period, including the establishment of a naval base in Chŏlla in 1479, clearly marked on the map's south-western coast; in contrast, it makes little attempt to update its geography of the wider world, still showing China as it looked on early fourteenth-century Yuan maps, despite the availability of much more up-to-date ones. The Ryūkoku map may therefore not be just a simple copy of the lost, original 1402 Kangnido map, but an updated record of the rapid changes to the Chosŏn state. The late fifteenth-century copyists may have wished to convey that, while the rest of the world stood, the civic and military administration of the relatively new government was forging ahead.[65]

By choosing the 1402 map as its template, and retaining Kwŏn Kŭn's preface, the Ryūkoku Kangnido shows that the regime's interests in the 1470s otherwise remained close to those of the beginning of the century. Both versions were concerned to 'site' (to use a geomantic term) the Chosŏn kingdom within a wider world. In that shifting world it had to triangulate the imperial ambitions of Korea with those of China and Japan. But it was also a world in which, relatively free from absolute adherence to Chinese principles, the team of scholar-officials responsible for making the original map could project the 'barbarian' lands beyond East Asia. Though it was often seen as barbarian by the Chinese, Korea was also sufficiently independent to appreciate that the 'world was very wide', and to want to map its place and history within it independently, whatever might lie at its edges.

To modern Western eyes, the Kangnido map is a paradox. It *appears* to be a map of the world that is comparable with those found in *The Book of Curiosities*, or with the Hereford *mappamundi*. At the same time, Western viewers also sense that they are looking at a world picture produced by an alien culture with a very different method of understanding and organizing physical space. The idea of the world may be common to all societies; but different societies have very distinct ideas of the world and how it should be represented. Nevertheless, as the Kangnido map and its Chinese predecessors show, these very different world views are absolutely coherent and functional for those who make and use them. The Kangnido map is a particular cartographic response to one of the world's greatest classical empires, one shaped by Korea's perception of its own physical and political landscape. Both the Chinese and Korean experiences created maps that were concerned with so much more than accurately mapping territory: they were also effectively plotting structured relationships.[66] The Kangnido map and its copies were proposing a way in which a small but proud new dynasty could locate itself within the sphere of a much larger empire.

# 5

## Discovery

*Martin Waldseemüller, World Map, 1507*

*Hamburg, Germany, 1998*

Philip D. Burden is one of the most respected map-dealers in the United Kingdom, an expert in the cartography of the Americas, and author of *The Mapping of North America*. In the summer of 1998 he was approached by a London book-dealer on behalf of a client based in Hamburg who required his services to authenticate an antique map. Such approaches were not unusual in Burden's line of work, but his curiosity was piqued when told his expertise was needed as a matter of urgency, and that he would have to sign a confidentiality agreement before discovering the nature of the map involved. After signing the agreement, Burden later recalled, 'there followed a telephone conversation I will not quickly forget.'

The information Burden was given was sufficiently extraordinary for him to interrupt a family holiday to Disneyland in California and agree to fly straight back to London and then on to Hamburg. He was met by his client's representatives and driven to Hamburg's banking district. Ushered into a conference room in one of the district's banks, Burden was presented with the object he was being asked to authenticate: what was claimed to be the only surviving example of the printed world map attributed to the German mapmaker Martin Waldseemüller, entitled *Universalis cosmographia secundum Ptholomaei traditionem et Americi Vespucii aliorumque lustrationes*: 'A Map of the World According to the Tradition of Ptolemy and the Voyages of Amerigo Vespucci and Others', which is thought to date from 1507 and is generally accepted as the first map to name and describe 'America' as a continental land mass separate from Asia. Burden had spent years handling antique maps, and the distinctive feel of the paper on which this one was printed convinced him him that 'it was the genuine article, and not some elaborate fraud'. He was well aware that he was looking at one of the most important (and valuable) objects in the history of cartography. 'I believed', he later wrote, 'that it was, after the Declaration of Independence and the United States Constitution, the most important item of printed Americana in existence and the birth document of America, having named it so.'[1]

Burden spent four hours with the map before preparing a report for his client, a wealthy German businessman who had recently sold his computer software company and was interested in acquiring the map from its then owner, Count Johannes Waldburg-Wolfegg of Wolfegg Castle, in Baden Württemberg, southern Germany. Once it became known that the map was for sale, another buyer with a particularly compelling interest stepped forward: the US Library of Congress. Burden's original client lost interest, choosing to invest his money in another company rather than a map. With an asking price of $10 million, the Waldseemüller map was valued as the most expensive in the world. The library's representatives now posed Burden a different question: was the map really worth what many regarded as an exorbitant sum? Once Burden confirmed that he had at least two clients prepared to pay the asking price, the library's representatives moved to buy it in the summer of 1999. In drawing up the contract, the library listed a series of points which explained the map's importance for both cartographic and American history to justify the acquisition:

- The Map contains the first known use of the name 'America' as an original invention by Martin Waldseemüller to designate the new continent discovered by Christopher Columbus in the year 1492;
- The Map is the only existing copy of a woodcut made by Martin Waldseemüller, probably in the year 1507;
- The invention of the name 'America' by Martin Waldseemüller for a new continent that had previously been designated as 'terra incognita' bestows an historical identity upon the continent, and
- On this basis, the Map by Martin Waldseemüller represents a document of the highest importance to the history of the American people.

The document went on to say that a further 'objective behind selling the Map to the Library is to enhance the cordial relationship between Germany and the United States'.[2]

The origins of the map's sale went back to the beginning of the twentieth century. In 1901 Father Joseph Fischer, a German Jesuit priest and teacher of history and geography, discovered the map's only surviving copy in the archives of Wolfegg Castle. Fischer's discovery led to a series of efforts by American libraries and collectors to buy it, including the Library of Congress, which was first offered the map in 1912, but declined due to lack of resources. The library made subsequent efforts to buy the map over the next fifty years, but it was not until 1992, and the quincentenary of Columbus's first landfall in the Americas, that the fate of the map's future took a decisive turn. The celebrations designed to mark the anniversary included the Washington National Gallery's exhibition 'Circa 1492: Art in the Age of Exploration', which featured the rarely exhibited Waldseemüller map as its centrepiece. Keen to interest Count Waldburg-Wolfegg in selling it, the Library of Congress asked Daniel Boorstin, its Emeritus Librarian and Pulitzer-prize-winning author of *The Discoverers*, to write to him. Boorstin wrote that 'As the first map to contain the name for the American continent, the document signals the opening of the continuing relation between Europe and America and the pioneering role of European cartographers in the development of Western civilization'. The count, who, since inheriting his title had turned Wolfegg Castle and the family estate into a thriving health and golf resort, needed little persuading. He quickly let it be known to the Library of Congress that he was willing to sell the map, owned by his family for more than 350 years, later claiming in an interview that his decision to sell was based on a combination of 'a nobleman's awareness of tradition with modern entrepreneurship'. But before the count and the library could agree to a deal, they needed to overcome a serious political obstacle: the map was listed on the National Register of Protected German Cultural Property, and no item on the Register had ever been granted an export licence. When in 1993 the Library of Congress's representatives petitioned the then chancellor Helmut Kohl (a historian by training), their request was flatly denied.

Chancellor Kohl's defeat at the hands of Gerhard Schröder in the 1998 German national elections signalled a change in German-American cultural relations. Schröder's appointment of Dr Michael Naumann as the first Minister of Culture since 1933 (when the post was abolished by the Nazis) was key in deciding the map's future. Naumann, a former publisher for the Holtzbrinck Group, a multinational publishing group with holdings in the United States, was unsurprisingly a keen advocate of closer cultural ties, and perhaps trade relations, between the two countries. He strongly supported both the count and the Library of Congress in restarting negotiations with the German federal government, even going so far as to suggest that the recently amalgamated car corporation DaimlerChrysler might be interested in funding the map's acquisition as 'the perfect partner in this dramatic expression of German-American friendship'. Throughout 1999 Naumann deftly paved the way for an agreement granting the map an official export licence, while lawyers drew up the contract agreeing the terms of the sale.

On 13 October 1999 the count and the Library of Congress signed a contract agreeing the sale of the Waldseemüller map. Though the price was $10 million, the library could only for the moment afford a $500,000 down payment: the contract stipulated that they had just two years to find the balance, or face the humiliation of returning the map to the count. The Library went on a frantic fundraising effort to cover the extra cost. They consulted the Forbes 400 list of America's richest individuals, and approached individuals and corporations, from the Texan businessman and former presidential candidate Ross Perot, Henry Kissinger and Henry Mellon, to AOL and American Express. As the library solicited millions from the multinationals, more modest offers came in from the great American public. 'I'm not rich,' emailed Greg Snyder in October 2000, 'but I do have a few hundred dollars I would like to donate for the acquisition of the Waldseemüller map.' Despite this, initial efforts to secure the money were disappointing, and the library pursued other avenues. After deciding against a plan to offer rare books from its collection as part-payment, the library acquired $5 million from a House of Congress Committee towards the cost, on agreement that the money would only be forthcoming once matching funds had been secured from the private sector. The Committee justified its contribution by citing a bizarre precedent: that in 1939 Congress paid $50,000 for the 'Castillo Locket' – a gold and crystal crucifix containing 'fragments of the dust of Christopher Columbus'. The private half of the money was raised from a small group of wealthy private donors, including a substantial contribution from the Discovery Channel, which the library agreed to help develop a series of programmes entitled 'The Atlas of the World'. Not everyone was delighted with the purchase. The German academic Dr Klaus Graf had already complained in an online article that 'any attempt to buy cultural property which is officially listed in the very small catalogue of national cultural property is an act of immorality', and asking: 'Has the Library of Congress no sense of shame?' Commenting on the acquisition, the *New York Times* acerbically noted that the United States' relations with Germany had in fact recently plummeted, and that Congress's decision to provide so much money for the map was in stark contrast to the federal government's simultaneous cuts in funding public libraries.[3]

Finally, in June 2003, the library announced that the acquisition of the map was complete. On 23 July 2003, after more than a decade of negotiations, Waldseemüller's map was unveiled for the very first time as the property of the American Library of Congress in its Thomas Jefferson Building. Appropriately enough, it was displayed as a companion piece to an exhibition on the Lewis and Clark Expedition of 1803–6, the first state-sponsored mission to map North America systematically from the Mississippi to the Pacific. Led by Meriwether Lewis, William Clark and other members of the Corps of Discovery, the expedition began the epic process of surveying the 9.5 million square kilometres of the interior of a continent whose name and outline had apparently first been placed on a map by Martin Waldseemüller almost 300 years earlier.

The circumstances surrounding the library's purchase of the Waldseemüller map are not unusual to anyone working in the cultural industries. The traffic in historical artefacts between powerful nations and empires has invariably involved the development or resolution of larger diplomatic, political and financial interests. In this case, the library's acquisition and display of the Waldseemüller map says much about America's understanding of itself as a nation and its place in the wider world. When the sale was completed, the Library of Congress's website drew on Burden's evaluation, hailing the map as 'America's birth certificate', the first 'to depict the lands of a separate Western Hemisphere and with the Pacific as a separate ocean'. It was 'an exceptionally fine example of printing technology at the onset of the Renaissance', which 'reflected a huge leap forward in knowledge, recognising the newly found American land mass and forever changing mankind's understanding and perception of the world itself'.[4] The Waldseemüller map gave America what most nations crave: the legitimacy of a precise point of origin, usually tied to a particular event or document. In this instance it was a birth date of 1507, when, as

Waldseemüller showed, America was recognized as a continent in its own right.

Along with a birth certificate comes paternity, and America's was identified by the Waldseemüller map as unquestionably European. As Daniel Boorstin's letter to the count in 1992 had suggested, the map enabled America to see itself as intimately involved in the drama of the European Renaissance, the moment when Europe reinvented itself through the rediscovery of the values of the classical civilizations of Greece and Rome, leading to what the great nineteenth-century historian Jacob Burckhardt called 'the discovery both of the world and of man'.[5] By this interpretation, the rebirth (the literal meaning of the French word 'renaissance') of the classical past went hand in hand with the rise of Renaissance humanism, a new method of thinking about the individual self, as well as the 'discovery' of the individual's place within a rapidly expanding world, that anticipated the rise of Western modernity. And indeed, the map's bottom right-hand legend supports such an approach. 'Although many of the ancients were interested in marking out the circumference of the world,' it says, 'things remained unknown to them in no slight degree; for instance, in the west, America, named after its discoverer, which is now known to be a fourth part of the world.'[6] This sounds like the confident modernity of a newly discovered rationality, which draws on the classics only, ultimately, to discard them as a modern, European self-consciousness is shaped. It is this belief that permeates the Library of Congress's statements on the Waldseemüller map: that it represents a huge leap forward in knowledge; that it utilizes the revolutionary new technology of print; and that it changes our understanding not just of our world, but of our place within it. The map is, in other words, a quintessential document of the European Renaissance.

The map certainly represents a totally different world to that of the Hereford *mappamundi*, the previous map produced in Europe which we examined. In the 200 years that separate the maps, the entire representation of the world, its intellectual and practical creation, even the term used to describe both objects, were transformed (although *mappaemundi* continued to be made well into the sixteenth century and displayed alongside newer maps showing recent discoveries). In 1290 the Hereford *mappamundi* is called an 'estorie', or a history; by 1507 the Waldseemüller map calls itself a *cosmographia* (cosmography) – a science describing the earth and the heavens. Gone is the *mappamundi*'s eastern orientation, its religious apex and monstrous margins. They are replaced on Waldseemüller's map by a north–south orientation, with a representation of recognizable coastlines and land masses, scientific lines of longitude and latitude, and a series of classical motifs. Bringing together Ptolemy's recommendation that maps be oriented with north at the top, and the development of navigational methods which drew on the use of compass bearings that privileged north as their prime direction, most late fifteenth- and early sixteenth-century European world maps like Waldseemüller's gradually replaced the east with the north as the basic point of orientation. Both maps display their classical learning, but in very different ways. Where the Hereford *mappamundi* drew on Roman and early Christian authors to confirm its religious understanding of creation, the Waldseemüller map reaches even further back to the Hellenistic world of Ptolemy, and his geometrical perception of the terrestrial and celestial worlds. Where the apex of the Hereford *mappamundi* portrays Christ in His majesty, the top of the Waldseemüller map enshrines a classical geographer and a contemporary navigator. While the Hereford *mappamundi* exhibits little or no interest in explicitly learning from other maps, the Waldseemüller map announces its debt to a whole world of earlier mapmakers – both the theoretical, academic maps and projections of Ptolemy, and the more practical portolans, sailing charts and maps produced by contemporary pilots and navigators trying to work out how to sail beyond Europe's shores from the early fifteenth century.

It was sea charts like the so-called Caveri chart, made in 1504–5 by the Genoese mapmaker Nicolo Caveri (or Canerio), that began slowly to map the lands discovered to the east, west and south of mainland Europe over the previous 100 years. The Caveri chart acknowledges the geographical world of the *mappaemundi*, with its tiny circular world picture placed at its heart, in central Africa, but this has been overtaken by the chart's elaborate network of rhumb lines (lines crossing a meridian at a constant angle) and compass roses that chart navigational lines of direction and bearing for pilots sailing out of sight of land.

This kind of chart had been used by sailors in the Mediterranean since at least the twelfth century and developed by pilots sailing beyond Europe in the fifteenth, including Christopher Columbus on his four voyages to the 'New World' of America that began in August 1492. By 1498, Columbus's third voyage led to the first known footfall of a

European on continental land in the western hemisphere when his crew landed on the Venezuelan coast on 5 August. Famously, Columbus never believed he was responsible for discovering a new continent: the Waldseemüller map's full title, and the legend in its bottom left-hand corner, celebrate another Italian explorer who would come to briefly eclipse Columbus as the 'discoverer' of the 'New World', but would enduringly provide the continent with its name. The legend describes the map as:

> A general delineation of the various lands and islands, including some of which the ancients make no mention, discovered lately between 1497 and 1504 in four voyages over the seas, two commanded by Fernando of Castile, and two by Manuel of Portugal, most serene monarchs, with Amerigo Vespucci as one of the navigators and officers of the fleet; and especially a delineation of many places hitherto unknown. All this we have carefully drawn on the map, to furnish true and precise geographical knowledge.[7]

The western voyages undertaken by the Florentine merchant and navigator Amerigo Vespucci at the close of the fifteenth century were, according to Waldseemüller's map, confirmation that the European voyages of exploration across the Atlantic had indeed discovered a new, fourth part of the world, unknown to the medieval world of the Hereford *mappamundi* and its tripartite world of Europe, Africa and Asia.

It was not just the map's geography that looked so dissimilar to that of the Hereford map. Its style and form came from a world that approached the business of mapping in a very different way to that of the makers and viewers of medieval *mappaemundi*. The Waldseemüller map was produced through an invention that was new to Europe: movable type. The idiosyncrasies of the manuscript scribe and illuminator were gone, replaced by the woodblock cutter, the printer and the compositor, who were responsible for transferring the original hand-drawn map onto the printing presses of early sixteenth-century Germany. Its ideas drew less on religious beliefs concerning the world's divine creation, and more on classical geographical texts like Ptolemy's *Geography*, assessed alongside modern sailing charts like Caveri's; these mapping practices were compared, contrasted, in some cases incorporated and in others discarded, in the creation of this new world picture. Although Ptolemy's name is included in the map's title, and his portrait stands at the top left of the map, it is directly contrasted with the newer discoveries of Vespucci, who is portrayed opposite his classical counterpart.

In some respects the *Universalis cosmographia* shattered the classical Ptolemaic world picture, introducing a fourth continent into Europe's geographical consciousness, and with it a whole set of new religious, political, economic and philosophical questions that would preoccupy scholars for generations to come. But we must qualify assessments of the map as a radical, even revolutionary description of a new world of geography. This is certainly not how it was first received when it was published – or even how it was first conceived. Neither Waldseemüller's name nor the supposed date of the map's first publication in 1507 is to be found in any of its legends or margins. It is in fact not even clear whether the Library of Congress's treasured map was printed in 1507, or if it really was the first map to name and represent America as a separate continent. In the book published to accompany the map, Waldseemüller and his associates hedged their bets about the nature of the new discoveries in the west, arguing (as we shall see later) that America was not necessarily a new *continent*, but was instead 'an island', a cautious qualification which suggested they would be prepared to revise their assumptions if future voyages to and discoveries of this 'New World' persuaded them they should. The map is also based on Ptolemy's 1,300-year-old map projections, reproducing many of the Greek geographer's errors, and adhering to a geocentric view of the universe that would only be challenged with the publication of Copernicus's *On the Revolutions of the Celestial Spheres* in 1543. These were hardly signs of a challenging modernity.

Waldseemüller went on to create a series of maps up until his death around 1521, but he never again used the name 'America' on any other map showing this 'New World'. The mapmaker appears to have held serious reservations about the wisdom of naming the new continent 'America' in 1507, and it took another generation before the name 'America' was universally accepted on world maps and in atlases. Although an extraordinary amount of publicity surrounded the acquisition of the map by the Library of Congress, the *Universalis cosmographia* received little public attention on its first and subsequent publications, and within just a few decades all copies of the map (of which no more than 1,000 were printed) were believed lost.

The history of the *Universalis cosmographia* shows that defining origins and establishing a moment of singular geographical discovery are far more complicated than we might imagine. The origins of America as a continent, just like those of this particular map, are shot through with competing claims and disputed beginnings made by a range of explorers, mapmakers, printers and historians. With the benefit of hindsight it is easy to look back on this period of global history as the 'Great Age of Discovery', and the *Universalis cosmographia* as commensurate with the scale and drama of such events. Certainly the achievements of both the Portuguese and Spanish empires from 1420 to 1500 are extraordinary. In this period the Portuguese sailed into unknown space, making landfalls right down the Africa coast, and colonizing the Azores, the Canary Islands and the Cape Verde islands. By 1488 they had established trading posts in West Africa and rounded the continent's southernmost tip, and by 1500 they reached India and Brazil. Spain's financial support for Columbus's first voyage to the New World in 1492 was the first of three such ventures that would bring the Caribbean islands and Central America to the attention of Europeans, followed by subsequent voyages encountering unknown stretches of the Northern and Southern American coastlines. All these discoveries are recorded on the *Universalis cosmographia*, which shows a world more than twice the size of Ptolemy's *oikoumenē*.

Nevertheless, the most difficult of all terms to explain in relation to this particular map is the one repeatedly used whenever it is mentioned: *discovery*. Today, we think of discovery as a straightforward concept, which involves learning about or revealing something previously unknown, especially when associated with travel and the 'discovery' of hitherto unknown places. The Waldseemüller map might at first appear to represent a defining 'discovery' of 'new worlds' in the history of Western mapmaking, but its use of the term indicates that it took a rather more circumspect approach to the 'new' lands it portrayed.

For people in the early sixteenth century, the discovery of new places, even new worlds, was regarded with caution, even suspicion. It challenged the foundations of knowledge inherited from classical writers like Aristotle and Ptolemy, and even questioned biblical authority: if the new world of America and its inhabitants really existed, why were they not mentioned in the Bible? The problem was compounded by the variety of inconsistent and often contradictory meanings associated with the word 'discovery' and the contemporaneous rise of European vernacular languages. In English the word only became common currency in the later sixteenth century, where it has at least six different meanings, including 'to uncover', 'disclose', or simply 'reveal'. In Portuguese, one of the first languages to record the new seaborne 'discoveries' from the early fifteenth century onwards, the term *descobrir*, usually translated as 'to discover', was regularly used to mean 'exploring', 'uncovering', but also 'finding by chance', and even simply to 'pick up'.[8] In Dutch, 'discovery' is usually translated as *ontdekking*, meaning to uncover, to find out the truth, or to detect a mistake. 'Discovery' was therefore as much about describing an encounter with territory and lands that were already known, through myth or classical learning, as it was about the revelation of 'new worlds' for the very first time. Even the term 'new world' is studiedly vague: the Portuguese refer to the rounding of the Cape of Good Hope in 1488 as the 'discovery' of a 'new world', even though maps of the time represented a version of the Indian Ocean and its related territories. Renaissance scholars were not as excited by the shock of the new as we are today, and invariably tried to assimilate this kind of 'discovery' into classical geographical knowledge. As a result, landfalls in places like Cuba or Brazil might be labelled 'discoveries' of 'new worlds', but the descriptions of explorers and mapmakers shows that they were often wrongly identified as existing places – Cuba could be named Japan, Brazil China, and so on.

We see maps of the Renaissance as embracing the 'discovery' of new lands, but their creators were trying, rather, to reconcile new information with established classical models of the world produced by writers like Ptolemy and Strabo: empirical reports often differed from learned authority, and mapmakers were reluctant to give up on revered classical texts unless they had compelling grounds to do so. Such information as they received was piecemeal and often contradictory, a problem noted by writers and mapmakers such as al-Idrīsī and even Herodotus, and assessing it alongside classical geographical models that seemed perfectly adequate was a delicate process. Mapmakers also needed to balance their desire for comprehensiveness and accuracy with a new imperative introduced through the new medium of print, and hitherto unknown in mapmaking: the need to sell maps and make money. Printing was a commercial industry that needed to make a profit, as well as providing a new way of making maps. The delicate balancing act of meeting all these objectives is central to the *Universalis cosmographia*'s creation. To celebrate the Waldseemüller map as a central object in the history both of Europe's discovery of itself and of America misunderstands the practical and intellectual development of geography in the early sixteenth century. To understand this development, a good place to start is with the map's putative creator.

Martin Waldseemüller (c. 1470–c. 1521, also known by the Hellenized version of his name, Hylacomylus or Ilacomilus) was born in the village of Wolfenweiler, near Freiburg im Bresgau, in what is today the state of Baden Württemberg in south-western Germany. The son of a butcher who rose to a position on the city council, Martin enrolled at the University of Freiburg in 1490, where he studied (presumably theology) under the renowned Carthusian scholar Gregor Reisch. Waldseemüller would have pursued the study of subjects advocated by Martianus Capella in his fifth-century book *The Marriage of Philology and Mercury*: the *trivium* of grammar, logic and rhetoric, and the *quadrivium* of arithmetic, music, geometry and astronomy. The geometrical and astronomical elements of the *quadrivium* introduced him to writers such as Euclid and Ptolemy, giving him a basic grounding in the principles of cosmography. In the late 1490s Waldseemüller moved to Basle, where he came into contact with the renowned printer Johannes Amerbach, an associate of Reisch. Amerbach was part of a second generation of printers who were starting to refine the original development of printing using movable type, publishing a mixture of biblical, devotional, legal and humanist books for a growing community of literate readers. It was probably here that Waldseemüller began to learn how to translate his humanist education in cosmography and mapmaking into the kind of printed map for which he would become famous.

The development of movable type in Germany around 1450 postdated its invention in China by around 400 years. Nevertheless, it is arguably the most important technological innovation of the European Renaissance. The first printing press is believed to have emerged from a partnership in Mainz in the 1450s between Johann Gutenberg, Johann Fust and Peter Schöffer. By 1455 Gutenberg and his team had printed a Latin Bible and by 1457 an edition of the Psalms. By the end of the fifteenth century, printing presses were established in all the major cities of Europe, and it is estimated that these presses were responsible for printing between 6 and 15 million books in 40,000 different editions – more than had been produced in total in manuscript since the fall of the Roman Empire (the European population in 1500 has been estimated at 80 million).[9] Those who experienced this first wave of mass printing were quick to grasp its significance: the German humanist Sebastian Brant noted, with only mild exaggeration, that 'by printing, one man alone can produce in a single day as much as he could have done in a thousand days of writing in the past'.[10]

In recent years scholars have questioned the revolutionary impact of the printing press as what Elizabeth Eisenstein has called 'an agent of change', but there is little doubt that the new invention (or re-invention) transformed knowledge and its method of communication.[11] Print promised speed, standardization and accurate reproducibility in the publication and distribution of books of all kinds. The reality of the working of printing houses and the technological and financial pressures they faced meant that these promises were not always fulfilled, but the ability of printed texts to introduce relatively consistent pagination, indexes, alphabetical ordering and bibliographies – all of which were virtually impossible in manuscripts – allowed scholars to approach learning in exciting new ways. Two geographically separated readers who owned, say, the same printed edition of Ptolemy's *Geography*, could now discuss and compare the book, right down to a specific word (or map) on a particular page, knowing that they were looking at the same thing. The idiosyncrasies of manuscript culture, which was so reliant on the hand of the particular scribe, could never allow for such uniformity and standardization. This new process of exact duplication also gave birth to the phenomenon of new and revised editions. Printers could incorporate discoveries and corrections into the work of a writer or a particular text. New reference books and encyclopedias were published on subjects like language, law and cosmography, which claimed the ability to produce precise definitions, comparative study, and the classification of knowledge according to alphabetical and chronological order.

The impact of the new printing presses also affected visual communications – particularly mapmaking. Part of the importance of printing was that it allowed for what one critic has famously termed 'the exactly repeatable pictorial statement'.[12] The new presses allowed mapmakers to reproduce and distribute identical copies of their maps in their hundreds, even thousands, at a level of precision and uniformity that was hitherto unimaginable. By 1500 there were approximately 60,000 individual printed maps in circulation within Europe. By 1600 this number had risen to a staggering 1.3 million.[13] These figures are all the more extraordinary when we remember that transforming manuscript maps into printed versions presented fifteenth-century mapmakers and printers with a series of enormous technical challenges.

It was with some awareness of the problems and opportunities afforded by the printing press that Martin Waldseemüller arrived in the town of Saint-Dié in the duchy of Lorraine in 1506. Now known as Saint-Dié-des-Vosges, close to the border with Germany, the town's geographical location at the confluence of so many aspects of European culture has decisively shaped its history. From the Middle Ages the duchy of Lorraine was on the axes of trade routes from the Baltic in the north to the Mediterranean in the south, and from Italy in the east to the markets of the Low Countries in the west. It was also sandwiched between the rival states of the French, Burgundian and Imperial rulers, and easily became entangled in their political and military conflicts. This made for a tense, but extremely cosmopolitan atmosphere. By the late fifteenth century the duchy was under the control of René II, duke of Lorraine, who in 1477 fought and won the Battle of Nancy over his rival Charles the Bold, duke of Burgundy. The victory gave René the political autonomy and military security he craved, and he set about establishing Saint-Dié as a centre of learning to rival those of the French, Burgundian and Habsburg courts that surrounded his duchy.

René entrusted Gaultier (or Vautrin) Lud, his personal secretary and a canon of Saint-Dié, with the task of establishing a humanist academy, known as the Gymnasium Vosagense, for the pursuit of his personal glory rather than financial gain. To ensure that the academy's ideas could be successfully disseminated, Lud drew up plans (on René's orders) to establish Saint-Dié's first printing press, using the expertise of printers based in Strasbourg, just 60 kilometres away, already one of northern Europe's biggest printing centres, and by the later sixteenth century home to more than seventy printers. Lud was looking for a Strasbourg-based cosmographer, and identified Martin Waldseemüller as 'the most knowledgeable man in these matters'.[14] Like Lud, Waldseemüller was a theologian with an interest in cosmography, as well as in the new techniques of representing them in print. By 1506 he was one of the earliest and most important members of the Gymnasium Vosagense.

Waldseemüller was joined at the academy by a handful of other humanist scholars, in particular two who were to become intimately connected with the production of the *Universalis cosmographia*. The first was Matthias Ringmann (also known by his Hellenized name of Philesius). Born in Alsace around 1482, Ringmann studied in Paris and Heidelberg before working for various Strasbourg printing shops as a corrector, proofreader and scholarly adviser. Like Lud, Ringmann was involved in the printing of books on Portuguese and Spanish voyages of exploration, which probably explains his involvement in the Gymnasium. The second was Jean Basin de Sendacour, another theologian with an expertise in Latin, who would prove indispensable in translating classical and contemporary texts.

Waldseemüller's arrival in Saint-Dié in 1506 provided the catalyst for work on an ambitious geographical project intended to put the Gymnasium at the heart of northern European intellectual life, but not initially designed to produce a world map depicting the discovery of America. Instead, the trio of Waldseemüller, Ringmann and Sendacour began with the intention of producing a new edition of Ptolemy's *Geography*. It may now seem surprising for such a group to turn to Ptolemy's 1,300-year-old book just when its geographical knowledge was being undermined by seaborne voyages to the west and east of mainland Europe, but it was in fact a logical choice. Although Ptolemy's book was mentioned by scholars from at least the sixth century, it was not until the fourteenth century that manuscripts of the Greek text found their way to Italy for serious study and translation. In 1397 the Greek scholar Manuel Chrysoloras was invited to travel from Constantinople to Florence to teach Greek to the humanist circle surrounding one of Italy's leading scholars, Coluccio Salutati. Chrysoloras's Florentine colleagues were so eager to learn Greek that they also paid for manuscripts to be sent from Constantinople, and these included copies of Ptolemy's *Geography*. Chrysoloras began work on the first Latin translation, completed by another Florentine humanist, Jacopo Angeli, around 1406–10. Angeli gave an indication as to how the early Italian humanists regarded Ptolemy's book by translating its title as *Cosmography*, rather than *Geography*, a decision that would influence mapmakers and their maps for the next two centuries. Cosmography, as we saw in Chapter 1, describes the features of the universe by analysing both the heavens and the earth. For the Renaissance, with its belief in a divinely created geocentric universe, this involved providing a mathematical description of the relations between the cosmos and the terrestrial earth. Cosmography therefore included a comprehensive (if somewhat vague) description of the activities of what we would today ascribe to a geographer, all overlaid with a veneer of classical authority through its evocation of Ptolemy and his own celestial-terrestrial methodology.[15]

For Angeli and his Florentine friends, the translation of Ptolemy's *Geography* as *Cosmography* was of more interest in resolving celestial and astrological matters than making scientific claims to project the terrestrial sphere on a plane surface. Many Italian humanists consulted the text for philological reasons, checking the ancient topographical nomenclature against that of modern place names. Angeli's translation produced a garbled and truncated version of Ptolemy's complex mathematical projections, and as a result it was read far more prosaically throughout the fifteenth century than many scholars since have believed. It did not launch the revolution in Renaissance mapmaking that is often claimed, because its innovative methods were poorly understood and ignored by most of its readers.[16] Even with the publication of Ptolemy's text through the new medium of print, most of the newly designed and updated maps which accompanied it were printed without a network of mathematical coordinates, showing that there was only a limited understanding of Ptolemy's scientific methods of projecting the earth onto a map. The challenge of simply printing maps was quite enough to occupy most printers and scholars.

By the time Waldseemüller and his colleagues began work on their map, no fewer than five new printed editions of Ptolemy's text had been published. The first, printed in Latin in Vicenza in 1475, lacked any maps, but this was quickly followed in 1477 in Bologna by the first to reproduce regional and world maps (and which is therefore regarded as the first ever printed atlas, although it did not use the name). The following year another edition was printed in Rome, and then a loose translation of Ptolemy's text into Italian, complete with maps, was published in Florence in 1482. In the same year the first German edition of Ptolemy was published in Ulm. While woodcut flourished north of the Alps and was used in the Ulm edition, all these early Italian maps were printed using the technique of copperplate engravings. This was more time-consuming, in that, unlike a woodcut, a copperplate engraving plate could not be set alongside movable type, but it held the advantage of a finer and more versatile use of line, which by the later sixteenth century would allow it to supercede woodcut maps.

The recovery and publication of Ptolemy's *Geography* in the fifteenth century did more than just satisfy the philological curiosity of humanist scholars. Ostensibly, Ptolemy's account of the world looked increasingly outdated in the face of the Portuguese and Spanish seaborne voyages of exploration. The early Portuguese voyages down the coast of West Africa revealed that, contrary to Ptolemy's belief, it was possible to circumnavigate Africa, and that the Indian Ocean was not landlocked. Even more significantly, Columbus's voyages into the western Atlantic proved the existence of land masses apparently unknown to Ptolemy and the Greeks, and had profound consequences for Ptolemy's overall calculations about the scope and shape of the known world. But at the same time that these voyages were undermining Ptolemy, his texts were proving more popular than ever. New editions of the *Geography* were published following Columbus's return – by 1500, of the 220 recorded maps in print, over half were based directly on Ptolemy – but those printed after 1492 contained little or no acknowledgement of Columbus's findings.[17]

Rather than discarding Ptolemy, Renaissance scholars adopted a more accumulative approach in their attempt to unite classical with modern geographical knowledge. Ptolemy's tables and written descriptions alongside medieval *mappaemundi* were the only comprehensive models of the world available to scholars and navigators like Columbus whose approach was therefore to try to reconcile their discoveries with these classical and medieval paradigms, even where the models apparently contradicted what they had found. Although it was still poorly understood by many, Ptolemy's *Geography* did explain how to draw a geographical projection of the known world using spaced parallels and converging meridians within which navigators and scholars could try and plot their new discoveries. The results were often puzzling and contradictory, but they stimulated further physical and intellectual exploration. They can be seen in the early printed editions of Ptolemy, which increasingly incorporated the new discoveries to the point that Ptolemy's original description appeared almost unrecognizable.

By the early sixteenth century, the leading innovations in print were taking place in German city states like Nuremberg and Strasbourg (which would play a part in the publication of the Waldseemüller map), with their active interest in classical learning and seaborne discoveries.

Both cities were closely tied to intellectual developments in Renaissance Italy and seaborne exploration in the Iberian peninsula through trade and finance. The first known terrestrial globe of the world was created in Nuremberg in 1492 by the merchant Martin Behaim, who had financed and taken part in a Portuguese trading voyage down the coast of West Africa in the 1480s. Cities like Nuremberg were also acknowledged centres of excellence in the production of not just print but also scientific instruments used in mapping and navigation.

Writing to a friend in 1505, Matthias Ringmann revealed that the original plan of the Gymnasium printers was to publish a new edition of Ptolemy to eclipse both the Italian and the first German edition published in Ulm. But as work on the edition began, the group was confronted with texts that appeared to describe a new and very different world to the west of Europe from that envisaged by Ptolemy. These were printed translations of the Florentine merchant and traveller Amerigo Vespucci's letters describing a series of voyages undertaken between 1497 and 1504 in which he claimed to have discovered a new continent. In the same letter Ringmann explained the two main elements that would come to influence the *Universalis cosmographia* on its publication just two years later:

> The book itself of Americus Vespucius has by chance fallen in our way, and we have read it hastily and have compared almost the whole of it with the Ptolemy, the maps of which you know we are at this time engaged in examining with great care, and we have thus been induced to compose, upon the subject of this region of a newly discovered world, a little work not only poetic but geographical in its character.[18]

In 1503 a Latin translation of a letter ostensibly written by Vespucci to his Florentine patron, Lorenzo de' Medici, was published under the sensational title *Mundus Novus*. Describing a voyage to the east coast of South America, this short letter described 'those new regions which we searched for and discovered', and which 'can be called a new world, since our ancestors had no knowledge of them'.[19] For the first time, the discoveries in the western hemisphere were regarded as a new continent. The publication of Vespucci's letter seems to have been a deliberate attempt to rival Columbus's earlier letter to Luis de Santángel, published in 1493, which described his momentous landfalls in the Caribbean during his first voyage between August 1492 and March

1493. By claiming the discovery of a 'new world' (in contrast to Columbus's belief that he had landed in Asia), and adding some lurid accounts of the sexual and dietary customs of the natives, the success of the *Mundus Novus* was assured. Within weeks it was rushed into print in Venice, Paris and Antwerp, and by 1505 there were at least five printed editions published in German, including a version edited by Matthias Ringmann.

In the same year another letter attributed to Vespucci was published, entitled *Lettera di Amerigo Vespucci delle isole nuovamente trovate in quattro suoi viaggi* ('Letter of Amerigo Vespucci Concerning the Isles Newly Discovered on his Four Voyages'), addressed to a 'Magnificent Lord', believed to be Piero di Tommaso Soderini, the then head of the Florentine Republic, and describing four voyages undertaken by Vespucci for the Spanish and Portuguese crowns between 1497 and 1504. Although the letter lacked the sensationalism of the *Mundus Novus*, it dramatically claimed that on Vespucci's first voyage, between May 1497 and October 1498, the Florentine 'discovered many lands and almost countless islands', 'of which our forefathers make absolutely no mention'. From this the writer concluded 'that the ancients had no knowledge of their existence'.[20] The account goes on to describe a series of landings along the coast of Central and South America, which predated Columbus's first recorded landfall on the continent in Venezuela in August 1498 by nearly a year.

Both printed letters were forgeries, or at least inflated and sensational versions of Vespucci's travels, as can be seen when they are compared to the more prosaic letters he actually wrote, which were only discovered in manuscript in the eighteenth century. These letters proved that Vespucci's first continental landfall was in 1499, a year later than that of Columbus, and that it was not Vespucci but his over-zealous publishers who pushed for his claims to be the first to 'discover' America. By the time Vespucci's letters were discovered, national interests had already downgraded his achievements: from the mid-sixteenth century, Spanish writers eager to celebrate Columbus and his Spanish-sponsored voyages poured scorn on the claims made by Vespucci's printers, even going so far as to call for the suppression of all maps using the name 'America'.

In Saint-Dié in 1505–6 the members of the Gymnasium Vosagense were unaware of how Vespucci's travels were being manipulated and sensationalized. They had no alternative but to rely on the trickle of information which reached them about Vespucci's voyages, which meant the *Mundus Novus* and, more recently, the four voyages letter, with its claim that Vespucci reached the new continent before Columbus. As Ringmann's 1505 letter shows, Vespucci's letters transformed the Gymnasium's project. They now embarked on an even more ambitious project than just editing the *Geography*: the creation of a world map comparing Vespucci's geographical information with Ptolemy's, and to publish alongside the map a geographical description describing their reasons and methods for departing from Ptolemy's *Geography*.

The Gymnasium worked remarkably quickly, and by the spring of 1507 their endeavours were complete. Their project was published in three parts. The first, the *Cosmographia introductio*, was published in Saint-Dié on 25 April 1507. It was a short, forty-page theoretical introduction to cosmography, followed by a further sixty pages containing a Latin translation by Jean de Sendacour from a French printed text of *The Four Voyages of Amerigo Vespucci*. The *Cosmographia introductio*'s full title announced the other two parts of the project: 'Introduction to Cosmography: containing the requisite principles of geometry and astronomy beside the four voyages of Amerigo Vespucci, and a proper representation of the whole world, both as a globe and a map, that includes remote islands unknown to Ptolemy recently brought to light.'[21] It was hardly snappy, but it indicated the scale and ambition of the project, as did its dedication to 'Maximilian Caesar Augustus', the Habsburg prince and Holy Roman Emperor Maximilian I (1459–1519). Ringmann dedicated a poem to Maximilian, which was followed by Waldseemüller's prose dedication, in which he gave a brief account of the Gymnasium's labours. 'I have studied', he began, 'with the help of others the books of Ptolemy from a Greek manuscript and, having added the information from the four voyages of Amerigo Vespucci, I have drawn a map of the whole world for the general education of the scholars as a way of introduction to cosmography, both as a globe and as a map. These works', he concluded, 'I am dedicating to you, since you are the lord of the known world.'

Subsequent chapters provided a fairly orthodox account of cosmography closely based on Ptolemy, explaining the principal elements of geometry and astronomy, and their application to geography. The first mention of Vespucci's discoveries comes in chapter 5, which describes the division of the earth into five zones in line with Ptolemy and other classical geographers. Describing the 'torrid' zone situated south of the equator between the tropics of Cancer and Capricorn, the chapter explains: 'there are many peoples who inhabit the hot and dry torrid zone, such as the inhabitants of the Golden Chersonese [the Malay peninsula], the Taprobanenses [Sri Lanka], the Ethiopians, and of a very large part of the earth that for all time was unknown, but has recently been discovered by Amerigo Vespucci.'[22] In this account, Vespucci's putative discoveries in the western hemisphere are easily incorporated into Ptolemy's classical zones, and are seen as contiguous from east to west with other inhabitants of countries within the same parallel. Two chapters later, refining this description of climatic zones that divide the earth, the *Cosmographia introductio* describes seven zones north and south of the equator, again drawing on Ptolemy. Almost in passing, the chapter explains that 'the farthest part of Africa, the islands of Zanzibar, the lesser Java, and Seula and the fourth part of the earth are all situated in the sixth climate towards Antarctica', south of the equator.

The passage which follows is one of the most important statements in early European exploration. 'The fourth part of the earth, we have decided to call Amerige, the land of Amerigo we might even say, or America because it was discovered by Amerigo.'[23] This is the first recorded mention of the naming of America after Vespucci, but remarkably, the passage is made to fit almost seamlessly within the classical understanding of the earth divided into climatic zones. Vespucci's discoveries in the Americas are incorporated into the same zone running from east to west that includes southern Africa and the islands of the southern Indian Ocean. As a result, according to the *Cosmographia introductio*, Vespucci's 'discoveries' strengthened, rather than eroded, Ptolemy's world picture.

Finally, in chapter 9, the *Cosmographia introductio* provides a general description of the earth. It begins: 'There is at this time a fourth part of this small world barely known to Ptolemy and inhabited by beings like ourselves.' It goes on to describe Europe, Africa and Asia, before returning to the new territories and repeating the idea for naming them:

Today these parts of the world have been more extensively explored than a fourth part of the world, as will be explained in what follows, and that has been discovered by Amerigo Vespucci. Because it is well known that Europe and Asia were named after women, I can see no reason why

anyone would have good reason to object to calling this fourth part Amerige, the land of Amerigo, or America, after the man of great ability who
discovered it.

In concluding, the chapter states: 'The earth is now known to be divided
into four parts. The first three of these are connected and are continents,
but the fourth part is an island because it has been found to be completely surrounded on all of its sides by sea.'[24] At the same time as
celebrating the new discoveries, the text tells its readers that Ptolemy
'barely' knew the fourth part of the world – which is very different from
saying he did not know it at all. The impact of new geographical information and maps can be detected in the phrases 'now known' and 'been
found', but with it comes the ultimate quibble regarding the new-found
land's status as either an island or a continent. Renaissance mapmakers
understood islands and 'parts' of the world based on classical 'zonal'
maps, but 'continents' were more difficult to define. The cosmographer
Peter Apian defined it in 1524 as 'firm or fixed land which is neither an
island nor a peninsula nor an isthmus',[25] which was hardly helpful.
Europe, Asia and Africa were understood to be 'continents', but Waldseemüller and his colleagues were understandably reluctant to give the
new land of America such important status in 1507 without additional
verification of its shape and size. As a result, it remained an island until
further notice.

The second part of the publication was, as the dedication promised,
a small woodcut map just 24 × 39 centimetres, composed of map gores –
strips with curved sides tapering to a point which when pasted together
onto a small sphere made up a complete terrestrial globe. These were
the first known printed gores for a terrestrial globe ever made, and
include a western hemisphere, with South America labelled 'America'.
The globe gores were closely related to the final, and most ambitious
element in the whole project, the enormous twelve-sheet world map,
*Universalis cosmographia*, the first printed wall map.

Although the printing of the whole *Cosmographia introductio* was a
reasonably straightforward task for the small Saint-Dié press, the scale
and detail of the *Universalis cosmographia* was beyond its limited
means, and printing was moved to Strasbourg, where it was probably
finished in the printing house of Johann Grüninger. Even by today's
standards, its printing was an extraordinary technical achievement. It is

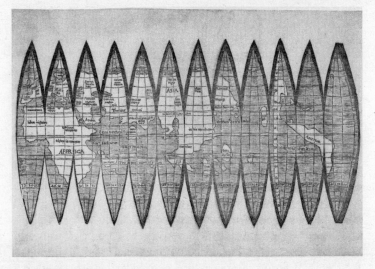

Fig. 13 Martin Waldseemüller, globe gores, 1507.

composed of twelve separate sheets of individual woodcuts, printed on handmade rag paper, each measuring 45 × 60 centimetres. Once all twelve sheets are assembled, the map measures a massive 120 × 240 centimetres (approximately 34 square feet). It is all the more staggering in the light of the kind of practical problems it presented to its printers.

The map was made using the relief woodcut technique, which was common well into the sixteenth century. Towns and cities like Strasbourg, Nuremberg and Basle, with their strong tradition in craftsmen, and easy access to wood, paper and water, were perfectly positioned to develop printed woodcuts. The woodcut method involved fashioning a block from a plank of wood; the craftsman (in German known as a *Formschneider*) carved out the non-printed areas (white in the final printed version) with knives and chisels, to leave the linear design of the map in relief, which then received the ink and produced the impression of geographical features. This was a far more laborious and skilled process than setting type for a short written text like the *Cosmographia introductio*, and prescribed the visual vocabulary of the final printed object. The woodcut technique was limited in its capacity to reproduce gradations of tone and fineness of line and detail, all of which are essential in territorial representation. Where geographical information was limited, the woodcut was left flat, creating no impression on the surface of the paper. The large, blank areas in Africa and Asia on the *Universalis cosmographia* are therefore a result as much of the printing process as the limitations of geographical knowledge.

Another problem the printers faced was lettering. Maps need to combine text with line, and this led early printers to cut letters directly onto the block alongside the map's visual detail, which then carried the distinctive square, severe Gothic lettering produced by flat-bladed knives, but the Waldseemüller map was being produced at exactly the moment when the older technique was giving way to the more elegant Roman type favoured by Italian humanists. It is a sign of the speed with which the map was put together that it uses both Gothic and Roman lettering, although this led to various inconsistencies in letter size and shape. In fact, the map shows two ways of reproducing the letter forms. The first was to cut them directly onto the block, although this was time-consuming. Another was to chisel a slot in the woodblock and wedge in type using glue. This also presented the printer with problems, as mistakes could easily creep in, and multiple insertions of type led to the block looking like a honeycomb, which could produce warping or even splitting. Setting one masterforme (the two-sided frame on which type was set to print both sides of a folio sheet) could be at least one day's work by two compositors, just for the text. This did not include carving intricate geographical outlines onto a woodblock, and then setting it with type, which would have taken much longer, stretching over weeks rather than days. Multiplying this kind of specialized labour by twelve (the number of sheets which made up the *Universalis cosmographia*) gives some idea of both the daunting nature and the remarkable speed with which the Gymnasium's project was executed throughout 1506–7.[26]

A further difficulty was reconciling the use of woodcut illustrations with type. Printers would often take as many 'pulls', or imprints from the woodcut map illustration as they felt were required for a particular edition, then put them to one side, while they broke up the valuable type for use in the printing of other books. When the maps were reassembled for another print run, the type needed to be reset, at which stage minor corrections could be made – or new errors could appear. This might have important consequences for the surviving Waldseemüller map. Many other apparently 'identical' printed maps from the early sixteenth century still exist in different editions with noticeably different lettering, giving the lie to the belief that printed maps are always exact copies of an original.[27] These problems of reproduction led many readers and scholars to temper the enthusiasm for print expressed by the likes of Sebastian Brant; one of Brant's contemporaries warned that such errors by careless printers turned the medium of print 'into an instrument of destruction when they, completely devoid of judgment, do not print well-emended books, but ruin them by bad and careless editing'.[28]

A final problem faced by the Strasbourg printer was how to transfer the enormous cartographic design (presumably drawn by Wald-seemüller) onto the woodblocks. With prime responsibility for the original manuscript map, Waldseemüller also had to supervise its transfer onto the twelve blocks, either by drawing on the blocks in reverse or by pasting the original manuscript maps onto the block and cutting through it before it was carved in relief. This second method would have involved varnishing the map on the back to allow the image to come through, and then cutting through and down into the block. The major drawback to this process was of course that the original map was destroyed, although it might also explain why Waldseemüller's hand-drawn map has not survived (like many maps from this period that went into print). With simple, diagrammatic maps like the first known

Fig. 14 T-O map from Isidore of Seville, *Etymologies*, 1472.

printed example, a T-O map used to illustrate an edition of Isidore of Seville's *Etymologies*, printed in Augsburg in 1472, many of these problems were relatively straightforward. But with maps printed on the scale of the *Universalis cosmographia*, the logistical problems involved were immense.[29]

We do not know exactly whether the tripartite publication was sold as a package, or if its elements were sold separately. They were certainly very different: each of the wall map's twelve sheets was nearly twice the size of the dimensions of both the introductory text and the globe gores. But taken together this was an ambitious declaration of the classical and modern state of cosmography and geography in all its dimensions. These texts collectively represented an irrevocable departure from medieval *mappaemundi*. The causes were obvious: the impact of printing, which produced the map's utterly different appearance; the influence of Ptolemy's *Geography*; and the effect of contemporary geographical discovery, most noticeably that of Vespucci in the 'New World' of the Americas. The Gymnasium's achievements were not confined to changing the geographical representation of the world: they were also part of a new approach to geography as an intellectual discipline, in terms of both the ways it was produced and how it was used. Where the Hereford *mappaemundi* provided answers to the world's divine creation and the afterlife, the *Universalis cosmographia* tried to unify classical, medieval and modern representations of the world in line with Renaissance humanist thinking, and made possible the circulation of multiple copies of roughly the same image to a range of individuals – scholars, navigators and diplomats – all with very different interests in this emerging 'new world'.

The *Universalis cosmographia* neatly divides the world into two halves, a western and an eastern hemisphere (although they are not named as such), oriented with north at the top. To the right, the six sheets run down from north to south through the Caspian Sea, the Arabian peninsula and the east coast of Africa. Although the orientation and shape of the medieval *mappaemundi* has gone, much of the descriptive detail on the map still derives from medieval and classical geography. The depiction of central and eastern Asia is mainly drawn from Marco Polo's late thirteenth-century travels, and the rest of the region reproduces Ptolemy's erroneous geography. Although the map draws on Caveri's sea chart with traces of early Portuguese voyages to the Indian narrow isthmus at approximately 30° N. To the north, the continent ends abruptly with a right-angled line drawn at 50° N; to the west are mountains and a legend stating 'Terra ultra incognita' (the land beyond is unknown). It is a highly abbreviated version of modern North America, but with intriguing elements, including what look like the Florida peninsula and a Gulf coast. The Caribbean Islands, including 'Isabella' (Cuba) and 'Spagnolla' (Hispaniola), are shown off the eastern coast, in a sea labelled for the first time as 'Oceanus Occidentalis', or Western Sea. The continent supports Spanish claims to the region by flying the flag of Castile, but is not given the name of America. Instead, in its southern regions, it is named 'Parias' in capital letters. So the great birth certificate of America actually calls North America 'Parias', a word taken from Vespucci's account of his meeting with the local inhabitants who used it to designate their homeland.

The map reserves the name 'America' to describe the southern land mass, and is placed in the location of modern-day Brazil. This southern region is far more extensive and detailed than its northern neighbour. Although the southernmost point is cut off at 50° S (conveniently eliding questions of its possible circumnavigation), the region bears the imprint of fifteen years of intensive Spanish and Portuguese exploration of the coastline. To the north, a legend reads, 'This province was discovered by order of the King of Castile', and a legend above the Castilian flag flying off the north-east coast states that 'these islands were discovered by the Genoese admiral Columbus by commission of the Castilian king'. Although these legends give prior political claim to Spain, the legend off the south-east coast below the depiction of a Portuguese ship reads, 'the vessel was the largest of ten ships, which the King of Portugal sent to Calicut [in India], that first appeared here. The island was believed to be firm and the size of the previously discovered surrounding part was not known. In this place men, women, children and even mothers go about naked. It was to these shores that the King of Castile later ordered voyages to ascertain the facts' – a reference to Pedro Alvares Cabral's voyage of 1500.[30] Because Cabral sailed further out into the Atlantic than da Gama, he accidentally 'discovered' Brazil. Like Waldseemüller and his colleagues, he assumed it was an island, and left to sail on to India.

The map's representation of this new western continent was without precedent, but within the map as a whole it was hardly advertised as

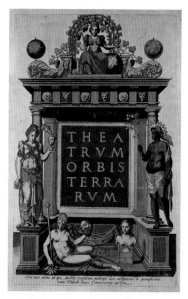

1. The earliest known world map: the Babylonian world map, from Sippar, southern Iraq, *c.* 700-500 BC.

2. The world as a theatre: the frontispiece to Abraham Ortelius, *Theatrum orbis terrarum* (1570).

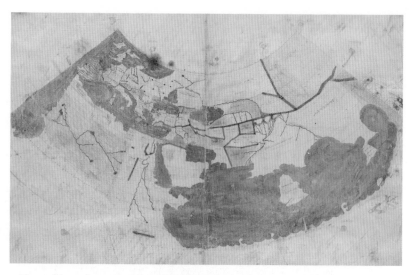

3. The world map from one of the earliest known copies of Ptolemy's *Geography*, written in Greek, thirteenth century. © 2012 Biblioteca Apostolica Vaticana (Urb. gr. 82, ff. 60v-61r)

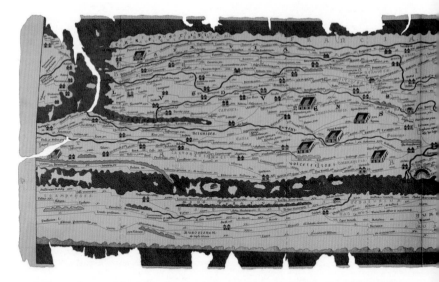

4a. A nineteenth-century facsimile of the Peutinger map (*c.* 1300), showing (from left to right) England, France and the Alps, and North Africa running along the bottom.

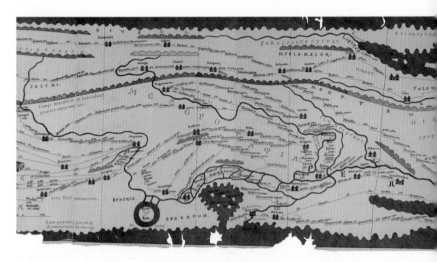

4b. The easternmost limits of the Roman world on the Peutinger map: Iran, Iraq, India and Korea.

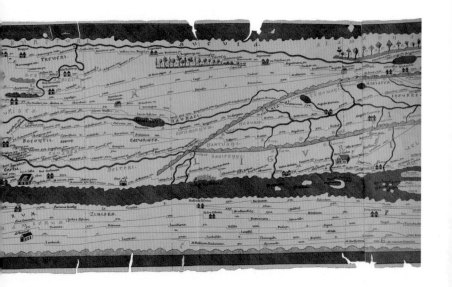

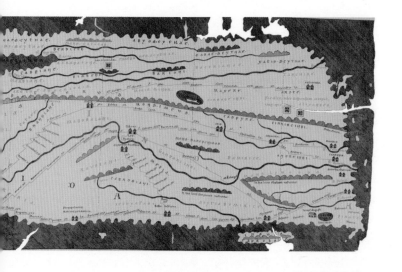

5. Twelfth-century Greek, Arab and Latin scribes working alongside each other in the chancery of King Roger II of Sicily.

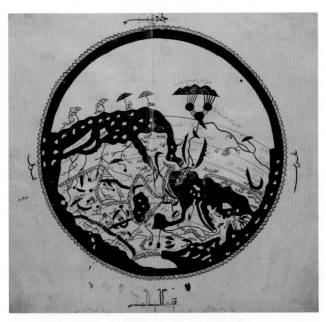

6. The circular world map from a sixteenth-century copy of al-Idrīsī's the *Entertainment* (1154), showing the convergence of Latin and Arabic geographical knowledge.

7. Suhrāb's diagram for a world map, in 'Marvels of the Seven Climates to the End of
Habitation' (tenth century), with a diagrammatic map showing the earth's seven climates.

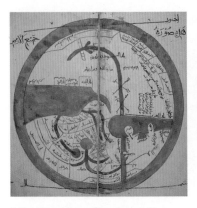

8. Ibn Hawqal's world map (1086), oriented with south at the top.

9. The circular world map from the anonymous *Book of Curiosities*, almost identical to the world map found in al-Idrīsī's *Entertainment*.

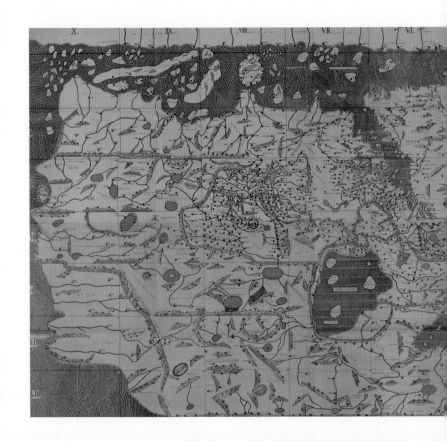

10. The unique rectangular world map from the *Book of Curiosities*, from a thirteenth-century copy, oriented with south at the top and with a scale bar.

11. A reconstruction of the world map combining the seventy regional maps drawn in al-Idrīsī's *Entertainment*.

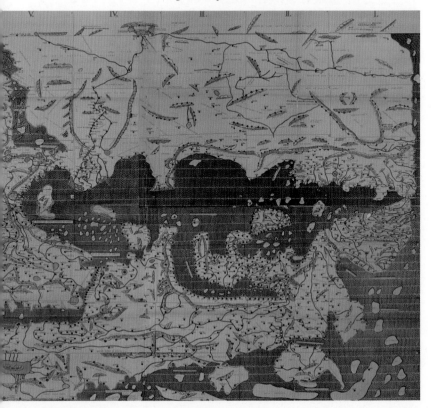

12a. The Hereford *mappamundi* (*c.* 1300), with east at the top.

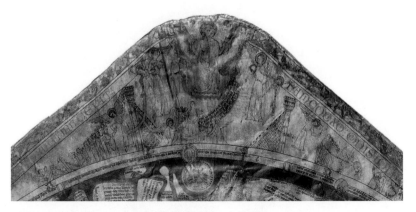

12b. Christ flanked by angels leading people to heaven and hell.

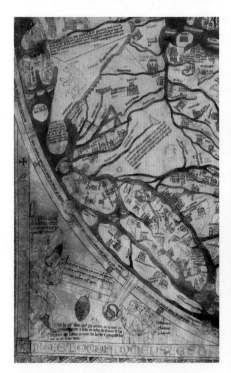

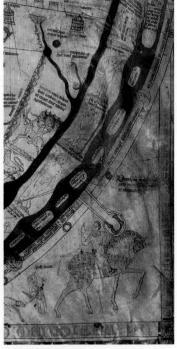

12c. The Roman emperor Augustus Caesar sending consuls to survey the earth. The British Isles are shown on the map directly opposite him.

12d. A rider gazes up at Africa and its 'monstrous' races, next to the words 'Go ahead'.

13. Zonal map from Macrobius' *Commentary on Scipio's Dream* (ninth century), showing the earth divided into temperate, frozen and 'torrid' zones.

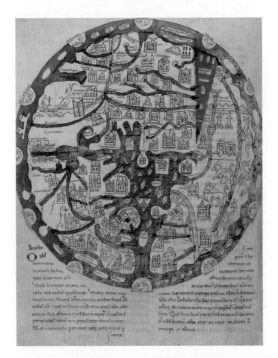

14. A twelfth-century world map illustrating Isidore's *Etymologies*. Despite a diameter of just 26 centimetres, it bears a striking resemblance to the Hereford *mappamundi*.

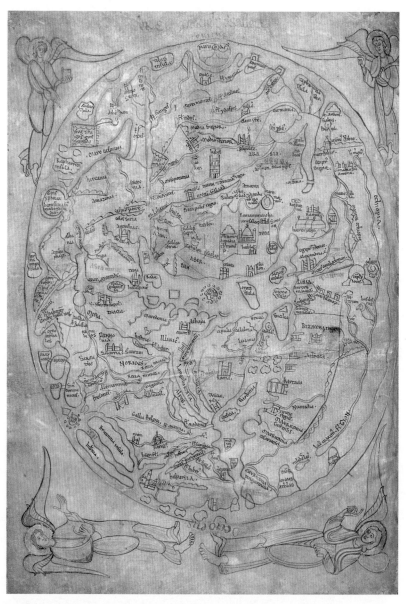

15. The Sawley map: the earliest known English *mappamundi* (1190), discovered in a Cistercian monastery in Yorkshire.

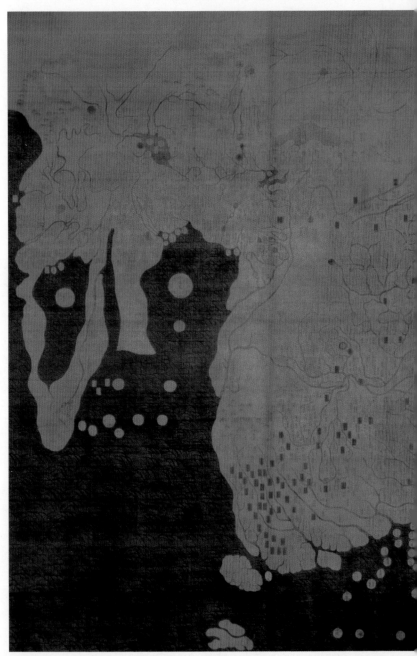
16. The Kangnido map (1470), the earliest known East Asian map to show the whole world, Europe, and Chosŏn Korea.

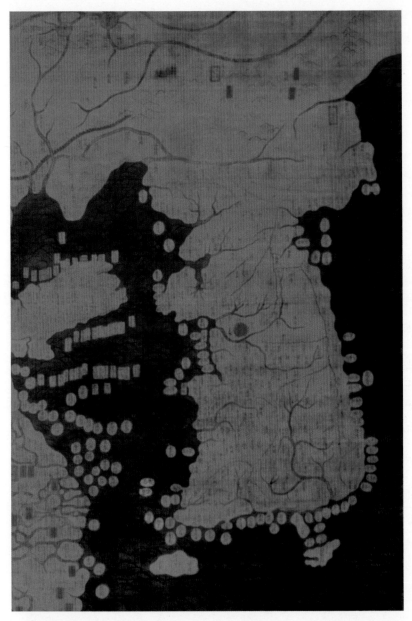

17. Detail of the Korean peninsula from the Kangnido map, showing key administrative and military sites.

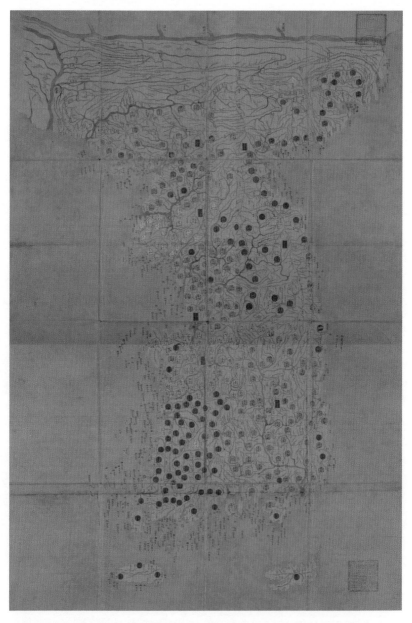

18. Copy of an official map of Korea by Chŏng Ch'ŏk (1390–1475), showing the influence of geomantic mapping, with colour-coded 'cosmic energy' flowing through the river systems (coloured blue) and mountain ranges (green), with district seats given other distinctive colours according to their provinces.

19. America's birth certificate: Martin Waldseemüller's map of the world (1507), the first to name and show America as a separate continent, bought by the US Library of Congress in 2003 for $10 million.

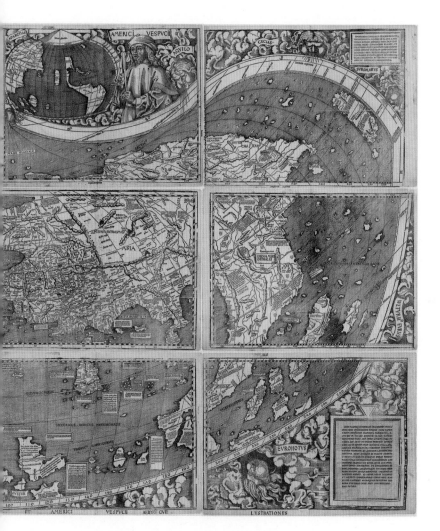

20. Nicolo Caveri's world chart (*c.* 1504–5), showing the new discoveries of the time, but still indebted to the *mappamundi* tradition, with Jerusalem at its centre.

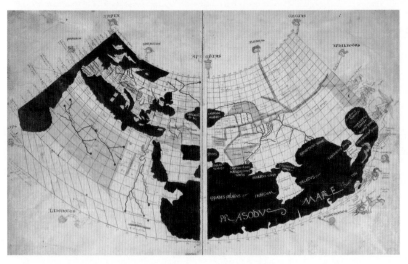

21. The earliest world map from Ptolemy's *Geography* in Latin (early fifteenth century), part of the European Renaissance's 'rediscovery' of classical learning.

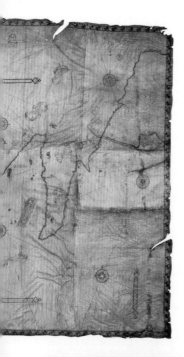

22. A change of mind? Martin Waldseemüller's map from his 1513 edition of Ptolemy's *Geography*, where 'America' has been replaced by 'Terra incognita'.

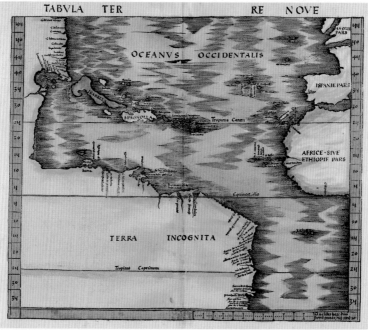

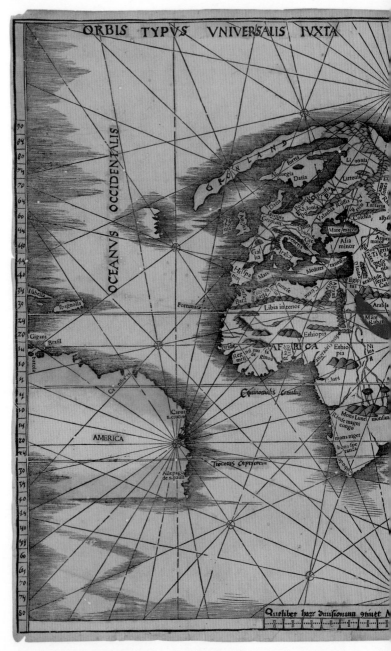

23. A world map attributed to Waldseemüller showing 'America' but (according to Henry N. Stevens) dated to 1506. Is *this* the first map to name the continent?

Balor
regio

indel
claud.

Tangat
p·uin.
Poliscu fl. fingul
puin.

Cathaya

ASIA
Auracithis
regio

Occndo z

Anuica fl

Bateita regio

tholo
uia
puin.

Quinfay
ciuit

Isfartisis

ua ucra
prouā

hircinū

Ostiastus

sacharum
regio

serica
regio

O re fl.

Māgi
puin.

Hircania ferg ula
Par aua

Scithia intra Imaū

India
superior

Perchia

Rudian

Cantaria

India   in tra

Ihicia
regio

Curmoba

Sinᵍ
Perfi

Catmai

Indus fl.   Ca

Cumba

India

Atitirea

Canalex
indiafcha

Pecini

gange

Falinga

Calicut

Sinᵍ gan
getic9

Murfuli
regiō

Tapobana

Mallaqa

Sinus magnus

Meabat
regniū

Lear regnū

Iaua maior

Regnū ac

Regnū
uia

Regnū

MARE   INDICVM

Saila

Iaua
minor

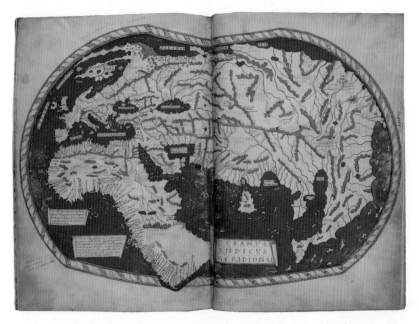

24. Henricus Martellus's world map (*c.* 1489). The discovery of the Cape of Good Hope breaches the Ptolemaic boundaries of the classical world map.

25. The Cantino Planisphere (1502), smuggled out of Lisbon by an Italian spy eager to learn about Portugal's commercially lucrative discoveries.

26. The earliest known terrestrial globe, made by Martin Behaim in 1492. Its underestimation of the size of the earth inspired Columbus and Magellan to embark on their eastward voyages.

27. Antonio Pigafetta's map of the Moluccas (1521), based on his first-hand experience of the spice-rich islands.

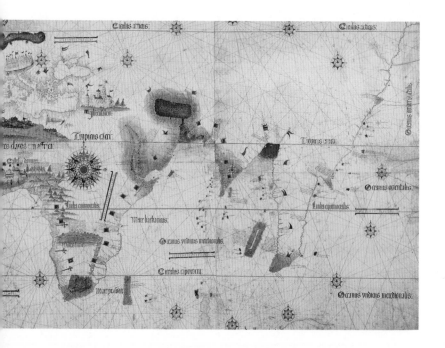

28. Nuño García's chart of the Moluccas (c. 1522), showing the islands in the Castilian half of the globe, east of the red dividing line agreed at Tordesillas (1494) which runs through Sumatra, where it intersects with the Equator.

revolutionary. Look again at the first known depiction of separate eastern and western hemispheres at the top of the map: to the left, Ptolemy is shown holding a quadrant, a symbol of his classical measurement of the stars and land. He stands next to an inset map of the classical *oikoumenē* of Europe, Africa and Asia, which is also the world on which his gaze falls as he looks at the larger map below. To the right stands Amerigo Vespucci, holding a pair of compasses, a more practical emblem of his modern navigational method, who is depicted next to an inset of the western hemisphere, which lacks any mention of 'America', and is simply designated as 'Terra incognita'. It does, however, show the first known image of the Pacific Ocean, with the geographically unfeasible straight line demarcating the west coast of North America improbably close to 'Zipangri', or Japan, and further to the west, Java. Like Ptolemy, he looks down on the half of the world with which his discovery is associated. The two men's eyes meet as they gaze across each other's respective spheres of influence, a suitable look of mutual admiration, as if to emphasize the map's interpretation of the world: it records the monumental discoveries of Vespucci and his forebears, including Columbus, and places them on a par with classical geography, but it also remains indebted to Ptolemy.

Much of the map's geographical detail of America relies on what it knows of Vespucci's voyages, but its frame remains Ptolemaic in ways that satisfied the Gymnasium's beliefs. The map's peculiar, bulb-like appearance and its pronounced graticule is the outcome of an attempt here to map the world by modifying the second projection described by Ptolemy in his *Geography*. Waldseemüller's decision to adopt Ptolemy's projection shows a mapmaker returning to classical models of representation to understand and then describe the contours of a newly emerging world. Prior to the geographical expansion of the known world in the fifteenth century, mapmakers were able simply to depict the particular hemisphere which they inhabited without seriously addressing the problems of projecting the circular globe upon a plane surface. Columbus's and Vespucci's voyages to the Americas presented mapmakers with precisely this problem of depicting both eastern and western hemispheres on a flat map, and their contemporaries quickly grasped the conundrum. Writing in 1512, the Nuremberg scholar Johannes Cochlaeus admitted that 'truly the dimension of the earth as now inhabited is much greater than these ancient geographers described it'.

He could have been describing the *Universalis cosmographia* when he continued:

> For beyond the Ganges the immense countries of the Indies stretch out, with the largest island of the East: Japan. Africa is also said to extend far beyond the Tropic of Capricorn. Beyond the mouth of the river Don there is also a good deal of inhabited land as far as the Arctic Sea. And what about the new land of Americus, quite recently discovered, which is said to be bigger than the whole of Europe? Hence we must conclude that we must now allow for wider limits, both in longitude and latitude, to the habitable earth.[31]

There were three possible geographical responses to this problem, each of them represented in the book, map and globe gores published by Waldseemüller and his colleagues. The first possibility was to depict both hemispheres, which is what we see at the top of the *Universalis cosmographia*. The second was to split the world into discrete parts, rather like the globe gores printed to accompany the map and its introductory textbook. The final possibility was to create a projection that tried to represent as much of the globe as possible on a flat map, while minimizing the distortion of land at its edges. On the *Universalis cosmographia* this is achieved by once again turning to Ptolemy, and reproducing a version of his second projection.

Ptolemy pointed out in the *Geography* that the second projection was more ambitious than his first because it 'was more like the shape on the globe than the former map', and therefore 'superior', although harder to draw than his first projection.[32] This second projection retained the illusion of the sphericity of the terrestrial globe by drawing the horizontal parallels as circular arcs, and the vertical meridians as curved. This created the impression of viewing the earth from space, where the eye in effect 'sees' a globular hemisphere. Looking straight on, the viewer perceives the great circle of the central meridian as a straight line, with the other meridians appearing on either sides of this meridian as equally balanced arcs, growing increasingly curved the further east and west they stretch. Similarly, the vertical parallels, which are in effect circles running right round the globe, are shown as concentric circular arcs.[33]

Waldseemüller and his colleagues adopted Ptolemy's second projection as the best model they knew to represent the world as a globe, but this required substantial modification of the projection and its global surface area. The Waldseemüller map extended Ptolemy's latitudinal parallels to 90° N and 40° S, allowing another 50° in which to represent the recent voyages of exploration from north to south, particularly those down the coast of Africa and into the Indian Ocean. This was significant enough, but along its east–west longitudinal axis Waldseemüller's departure from Ptolemy was even more innovative. Although the map retained Ptolemy's prime meridian running through the Canary Islands, it doubled Ptolemy's breadth of the known world, increasing it to 270° E and 90° W. This allowed for the portrayal of North and South America in the west and Japan in the east, but it also led to serious distortion at its furthest longitudinal limits.

The results of this turn to Ptolemy were not always successful, but even their limitations suggest some intriguing puzzles. As Waldseemüller and his team were unable to use modern mathematical equations to plot their graticule, their solutions were uneven and discontinuous, which is perhaps why the meridians on the map seem to be segmented rather than smooth arcs running south from the equator, especially at their eastern and westernmost extents (although another more prosaic possibility is that the bottom left- and right-hand woodblocks were simply too small to retain the smooth curvature of the meridians, leading to the abrupt change in angle). Similar problems are also visible in the depiction of North and South America, with their unrealistic, angular coastlines. Until recently, scholars have assumed that they simply represent an inability to project land any further. Recent 'cartometric' analysis by John Hessler in the Geography and Map Division at the Library of Congress, uses methods of computation to assess the map's depiction of terrain, and claims that these regions of the map look as they do, not because of lack of geographical information, but because of the serious distortion caused by the partial adaptation and elongation of Ptolemy's second projection.[34] Hessler shows that if we take into account the distortion caused by Ptolemy's projection, the map's representation of America and in particular its western Pacific coast is startlingly accurate. This is all the more perplexing given that the map predates Vasco Núñez de Balboa's first European sighting of the Pacific in 1513, and Magellan's crossing of the Pacific in 1520. Hessler can only conclude that the Gymnasium had access to maps and geographical information that have since been lost, although why they would wish to remain silent about their sources for describing a whole new continent and ocean remains a mystery.

In its written descriptions, its accompanying introductory textbook and its incorporation of Ptolemy's second projection, the Gymnasium's publications accommodated the slow and contradictory information about the 'discovery' of new lands within the prevailing classical theories of the known world. The result was an impressive publication, but one which implicitly conceded that it was only offering a snapshot of how the rapidly evolving world of 1507 looked. Its various facets – world map, globe gores, textbook – offered different perspectives on how to look at and understand this changing world. Waldseemüller boasted that the map was 'scattered throughout the world not without glory and praise'.[35]

The map's subsequent impact was certainly 'scattered', but also decidedly mixed. Waldseemüller later claimed that 1,000 copies of the *Universalis cosmographia* were printed. This was not an unusual figure for the time, but certainly a large one for such a complicated printing job. However, only one allusion to an acquisition of the map survives, and even that cannot be definitely said to refer to the *Universalis cosmographia*. In August 1507 the Benedictine scholar Johann Trithemius wrote that he had recently 'purchased cheaply a handsome terrestrial globe of small size lately printed at Strasbourg, and at the same time a large map of the world containing the islands and countries recently discovered by the Spaniard [*sic*], Americus Vespucius, in the western sea'.[36] If this is the *Universalis cosmographia*, it is hardly celebrated as a revolutionary artefact: Trithemius seemed more pleased with his cheap, novel globe. Other mapmakers copied the map and adopted its naming of America, including Peter Apian's 1520 world map (which dates the continent's discovery to 1497), and Sebastian Münster, who called the region 'America' and 'Terra nova' on his world map of 1532, and then 'America or the island of Brazil' in a subsequent 1540 map. It was only in 1538 that Gerard Mercator first applied the term to the entire continent, but he dropped the name when it came to plotting his famous world map in 1569 (see Chapter 7). By the end of the sixteenth century the name finally acquired universal geographical and toponymical status, thanks to German and Dutch mapmakers who needed a name to describe the continent and one which avoided ascribing it to a particular empire (some maps referred to it as 'New Spain') or religion (other maps labelled it 'Land of the Holy Cross'). In the end, the name 'America' endured, not because of any agreement as to who discovered it, but because it was the most politically acceptable term available.

Even Waldseemüller himself had second thoughts about using the term 'America'. Following the publication of the *Cosmographia introductio* and the *Universalis cosmographia*, he and Ringmann continued with the project to complete a new edition of Ptolemy's *Geography*. Despite Ringmann's death in 1511, Waldseemüller carried on with the edition, which was published by the printer Johannes Schott in Strasbourg in 1513. Here, the region previously labelled 'America' becomes an enormous 'Terra incognita', poised ambiguously between an island and a continent, noticeably denied a western coastline in case subsequent voyages re-established a connection with Asia. Not only is 'America' wiped from the map, so is Vespucci: the map's legend reads, 'This land with the adjoining islands was discovered through the Genoese Columbus by order of the King of Castile.'[37]

Perhaps Ringmann was the driving force behind the decision to put 'America' on the 1507 *Universalis cosmographia* all along (he initially edited Vespucci's *Mundus Novus* and, it has been argued, took primary responsiblity for writing the *Cosmographia introductio*). Maybe his death in 1511 liberated Waldseemüller from having to reproduce a region and nomenclature in which he never really believed.[38] But it is more likely that Waldseemüller's decision to drop the term 'America' from all his subsequent maps was a response to the publication of another collection of travel narratives which he consulted, entitled *Paesi novamenti retrovati* ('Lands Recently Discovered'). This collection was published in Vicenza in 1507, but only reached Germany in 1508, when it was translated as *Newe unbekanthe landte* ('New Unknown Lands'). The book arrived too late to change the primacy of Vespucci's discoveries on the *Universalis cosmographia*, but it did allow Waldseemüller to adopt its chronology of discovery in all his subsequent maps. The *Paesi* argued for Columbus's first voyage of 1492 as the prime moment of discovery, followed by Pedro Cabral and his landing in Brazil in 1500, and then Vespucci, whose first landfall was dated 1501, rather than 1497.[39] In his later work Waldseemüller seems to have continued to rely on Ptolemy's geographical frame while cautiously introducing new information whenever it reached him until his death some time between 1520 and 1522. The irony is that, having originally been involved in putting 'America' on the map in 1507, Waldseemüller died having apparently retracted his belief in its name and its status as a separate land mass; and even the 1507 map kept its options open by referring to the continent as an 'island'.

There was one further moment of 'discovery' left to come. In the summer of 1901 the German Jesuit Father Joseph Fischer was granted permission by Count Waldburg-Wolfegg to examine Wolfegg Castle's collection of historical documents. As he sifted through the castle's archive, he came across an early sixteenth-century bound portfolio owned by the Nuremberg scholar Johannes Schöner (1477–1547). It contained a star chart by the German artist Albrecht Dürer, celestial globe gores made by Schöner (both dated 1515), Waldseemüller's 1516 world map, and the only known surviving copy of all twelve sheets of the 1507 world map. To find any one of these artefacts was exciting: to discover four at the same time made it one of the most important coups in the history of cartography. Fischer knew he had found one of the great lost maps of the Renaissance. He rushed an academic article into print on the subject, claiming that this was the lost map discussed in the *Cosmographia introductio*, and the very first printed version to come off the press. This was quickly followed by a facsimile edition of the newly discovered 1507 and 1516 maps, published in 1903 as *The World Maps of Waldseemüller (Ilacomilus) 1507 & 1516*.

Fischer's recovery of what he described as the first map of the new continent, attributed to Waldseemüller, did not meet with universal approval. By the late nineteenth century the provenance and originality of rare books and antique maps had become a lucrative business, particularly in North America, where wealthy philanthropists began to endow museums and cultural institutions in an attempt to turn the study of American history into an internationally respected discipline. One such figure was John Carter Brown (1797–1874), an avid collector who endowed a library named after him which is now attached to Brown University in Providence, Rhode Island, dedicated to the study of 'Americana'. Brown's most trusted adviser, with responsibility for the library's book and map acquisitions, was Henry N. Stevens. In 1893 Stevens acquired a copy of Waldseemüller's 1513 edition of Ptolemy. Although the world map it contained was similar in most respects to those reproduced in all the other copies of the 1513 Ptolemy, this one

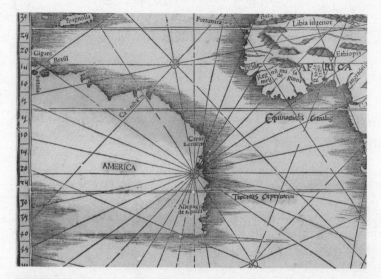

Fig. 17 Detail of America from Waldseemüller, world map dated 1506 by
Henry N. Stevens.

contained a particularly vital addition: the southern continent in the
western hemisphere was inscribed with the word 'America'. Stevens
believed that the map was by Waldseemüller, but made in 1506. He was
by implication claiming to have 'discovered' the long-lost world map
discussed by Waldseemüller and Ringmann in the *Cosmographia
introductio*.

Stevens's claim was prejudiced by the fact he was trying to sell his
map to the John Carter Brown Library for £1,000 (on which Stevens
also stood to earn a 5 per cent commission from the library). In the spring
of 1901 Stevens submitted a report explaining his reasons for dating the
map to 1506, based on examination of its paper, watermarks, type and
toponymy. He concluded that the map was inserted in a copy of the
1513 Ptolemy, and was an experimental design for the proposed edition
on which the Gymnasium Vosagense was working in 1505–6. The library
was satisfied that they were buying the first map naming America, and
purchased it in May 1901. It still resides there today. Just six months
later Fischer announced his discovery at Wolfegg, which he quickly
labelled 'the oldest map with the name America of the year 1507'. Stevens needed to move quickly to avoid professional embarrassment. His

response was to launch into correspondence with the John Carter Brown Library, offering to help buy the Wolfegg map, while also still asserting confidently that his map was earlier than Fischer's. He also persuaded various scholars and curators in the field to write apparently disinterested academic articles claiming that the map he had sold to the Carter Brown collection predated Fischer's discovery. Privately, Stevens revealed both his scholarly fears and a certain amount of national prejudice when he wrote at one point that 'I sincerely hope the Germans keep the darned thing. I wish they had never discovered it.'[40]

Just like his early twenty-first-century descendant, the incumbent prince, Max Waldburg-Wolfegg, initially expressed interest in selling his map, and in 1912 shipped it to London to be insured by Lloyds for £65,000, before offering it to the Library of Congress for $200,000 (worth $4 million in 2003). The library declined the offer. In 1928 Stevens returned to the fray to reassert the primacy of 'his' map, in a book reiterating his claims that the Carter Brown map was printed in 1506. It was based on his interpretation of the letters written by Waldseemüller and other members of the Gymnasium Vosagense in 1507 that a world map describing newly discovered regions of the world 'has been hurriedly prepared' for publication. Stevens concluded that 'his' map was printed in 1506, just before the far grander twelve-sheet *Universalis cosmographia*.

Subsequent debate was sceptical about Stevens's conclusions. Several scholars pointed out that the paper and type used in the making of Stevens's map were also used in books published as late as 1540; it was unlikely that the Gymnasium Vosagense would have produced a map in 1506 more geographically accurate than the supposedly later *Universalis cosmographia*. Writing in 1966, the distinguished map historian R. A. Skelton conceded that the Stevens map was probably printed in the same year as the *Universalis cosmographia*, but that no amount of technical analysis of paper, type or other such technical specifications would ever definitely resolve the debate over their exact chronology. There was one final intriguing twist in 1985, when the curator Elizabeth Harris conducted a detailed typographic analysis of the Wolfegg map rediscovered by Fischer. Harris analysed the map's paper, watermarks and woodblocks, which showed splits. This was usually a sign of repeated printing, which displayed noticeable blurring in the lettering. Harris concluded that the Wolfegg map was not the first printed version of

1507, and that it was in fact a later version, using the original wood-blocks, but printed no earlier than 1516, and possibly much later.[41]

If true, Harris's conclusions reveal that the only known copy of the *Universalis cosmographia* was actually printed at least nine years later than the date of the original woodblocks. This does not necessarily cast doubt on its original creation in 1507, but it does mean that the Library of Congress possesses a map physically printed around 1516, possibly later than the first printing of Stevens's map. Such conclusions further complicate any attempts to claim primacy or originality when it comes to printed maps. Rather like the debate over whether Columbus or Vespucci first 'discovered' America, the controversy over which map first named America as a continent is ultimately a matter of interpretation. Having lost the original woodblocks and first impression of the *Universalis cosmographia*, should Stevens's map in the John Carter Brown Library be given preference as the 'first' map to name America, even if scholarship remains unable to definitively date its creation to 1506, or even 1507? Both the John Carter Brown Library and the Library of Congress retain vested institutional and financial interests in the primacy of their maps: the US taxpayer would presumably be unimpressed if it knew that half the cost of a map acquired for $10 million by the national library with public money was predated by another map in a private library in Rhode Island, which cost just $1,000 in 1901.

The so-called Waldseemüller map of 1507 takes us a long way from the *mappaemundi* that preceded it, and the debates around 'discovery' that defined it ever since its creation in Saint-Dié between 1505 and 1507. It *does* represent a shift in the mentality of mapping and its makers which can be seen as representative of European Renaissance cartography. Mapping now drew on classical geography, and in particular Ptolemy, more confidently than ever before. It proposed a new role for itself as cosmography, the science of describing the earth and the heavens in a harmonious, universal whole. As well as drawing on classical geography to describe the world, maps like Waldseemüller's incorporated contemporary maps and charts showing navigational breakthroughs and explorations of places unknown to Ptolemy and his predecessors. This approach to knowledge was cumulative. It did not represent a revolutionary break with previous geographical beliefs. The map and its makers cautiously proposed changes to their classically inspired world, and where evidence conflicted, they were as likely to fall back on the old rather than accept the new.

In the new era of print, Waldseemüller and his colleagues worked with what little information on exploration and discovery they had in front of them, and made decisions accordingly. Naming a new region 'America' in 1507 was a highly provisional decision, and dependent on the ability of the printing press to circulate sensational but unverified news of the 'discoveries' of Columbus, Vespucci and others. For the scholars in Saint-Dié, this resulted in calling a continent an island, and then withdrawing the name they had given to one part of it, America, as subsequent publications cast doubt over their initial findings.

Ultimately, printing changed the whole tenor of our understanding of the *Universalis cosmographia*, and so many of the other maps surrounding it. This was not only because printing increased the possibility of exact reproducibility, standardization and preservation of maps and books, but also because it gave rise to piracy, forgery, misprinting, and the financial interests of printers, typesetters, compositors and editors in any attempt to describe what really happened in the creation of these maps. Printing introduced a whole new dimension to the making of maps that was unknown to medieval manuscript mapmakers, where the mapmaker alone, sometimes with a scribe and illuminator, had been responsible for creating a map. It added a new layer of personnel into the process of making a map, which is why identifying Waldseemüller, or Ringmann, or a particular printer as the author of a map becomes virtually impossible. Printing transformed how a map looked, including its depiction of geographical relief, shading, symbols and lettering; and it altered the purpose of a map, which became tied to money, and a new, humanist scholarship that saw it as a device for understanding the expansion of the world beyond Europe's borders.

The history of the Waldseemüller map remains in many respects a mystery. Questions about it remain unanswered, from its depiction of the Pacific and that peculiar wedge of America, to its almost immediate disappearance from historical records. But what it inadvertently shows is that the discovery of origins – of America and of the chronological primacy of one map over another – is a chimera. What is found at the historical moment of the creation of any world map is not the inviolable identity of its origin but the dissension of disparate stories, competing maps, different traditions. The French philosopher Michel Foucault's criticism of the belief in the certainty of origins could equally describe the history of the *Universalis cosmographia*: 'devotion to truth and the precision of scientific methods arose from the passion of scholars, their reciprocal hatred, their fanatical and unending discussions, and their spirit of competition.'[42] The dynamic complexity of early printing means that, for all its beautiful execution and years of scholarly labour, we shall probably never know for sure if the *Universalis cosmographia* can be called the 'first' map to properly describe and name America.

# 6

# Globalism

*Diogo Ribeiro, World Map, 1529*

*Tordesillas, Castile, June 1494*

In the summer of 1494, delegations representing the Castilian and Portuguese crowns met in the small town of Tordesillas near Vallodolid in central Castile. Their purpose was to resolve the diplomatic and geographical dispute caused by the return of Columbus from his first voyage to the New World in March 1493. Since the early decades of the fifteenth century the Portuguese had navigated their way down the African coast and out into the relatively unknown Atlantic, until Castile demanded clarification of the limits of the Portuguese sphere of possession. In 1479 the Treaty of Alcáçovas stipulated that Portuguese influence extended to territories 'in all the islands hitherto discovered, or in all other islands which shall be found or acquired by conquest from the Canary Islands down toward Guinea'[1] – a vague compromise which required immediate reappraisal upon news of Columbus's discoveries in 1492. The rulers of Castile and León, Queen Isabella I and her husband, the Aragonese-born King Ferdinand V, petitioned Pope Alexander VI (a native of Valencia) to uphold their claims to the newly discovered territories. Much to the anger of the Portuguese, the pope agreed in a series of bulls issued throughout 1493, which prompted the Portuguese king John II to demand a new round of negotiations.

The result was the Treaty of Tordesillas, signed on June 7 1494. In one of the earliest and most hubristic acts of European global imperial geography, the two crowns agreed that 'a boundary or straight line be determined and drawn north and south, from pole to pole, on the said ocean sea, from the Arctic to the Antarctic pole. This boundary or line shall be drawn straight, at a distance of three hundred and seventy leagues west of the Cape Verde Islands.'[2] Everything to the west of this line, including the territories discovered by Columbus, fell under the control of Castile, and everything to the east, including the entire African coastline and the Indian Ocean, was allocated to the Portuguese. The world was divided in half by two European kingdoms, using a map to announce their global ambitions.

The exact map used to demarcate the kingdoms' relative spheres of influence has not survived, but some world maps of the period reproduce the newly agreed meridian running west of the Cape Verde islands. The results of the partition were immediate: Spain took the opportunity to push on with voyages to the New World, while the Portuguese realized that if they were to capitalize on their control of the sea routes eastwards, they needed to reach India. The Portuguese king John II had informed Pope Innocent VIII in 1485 that he was confident 'of exploring the Barbarian Gulf [the Indian Ocean]', and that 'the farthest limit of Lusitanian maritime exploration is at present only a few days distant' from this ocean, 'if the most competent geographers are but telling the truth'.[3] John's claims may have been exaggerated, but by December 1488 the Portuguese pilot Bartolomeu Dias returned to Lisbon from a sixteen-month voyage during which he sailed down the African coast and became the first European to round the Cape of Good Hope.

Henricus Martellus's 1489 world map was one of the first to depict Dias's voyage. The mapmaker broke the frame of what is otherwise a typically Ptolemaic map to show that the southern tip of Africa was circumnavigable, a decision that Waldseemüller followed in his own attempt to represent the impact of the Portuguese voyages on his 1507 world map. By the late 1490s, with the route open into the Indian Ocean and the terms of the Treaty of Tordesillas barring Portuguese expansion westwards into the Atlantic, King John's successor Manuel I turned his attention to supporting an expedition to reach India.

The motives for such an expedition might have been couched in the language of religious conversion, but were equally concerned with breaking into the fabled spice trade. By the fifteenth century pepper, nutmeg, cinnamon, cloves, ginger, mace, camphor and ambergris were starting to trickle into Europe from the East, expensive and aspirational condiments enabling Christian courts to imitate exotic Arab recipes, as well as curing a variety of real and imagined ailments, and providing the constitutive elements for a range of perfumes and cosmetics. Until the late fifteenth century, it was Venice, the fabled 'Gateway to the East', which controlled all spice imports into Europe. From their harvest in South-east Asia, spices were sold to Indian merchants, who transported them back to the Indian subcontinent, where they were in turn sold on to Muslim merchants, who shipped them on via the Red Sea to Cairo and Alexandria. From there, they were bought by Venetians and shipped back to their native city, then sold to merchants from across Europe. The sheer length of time involved, and the customs duties imposed on transporting these precious commodities the thousands of kilometres from their origin, meant that by the time they reached Europe their price was high, but their freshness low.

The arrival of da Gama's fleet in Calicut on the south-western coast of India in May 1498 threatened to alter completely the balance of commercial power in Europe and in the Indian Ocean. Having successfully traded with the local merchants to obtain a cargo of pepper, spices and a variety of precious woods and stones, da Gama proved it was possible to circumvent the slow and expensive overland trade routes between Europe and Asia by transporting low-volume, luxury commodities like spices by sea back to Lisbon via the Cape of Good Hope. Manuel I quickly understood the consequence of da Gama's voyage for his kingdom's standing in European imperial politics. Writing to his Castilian counterpart following da Gama's return, Manuel piously hoped that 'the great trade which now enriches the Moors of those parts, through whose hands it passes without the intervention of other persons or people, shall, in consequence of our regulations, be diverted to the nations and ships of our own kingdom'. He solemnly concluded that 'henceforth all Christendom, in this part of Europe, shall be able, in a large measure, to provide itself with these spices and precious stones'.[4] Cloaking his delight at beating Castile to India in the rhetoric of Christian solidarity, Manuel knew that the principal kingdom in Christendom to benefit from da Gama's voyage would be Portugal.

It was not only Castile which felt eclipsed by news of da Gama's voyage: the Venetians were horrified at what they saw as a direct challenge to their control of the spice trade. Writing in his diary in 1502, the Venetian merchant Girolamo Priuli wrote that 'all the people from across the mountains who once came to Venice to buy spices with their money will now turn to Lisbon because it is nearer to their countries and easier to reach; also because they will be able to buy at a cheaper price'. Priuli understood that Venice could not compete with a situation where 'with all the duties, customs, and excises between the country of the [Ottoman] Sultan and the city of Venice I might say that a thing that cost one ducat multiplies to sixty and perhaps to a hundred'. He concluded that 'in this, I clearly see the ruin of the city of Venice'.[5]

Such predictions of Venice's demise turned out to be premature, but da Gama's voyage, and the subsequent establishment of the *Carreira da India*, the annual Portuguese commercial fleet sailing to India, transformed the emerging global economy. At its height in the mid-sixteenth century the Portuguese empire was dispatching more than fifteen ships a year to Asia, returning with an annual average of over 2,000 tons of cargo, rising to nearly double that towards the end of the sixteenth century. Nearly 90 per cent of Portugal's imports were made up of spices from the Indian subcontinent; pepper accounted for over 80 per cent of these spices. By 1520 the revenue from these imports represented nearly 40 per cent of the Portuguese crown's total revenue, although even that did not include the money collected from customs duties on trade moving in and out of Portugal's overseas possessions throughout the Indian Ocean.[6] The wealth that flowed into Lisbon and the Portuguese crown's revenues enabled the kingdom to transform itself into one of Europe's richest empires. Portugal's wealth and power now lay not in the possession of territory, but in the strategic control of commercial networks that lay thousands of kilometres from the imperial centre. Unlike earlier empires built on the acquisition and control of land, this was a new kind of empire built on water.

Without predominantly Portuguese scientific innovations in long-distance seaborne navigation developed throughout the late fifteenth century, the establishment of a regular fleet to the markets of South-east Asia would have proved hazardous at best. In such a climate, possession of geographical information became more precious than ever, and both crowns jealously guarded their cartographic secrets. In August 1501, at the height of Portugal's rivalry with Venice over control of the spice trade, Angelo Trevisan, secretary to the Venetian ambassador to Castile, wrote to his friend Domenico Malipiero, explaining the difficulties of obtaining Portuguese maps of India:

> We are daily expecting our doctor from Lisbon, who left our magnificent ambassador there; who at my request has written a short account of the

[Portuguese] voyage from Calicut, of which I will make a copy for Your Magnificence. It is impossible to procure the map of that voyage because the king has placed a death penalty on any one who gives it out.

Less than a month later, however, Trevisan wrote to Malipiero again, with a very different story:

If we return to Venice alive, Your Magnificence will see maps both as far as Calicut and beyond there less than twice the distance from here to Flanders. I promise you that everything has come in good order; but this, Your Magnificence may not care to divulge. One thing is certain, that you will learn upon our arrival as many particulars as though you had been to Calicut and beyond.[7]

Trevisan had somehow managed to obtain Portuguese maps whose circulation was, according to the Venetian, forbidden on pain of death. The maps offered invaluable information on the Portuguese sea route to India, but Trevisan was also interested in the more intangible, almost magical power of a map: the ability to allow its owner to imagine the territory itself. Trevisan rhetorically assures Malipiero that the map has the power to simulate the experience of actually being in Calicut – but safely insulated in his Venetian study from the dangers and hardships of months of life-threatening seaborne travel.

Although we do not know which maps the Venetian smuggled home, there is an example of a similar process of cartographic espionage, again at Portugal's expense, which took place the following year. The beautifully illustrated map known as the 'Cantino planisphere' is named not after the unknown Portuguese mapmaker who made it, but the Italian who stole it. In the autumn of 1502, Ercole d'Este, the duke of Ferrara, sent his servant, Alberto Cantino, to Lisbon, ostensibly to trade in thoroughbred horses. Instead, Cantino paid a Portuguese mapmaker twelve gold ducats to make a world map, which was smuggled out of Lisbon and sent back to Ferrara, where it was installed in Ercole's library.

The map remains to this day in northern Italy, in a library in the former Este residence of Modena, and shows the ferment of geographical knowledge at the beginning of the sixteenth century in gorgeous hand-illuminated colour. America is still undefined as a continent, with only a fragment of the Florida coast represented, dwarfed by the recently discovered Caribbean Islands. The interior of Brazil is also indeterminate, showing the Portuguese discovery of its eastern coastline in 1500. India and the Far East are only vaguely sketched in, reliant on da Gama's still relatively recent landing in Calicut in 1498. The map's detail is reserved for what mattered to the Portuguese crown: its trading stations in West Africa, Brazil and India, supplemented by a series of legends describing the commodities available in this newly emerging world. Ercole was uninterested in exploiting the map's navigational information about how to reach India: Ferrara was too small and geographically landlocked to consider itself a potential seaborne power. He was interested, rather, in displaying his access to the arcane knowledge which described how the shape of the sixteenth-century world was changing before the eyes of its rival kingdoms and empires.

In the western Atlantic, the Cantino planisphere reproduced the key feature of the Treaty of Tordesillas: a vertical line running from north to south, to the east of the Caribbean Islands, and bisecting Brazil. This partition appeared straightforward enough when projected upon a flat, plane map like this one, but it begged one monumental question: as the Portuguese sailed ever further east in the early years of the sixteenth century, and Castile pushed further into the New World, where would it fall if it ran right round the globe? A flat map conveniently avoided answering such a politically divisive question, but subsequent events would require both Europe's empires and its mapmakers to start to imagine the world globally, projected onto a sphere, rather than flat on a map attached to a wall or spread out on a table.

In 1511 the Portuguese captured Malacca on the southern tip of the Malaysian peninsula, one of the great distribution centres for spices arriving from the nearby Moluccas. The Portuguese realized that they were in touching distance of capturing the islands, and with it global domination over the spice trade. Then, just two years later, in 1513, the Castilian adventurer Vasco Núñez de Balboa crossed the isthmus of Darien in present-day Central America and became the first known European to see the Pacific Ocean. For Balboa, the discovery of the Pacific represented the possibility of claiming a whole new world for Castile. How far further westwards from Darien could Castile's claim to territorial possession extend? Where would the line drawn at Tordesillas fall when drawn through the Pacific? After taking Malacca in 1511, the Portuguese asked themselves the same question from the other direction. Could their influence extend as far eastwards as the Moluccas?

One man who thought that the Portuguese had reached the limits of their territorial claims under the terms of Tordesillas was one of the kingdom's most respected pilots: Fernão de Magalhães, better known today by his Hispanicized name, Ferdinand Magellan. Born around 1480 in Ponte da Barca in northern Portugal, Magellan joined the Portuguese fleet in 1505. By 1511 he participated in the Portuguese assault on Malacca, and it was at this point that he began to have doubts about Portugal's claim to territories any further to the east. Magellan himself never gave his reasons for doing so, but later writers gave fulsome explanations. Writing in 1523 after the return of Magellan's circumnavigation of the globe, the Habsburg adviser and scholar Maximilianus Transylvanus claimed:

> Four years ago, Ferdinand Magellan, a distinguished Portuguese who had for many years sailed about the Eastern Seas as admiral of the Portuguese fleet, having quarreled with his king who he considered had acted ungratefully towards him ... pointed out to the Emperor [Charles V] that it was not yet clearly ascertained whether Malacca was within the boundaries of the Portuguese or the Castilians, because hitherto its longitude had not been definitely known; but it was an undoubted fact that the Great Gulf [the Pacific] and the Chinese nations were within the Castilian limits. He asserted also that it was absolutely certain that the islands called the Moluccas, in which all sorts of spices grow, and from which they were brought to Malacca, were contained in the western, or Castilian division, and that it would be possible to sail to them and bring the spices at less trouble and expense from their native soil to Castile.[8]

As an adviser to Castile's ruler, the Habsburg emperor Charles V, it was in Transylvanus's interest to magnify Magellan's obscure dispute with his sovereign. Nevertheless, it seems that by October 1517 Magellan was convinced of the validity of Castile's claims to the Moluccas, because by then he was in Seville, working for Castile on his ambitious plans to capture the islands for Charles.

Of all the great early European voyages of discovery, none has been more misunderstood than Magellan's first circumnavigation of the globe, which in its ambition, duration and sheer depth of human endurance eclipses the achievements of Columbus's first voyage to the New World, or Vasco da Gama's voyage to India. There is no evidence that Magellan ever intended to circumnavigate the globe. His proposed expedition was a calculated commercial voyage aimed at outflanking the Portuguese control of the sea route to the Indonesian archipelago via the Cape of Good Hope by sailing not eastwards but westwards. Magellan was the first known navigator to recognize the possibility of sailing round the southern tip of South America and from there navigate his way through the Pacific to the Moluccas. Once there he would load his fleet with spices and sail back via South America, claiming the Moluccas for Castile and having, he hoped, established a quicker route to the islands.

The Castilian Dominican priest Bartolomé de las Casas (1484–1566), author of the *History of the Indies*, and a stern critic of the brutal behaviour of the Castilian adventurers in the Americas, recalled his conversations with Magellan in Valladolid in the spring of 1518, prior to his departure. Las Casas was unimpressed by the short, limping and undistinguished man he met, but he identified why Magellan was so convinced about Castile's claim. On his arrival in Seville, 'Magellan brought with him a well-painted globe showing the entire world, and thereon traced the course he proposed to take'. Las Casas went on:

> I asked him what route he proposed to take, he replied that he intended to take that of Cape Santa Maria (which we call Rio de la Plata), and thence follow the course south until he found the strait. I said, 'What will you do if you find no strait to pass into the other sea?' He replied that if he found none he would follow the course that the Portuguese took.

Presumably at this stage in his planning, Magellan maintained the official position that if he did not find a strait passing from the tip of South America into the Pacific, he would pursue the Portuguese route to the East via the Cape of Good Hope. But Las Casas knew better:

> according to what an Italian named Pigafetta of Vicenza, who went on that voyage of discovery with Magellan, wrote in a letter, Magellan was perfectly certain to find the strait because he had seen on a nautical chart made by one Martin of Bohemia, a great pilot and cosmographer, in the treasury of the King of Portugal, the strait depicted just as he found it. And, because the said strait was on the coast of land and sea, within the boundaries of the sovereigns of Castile, he therefore had to move and offer his services to the king of Castile to discover a new route to the said islands of Molucca.

The Italian Antonio Pigafetta, who sailed with Magellan, confirmed that the decision to sail west to reach the east was indeed based on Magellan's consultation of the geography of 'Martin of Bohemia', or Martin Behaim, the German merchant and globemaker who claimed to have participated in Portuguese voyages down the coast of Africa in the 1480s. If Behaim completed maps, as Las Casas and Pigafetta believed, none of them have survived, but Behaim did leave one object that guaranteed his lasting place in the history of cartography. In 1492, on the eve of Columbus's departure for the New World, Behaim completed his only surviving geographical work. This was not a map or chart, but what Behaim himself called an 'erdapfel', or 'earth apple', the earliest surviving example of a terrestrial globe made by a European. Although mapmakers since the Greeks had created celestial globes of the heavens, Behaim's is the first known globe depicting the earth.

Las Casas and Pigafetta grasped that Magellan's interest in Behaim lay in the revelation of a strait connecting the southern Atlantic with the Pacific – but an inspection of Behaim's globe reveals no such strait.[9] Perhaps Magellan saw other maps or charts made by Behaim which were subsequently lost or destroyed, or even later globes made by German cosmographers like Johannes Schöner. It seems more likely that Magellan consulted Behaim's terrestrial globe, not for a navigable route to the east via South America, but because it provided a global dimension through which to imagine his projected journey westwards to the east. Maps like the Cantino planisphere gave navigators general data on sailing across the Atlantic and Indian oceans, but their very nature as flat, two-dimensional maps prevented them from projecting a comprehensive picture of both western and eastern hemispheres with any reasonable accuracy. Terrestrial globes were little better. They were not used as navigational aids – their limited size meant that using them at sea when plotting seaborne voyages was all but useless. But for pilots like Magellan, the spherical projection of a terrestrial globe allowed him to think outside the geographical mentality of his time. While most princes and diplomats in Portugal and Castile continued to envisage the world on a flat map, with no real sense of the connection between the earth's western and eastern hemispheres, Magellan's planned voyage suggests that he was beginning to imagine the world as a global continuum.

There was one other vital aspect of the Behaim globe that seems to have inspired Magellan to embark on his voyage. Like many of his contemporaries, Behaim continued to imagine the world according to Ptolemy. Although his exploration of the coast of west and southern Africa led to minor revisions of the Greek geographer, where Behaim's first-hand knowledge ended he essentially reproduced Ptolemy's ideas about the size of the earth and the dimensions of the African and Asian continents. As we know, Ptolemy underestimated the circumference of the earth by one-sixth its actual length, but overestimated the breadth of South-east Asia. Having no concept of the Americas or the Pacific, Ptolemy's exaggeration of Asia meant that when Behaim came to plot the earth on a spherical globe, he corrected Ptolemy's belief in a land-locked Indian Ocean, disproved by Dias's rounding of the Cape of Good Hope in 1488, but still reproduced Asia according to the Greek geographer.

On a flat map, such exaggerations remained unremarkable to those familiar with Ptolemy, but reproduced on a terrestrial globe like Behaim's, their impact upon the eastern hemisphere was dramatic: the space between the west coast of Portugal and the east coast of China was just 130 degrees. The actual distance is nearer double that figure, 230 degrees. Looking at Behaim's globe would clearly have convinced Magellan that the voyage to the Moluccas via South America was shorter than the Portuguese sea route to Malacca. It was a mistake based on erroneous geography that would immortalize him for ever in world history; but it would also doom him and many of those who sailed with him.[10]

By the spring of 1518 Magellan was making preparations for his voyage. With the financial support of the emperor Charles V's creditors the German House of Fugger, he fitted out five vessels for the voyage with rigging, artillery, arms, provisions and pay for a crew of 237 at a cost of over 8 million *maravedís* (the voyage's sailors received 1,200 *maravedís* a month).[11] He also assembled a formidable team of Portuguese geographical advisers. They included Ruy Faleiro, an astronomer renowned for his attempts to solve the calculation of longitude, two of the most influential and respected of Portuguese mapmakers, the Reinel father and son team, Pedro and Jorge, and the pilot Diogo Ribeiro, who was appointed official chartmaker to the voyage. Faleiro, who was appointed chief pilot with responsibility for making charts and navigational instruments, drew more than twenty maps for use on board the fleet. The Reinels brought their practical knowledge of previous

Portuguese voyages, while Ribeiro, with his reputation for superb draughtsmanship, was responsible for collating and executing all the expedition's maps. Not surprisingly, considering all four men had defected from their employment by the Portuguese crown, Portuguese agents followed their every move while they were in Seville. One of the Portuguese agents, known only as Alvarez, wrote a letter to the Portuguese king in July 1519, informing him of the proposed voyage and the part played by the king's former mapmakers:

> The route which is said to be followed is from San Lucar directly to Cape Frio, leaving Brazil to the right side until after the demarcation line and from there to sail West ¼ North West directly to Maluco, which Maluco I have seen represented on the round chart made here by the son of Reinel; this was not finished when his father came there, and his father achieved the whole thing and placed the Moluccas lands. From this model have been made all the charts of Diego[sic] Ribeiro and also the particular charts and globes.[12]

Clearly versed more in espionage than geography, Alvarez reveals in his description the devastating political implications for Portugal of Magellan's proposed transgression of the 'demarcation line', the Treaty of Tordesillas: if successful, Magellan's voyage would challenge the Portuguese domination of the spice trade, and redraw the global map of European imperial politics.

Magellan's five ships and crew set sail from the port of San Lúcar de Barrameda on 22 September 1519. The events of the next three years have since passed into world history. Hunger, shipwreck, mutiny, political intrigue and murder punctuated Magellan's marathon voyage. From its outset the mainly Castilian crew were deeply suspicious of their Portuguese leader and his ambitious route to the Moluccas. Sailing down the coast of South America according to established Portuguese and Castilian navigation proved relatively unproblematic, but by the autumn of 1520 Magellan had reached uncharted waters at the tip of southern America. In November, after much searching and conflict over direction, Magellan found his way into the strait that still bears his name, and finally out into the Pacific Ocean.

Magellan named this new ocean *Mare Pacificum*, or 'peaceful sea'. It was to prove anything but peaceful. At just under 170 million square kilometres, the Pacific is the largest continuous body of ocean in the world, covering just under 50 per cent of the world's total water surface and representing 32 per cent of the total surface area of the globe. In 1520, of course, Magellan knew nothing of this, and had based his navigational calculations on Ptolemy and Behaim. The results of this miscalculation for Magellan's crew were formidable and for some, even fatal. Sailing westwards away from South America into uncharted open seas, it took the fleet more than five months to sight land in the eastern Philippines in the spring of 1521. Magellan landed on the island of Mactan in April. There he became embroiled in the island's local politics, and on 27 April, having sided with one of the island's tribal leaders, he led an armed party of sixty men in a skirmish with an opposing tribe. Vastly outnumbered and too far from the support of his remaining three ships, Magellan was singled out as the party's leader and killed.

Shocked and bewildered, the remaining crew set sail again, but now faced a series of fatal attacks from hostile local tribes, who took confidence from Magellan's death and the realization that his sailors were not invincible. Reduced to little more than 100 crew, with most of their high command dead and just two ships intact, the remaining officers split command of the fleet between three of their number, appointing the Basque pilot Sebastião del Cano as commander of the *Victoria*, despite his participation in an earlier mutiny against Magellan that left him in chains. The surviving fleet finally reached the Moluccas on 6 November 1521, where they managed to embark two shiploads of pepper, ginger, nutmeg and sandalwood. As the crew prepared to leave Tidore in the Moluccas, Antonio Pigafetta calculated in his journal that the island 'is in the latitude of twenty-seven minutes toward the Antarctic Pole, and in the longitude of one hundred and sixty-one degrees from the line of partition'; in other words, nineteen degrees within the Castilian half of the globe.[13]

After sailing across the Pacific for nearly a year, and reduced to just two ships, the fleet's officers were divided about which direction to take back to Castile: did they return via the Cape of Good Hope and complete the first known circumnavigation of the globe, or go back the way they came via Magellan's Strait? A decision was made that the *Trinidad* would retrace the fleet's treacherous path through the Pacific under the command of Gonzalo Gómez de Espinossa, while the *Victoria* would head for the African Cape, led by del Cano. Despite the horrors of the outward journey, returning via the Indian and Atlantic oceans seemed the riskier of the two options. The *Victoria* was already in terrible condition, and the likelihood of being captured by patrolling Portuguese ships was high. But while del Cano set off immediately, Espinossa vacillated as to his exact route. In May 1522 the *Trinidad* was captured and destroyed by a Portuguese fleet, and its crew imprisoned.

Meanwhile, on the other side of the Indian Ocean, the *Victoria* was successfully playing cat and mouse with the Portuguese all the way back to Europe. Finally, on 8 September 1522, after an eight-month return voyage, del Cano and his remaining crew reached Seville, completing the first recorded circumnavigation of the globe. Magellan was dead, four of his five ships lost, and of the 237 who had left Castile nearly three years earlier, just 18 survived to tell the tale of their extraordinary journey. In the first letter written to Charles V informing him of the voyagers' return, del Cano announced that 'we have discovered and made a course around the entire rotundity of the world – that going by the occident we have returned by the orient'.[14]

News of the return of the remains of Magellan's expedition reverberated across Europe. The papal nuncio to Germany, Francesco Chiericati, wrote to his friend Isabella d'Este in Mantua. Like her father Ercole (the owner of the purloined Cantino planisphere), Isabella was hungry for reports of the Castilian voyages of discovery, and Chiericati was happy to provide them. He told Isabella that Antonio Pigafetta 'has returned highly enriched with the greatest and most wonderful things in the world, and has brought an itinerary from the day he left Castile until the day of his return – which is a wonderful thing'. Describing the journey to the Moluccas, Chiericati reported that the surviving crew 'gained not only great riches, but what is worth more – an immortal reputation. For surely this has thrown all the deeds of the Argonauts into the shade.'[15]

For the educated elite of Renaissance Italy like Chiericati and Isabella, steeped in the classical past of Greece and Rome, the voyage indeed represented the eclipse of the great voyages of ancient myth, but for the diplomats of Portugal and Castile at the centre of a vital imperial dispute, the consequences were altogether more pragmatic. Del Cano's account of the voyage was very clear about his priorities. 'We discovered many very rich islands,' he reported, 'among them Banda, where ginger and nutmeg grow, and Zabba, where pepper grows, and Timor, where sandalwood grows, and in all the aforesaid islands there is an infinite amount of ginger.'[16] The Portuguese were horrified. In September 1522 King John III lodged a formal protest with the Castilian authorities over what he regarded as their infringement of Portuguese territory, and insisted that Charles V accepted Portugal's monopoly of all commercial traffic in and around the Moluccas. Charles V refused and instead claimed the Moluccas as lying within Castile's territorial dominion under the terms of the Treaty of Tordesillas. The Portuguese responded by refuting the claim, insisting that the voyage broke the terms of the treaty and maintaining that the Moluccas fell within their half of the globe. Charles countered again by offering to submit the matter to diplomatic arbitration, to which the Portuguese agreed.

Castile's initial diplomatic claims to the islands revolved around a fascinating, if slightly mendacious definition of 'discovery'. Charles's diplomats argued that even if Portuguese ships had seen and 'discovered' the Moluccas prior to Magellan's voyage, this did not technically represent imperial *possession*, and that Magellan's crew had extracted what they regarded as an oath of allegiance to the emperor from the island's native rulers, a standard Castilian practice when claiming newly discovered territory. Not surprisingly, the Portuguese refuted such semantic quibbles, arguing that the onus was on Castile to prove their possession of the islands according to geography. They also insisted that as negotiations continued, Castile should refrain from dispatching any further fleets to the Moluccas.

In April 1524 both sides agreed on formal negotiations aimed at resolving the dispute. They met on the border between the two empires at the towns of Badajoz and Elvas, high on the plains of Estremadura and separated by the Guadiana River. As the delegates arrived in the spring of 1524, they began to realize the magnitude of their task: this was no simple settlement of a territorial boundary dispute, but an attempt to split the known world in half. The Castilian delegation knew that if their claim was successful, their rule would stretch from northern Europe, across the Atlantic, and encompass the whole of the Americas and the Pacific Ocean. For Portugal, the loss of the Moluccas threatened to end the monopoly they had established over the spice trade, which had transformed the kingdom in less than a generation from a poor and isolated realm on the edge of Europe to one of the continent's most powerful and wealthy imperial powers.

It became apparent that maps would be the key to settling this global dispute, although, as one contemporary Castilian commentator wrote, geographical partisanship came in the most unusual of shapes:

> It so chanced that as Francis de Melo, Diego Lopes of Sequeira, and other of those Portugals of this assembly, walked up the river side of Guadiana, a little boy who stood keeping his mother's clothes which she had washed, demanded of them whether they were those men that parted the world with the Emperor. And as they answered, yea, he took up his shirt and showed them his bare arse, saying, come and draw your line here through the middest. Which saying was afterwards in every man's mouth and laughed at in the town of Badajoz.[17]

The story is probably apocryphal, a crude joke at the expense of the Portuguese delegation. But it shows that by the early sixteenth century even very ordinary people were beginning to be aware of the changing geography of the wider world.

Even before Magellan's circumnavigation, the realization that maps and charts facilitated better navigation and access to overseas markets led first the Portuguese then the Castilian crowns to fund institutions with responsibility for training pilots and collating geographical material relevant to seaborne exploration. The Portuguese Casa da Mina e India (the House of Mina – a fort on the West African coast, now in Ghana – and India) was created in the late fifteenth century to regulate trade and navigation with West Africa and India (once it reached there), and in 1503 Castile followed suit with the foundation of the Casa de la Contratación (the House of Trade) in Seville.[18] Portuguese navigation in the fifteenth century had shown that both an intellectual understanding of astronomy and a practical knowledge of sailing was required to map the Atlantic Ocean, and as a consequence both organizations aimed to unify the empirical data gathered from pilots and navigators with the inherited classical knowledge of educated cosmographers. Alexandria, Baghdad and even Sicily had seen the creation of centres of geographical calculation before, but they had usually aimed at creating a single map that would synthesize all known geographical knowledge and finally confirm what the world looked like. The maps and charts made by the Portuguese and Castilian trade organizations were different: they incorporated new discoveries, but were content to leave huge blanks across their surfaces, in anticipation of subsequent information that would be incorporated in later updated maps.

As the crowns of Portugal and Castile began to use these maps to resolve territorial claims and boundary disputes in the Atlantic and down the coast of Africa, they took on the status of legal authority. A map like that created under the terms of the Treaty of Tordesillas was regarded as part object, part document that the two political opponents accepted as legally binding because of its pivotal role in an internationally agreed treaty, confirmed by the pope. Such maps could settle disputes over places on the terrestrial globe that neither mapmakers nor their political paymasters had ever seen, never mind visited. They also laid claim to a new degree of scientific objectivity based on verifiable reports and logs of long-distance travel rather than hearsay and classical assumptions. Such claims were, as we shall see, somewhat dubious, and benefited the mapmaker as much as his political patron, but they allowed maps a new status, whereby early modern empires traded territory, and nowhere more decisively than with the Portuguese–Castilian conflict over the Moluccas, and its attempted resolution at Bajadoz-Elvas in 1524.

The changing perception of the role of the mapmaker and his maps can be gleaned from the composition of the negotiating teams that arrived in Badajoz-Elvas that spring. The Portuguese delegation consisted of nine diplomats (including the much maligned Francis de Melo and Diego Lopes of Sequeira), as well as three mapmakers, Lopo Homem and Pedro Reinel and his son Jorge. The Castilian delegation had more to prove than the Portuguese. It was their aggressive claim to Portugal's hold over the South-east Asian spice trade that brought the two delegations together, and they arrived with an equally impressive array of nine diplomats, including Sebastião del Cano, and no fewer than five geographical advisers drawn from across Europe. They included the Venetian Sebastian Cabot, head of the Casa de la Contratación. Cabot was one of the great navigators of his generation, said to have discovered Newfoundland in 1497 in the employ of King Henry VII before transferring his allegiance to the wealthier Castile. The team also included the Florentine mapmaker Giovanni Vespucci, nephew of Amerigo, as well as the Castilian mapmakers Alonso de Chaves and Nuño Garcia; Garcia was himself a former head of the Casa de la Contratación, and drew some of the maps for Magellan's circumnavigation prior to its departure. The final member of the Castilian team was neither Castilian nor Italian, but Portuguese: Diogo Ribeiro.[19]

Of all the Castilian team, we know least about Ribeiro. Born into relative obscurity in the late fifteenth century, Ribeiro joined the Portuguese fleets sailing to India in the first years of the sixteenth century, quickly rising to the position of pilot. Like many Portuguese mapmakers of his day, Ribeiro learnt to draw charts at sea rather than in the academy, which still privileged the knowledge of astronomy and cosmography over hydrography and mapmaking. By 1518, as we have seen, he was working for the Castilian crown in Seville, the centre of Castile's overseas imperial ambitions, and home of the Casa de la Contratación. By this time the Casa also included an office with sole responsibility for hydrography – the measurement of the seas for navigational purposes – founded to regulate the stream of new charts that came into Seville from the fleets returning from the New World and beyond. Ribeiro's success as a pilot led to his royal appointment as a cosmographer, and it was in this capacity that he took his place as an adviser to the Castilian negotiating team at Badajoz-Elvas, sitting opposite his Portuguese compatriots.[20] Despite his relative obscurity in comparison with his more distinguished colleagues, it was Ribeiro who, over the next five years, would offer the most compelling case in support of Castile's claim to the Moluccas, producing a series of beautiful, scientifically persuasive maps which would not only alter the course of those islands' history, but contribute to the change in global geography and mapmaking in the Renaissance.

The meeting at Badajoz-Elvas was preceded by weeks of intense espionage between the two imperial delegations. The Portuguese had managed to woo the Reinels back after their time working for Magellan in Seville, but as the delegations arrived, Pedro Reinel confessed to two of the Portuguese representatives that he had been 'invited together with his son to enter the emperor's service' for the substantial salary of 30,000 Portuguese *reis*, as had Simão Fernandez, another senior member of the Portuguese delegation.[21] Looking back on the dispute more than eighty years later in his book *Conquista de las islas Malucas* (1609), the Castilian historian Bartholomé Leonardo de Argensola summarized Charles V's diplomatic and geographical brief to his team. The emperor

urged, that by mathematical demonstration, and the judgment of men learned in that faculty, it appeared, that the Moluccas were within the limits of Castile, as were all others, as far as Malacca, and even beyond it. That it was no easy undertaking for Portugal to go about to disprove the writings of so many cosmographers, and such able mariners, and particularly the opinion of Magellan, who was himself a Portuguese ... Besides that, in relation to the article of possession on which the controversy depended, it was only requisite to stand by what was writ by and received among cosmographers.[22]

The Castilian delegation understood that this was a dispute that could only be resolved through the systematic manipulation of maps, the exploitation of national differences, the selective appropriation of classical geographical authority and, where necessary, bribery.

On 11 April both teams met on the bridge over the River Caya, right on the boundary between Portugal and Castile. Negotiations faltered almost immediately. The Portuguese protested against the presence of two Portuguese pilots amongst the Castilian delegation, Simón de Alcazaba and Esteban Gómez, who were quickly replaced. The Portuguese were also worried about the composition of the Castilian team of geographical advisers, and one man in particular. Days before negotiations began one of the Portuguese delegation wrote to King John in Lisbon, dismissing the authority of the Castilian geographers, with one exception. 'Their pilots are without any credit,' he claimed, 'except Ribeiro.' By this time Ribeiro's knowledge of the placement of the Moluccas seems to have been unrivalled. He knew the geographical claims of both sides, with privileged access to information about the islands before and after Magellan's voyage, and the Portuguese clearly feared that his contribution to the dispute could prove decisive.

The formal appointments of each team having been agreed, negotiations began in earnest. The lawyers quickly reached an impasse in establishing which side would act as the plaintiff, and it became clear that the geographers would be crucial to settling the claim. Both sides began by reiterating the terms of the Treaty of Tordesillas. The line of demarcation drawn at Tordesillas was 370 leagues west of the Cape Verde islands. This represented an unofficial prime meridian, from which Castile claimed all territories 180 degrees to the west, Portugal everything 180 degrees east. But the prize of the Moluccas was now so contested that both delegations even quibbled over from 'which of the said islands [in the Cape Verdes] they should measure the 370 leagues'.

In response, both sides requested maps and globes to ascertain the exact location of the demarcation line. On 7 May, 'the Portuguese representatives said that sea charts were not so good as the blank globe with meridians as it represents better the shape of the world.' For once, the Castilian delegation agreed, saying they also 'preferred a spherical body, but that the maps and other proper instruments should not be debarred'.[23] The thinking of both sides was by this stage noticeably global, although Castile still knew that the maps produced from Magellan's voyage would be crucial in supporting their claim. Not surprisingly, the Castilian team then argued for calculating the line of demarcation from San Antonio, the most westerly island in the Cape Verdes, which would allow them a greater portion of the Pacific, and by implication the Moluccas. The Portuguese predictably responded by insisting that the calculation begin from La Sal or Buena Vista, the easternmost points in the Cape Verdes. There were less than 30 leagues separating the two points – not enough to make the crucial difference in locating the Moluccas one way or another – but a sign of the delicacy of the negotiations.

The two sides reached stalemate, and from this point negotiations became almost comically adversarial. Maps were solemnly presented for inspection, fiercely attacked, then locked away, never to be revealed again. Both sides claimed map-rigging. God was brought into the argument on more than one occasion. On some days, when the claims became particularly heated, delegates simply feigned illness, or decided they were too tired to answer difficult questions. Castile responded to Portugal's claim to place the line of demarcation by saying 'they thought it best to pass beyond this question, and to locate the seas and lands on a blank globe'. The advantage of this was that at least 'they would not be standing still and doing nothing', and 'perhaps it would prove to whom the Moluccas belong no matter how the line be drawn'. Finally, both sides agreed to show the maps they possessed. The reason for their reticence was obvious: navigational knowledge was jealously guarded information. There was the added fear of presenting maps which might have been manipulated in the interests of a particular claim, and which could be exposed as fraudulent by experts from the opposing side.

On 23 May the Castilian delegation presented a map tracing Magellan's voyage to the Moluccas, from which they concluded that the islands lay 'one hundred and fifty [degrees] from the divisional line' in a westerly direction: thirty degrees within the Castilian half of the globe.

The maker of this map is unknown, but earlier maps locating the Moluccas point to Nuño Garcia, one of Castile's team of geographical advisers, who was involved in drawing Magellan's original maps. Garcia was responsible for a map dated 1522 which shows the eastern line of demarcation bisecting Sumatra, exactly where del Cano believed it to fall. The Castilian travel writer Peter Martyr regarded Garcia and Ribeiro as the most effective mapmakers on the Castilian team, 'being all expert pilots and cunning in making cards for the sea'. Capturing the tone of the negotiations, he described both men as presenting 'their globes and maps and other instruments necessary to declare the situation of the islands of the Moluccas about which was all the contention and strife'.[24]

That same afternoon the Portuguese responded by rejecting the map for its failure to depict key locations, including the Cape Verdes. Instead they 'showed a similar map on which the Moluccas were one hundred and thirty-four degrees distant [eastwards] from La Sal and Buena Vista, quite different from theirs', and forty-six degrees within the Portuguese half of the globe. Each side was claiming the authority to possess half of the known world, and yet their geographical knowledge apparently put them more than seventy degrees apart in locating the Moluccas on a world map. They were even still unable to agree on where they drew the meridian through the Cape Verdes, not that this made that much difference to the wider dispute. Five days later both delegations acknowledged that terrestrial globes represented the only way forward in trying to resolve their differences. As a result, 'both sides presented globes showing the whole world, where each nation had placed the distances to suit themselves'. The Portuguese gave little ground, estimating that the Moluccas lay 137 degrees east of the line of division, 43 degrees within their dominion. Castile then made a radical revision to its global estimate, claiming the islands lay 183 degrees east of the line – just 3 degrees within their half of the globe.

The Castilians reached for plausible but ever more complicated scientific arguments. At first, they argued for the accurate measurement of longitude in resolving the dispute. By the sixteenth century, pilots could calculate latitude quite accurately by taking measurements according to a relatively fixed point, the North Pole star. The absence of any such fixed referent when navigating across lines of longitude from east to west meant this was less of a problem when sailing across the open

Indian or Atlantic oceans, or down the coasts of Africa and America, but became one when the location of a group of islands on the other side of the world was under dispute. The only methods for calculating longitude were based on arcane and unreliable astronomical observations. The Castilians invoked the classical authority of Ptolemy to calculate longitude, claiming that 'the description and figure of Ptolemy and the description and model found recently by those who came from the spice regions are alike', and that as a result 'Sumatra, Malacca and the Moluccas fall within our demarcation'.[25] By now it was clear to everyone present that Magellan's use of Ptolemy's outmoded calculations inadvertently supported Castile's claims to the Moluccas. Trying to estimate the circumference of the earth was also dismissed as unreliable, because nobody could agree on the exact measurement of a league as a unit of distance. 'Much uncertainty is occasioned by this method,' according to the Castilians, especially when such measurements were taken at sea, 'for there are many more obstacles that alter or impede the correct calculation of them, such as, for instance, currents, tides, the ship's loss of speed' and a whole host of other factors.[26]

The Castilians therefore offered one final ingenious argument. Flat maps, they argued, distorted the calculation of degrees measured across the spherical globe. The Portuguese maps of the Moluccas and the 'lands situated along the said eastern voyage, placed on a plane surface, and the number of leagues being reckoned by equinoctial degrees, are not in their proper location as regards the number and quantity of their degrees'. This was because 'it is well known in cosmography that a lesser number of leagues along parallels other than the equinoctial, occupy a greater number of degrees'. There was some truth to this argument; most flat maps of the period portrayed the grid of latitude and longitude as straight lines intersecting at right angles, when geometrically they curved round the sphere, requiring complex spherical trigonometry to calculate the exact length of a degree. The Castilians therefore concluded that 'it will take a much greater number of degrees when they are transferred and drawn on the spherical body. Calculating by geometrical proportion, with the arc and chord, whereby we pass from a plane to a spherical surface, so that each parallel is just so much less as its distance from the equinoct is increased, the number of degrees in the said maps [of the Portuguese] is much greater than the said pilots confess.'[27]

Such arguments based on spheres were to no avail. Neither side was prepared to give way, and even the Castilians finally admitted in their closing remarks that they considered it 'impossible that one side can succeed in convincing the other by demonstrating that the Moluccas fall within his territory', without joint expeditions to agree on the size of a degree, and the correct measurement of longitude.[28] That was a hopelessly ambitious prospect, and by June 1524 negotiations were brought to a close without any resolution.

Throughout the conference, Diogo Ribeiro was closely involved in shaping Castile's geographical claim to the Moluccas, although rarely named in person. When Charles V took the diplomatic impasse over the Moluccas as an opportunity to dispatch fleets to the islands, Ribeiro was sent to La Coruña to act as the official cartographer of the recently founded Casa del la Especieria, established to challenge the Portuguese spice monopoly. Portuguese spies wrote to Lisbon from La Coruña, informing the crown that 'a Portuguese named Diogo Ribeiro is also here, making sailing charts, spheres, world maps, astrolabes and other things for India'.[29] Just five months after negotiations at Badajoz-Elvas foundered, Ribeiro was fitting out a new Castilian fleet with maps and charts, in an attempt to find a quicker westward route to the Moluccas. The fleet's commander, the Portuguese Esteban Gómez, was convinced that Magellan had missed a strait leading into the Pacific along the coast of Florida. After nearly a year of fruitless navigation, during which he reached as far as Cape Breton, Gómez returned to La Coruña in August 1525 with little to show for his efforts other than a group of native Americans kidnapped off Nova Scotia. Having welcomed back the fleet, Ribeiro took one of the Americans into his home. He baptized him 'Diego', acting as his godfather. Did he adopt the kidnapped Diego as an act of compassion and charity? Or did he spot an opportunity to acquire some local knowledge of the geography of the New World? It is a fascinating but ultimately elusive glimpse of the mapmaker's personality.

Gómez's voyage inspired the creation of the first of Ribeiro's series of world maps that provided a compelling case in support of Castile's claim to the Moluccas. The map, completed in 1525, can be seen as a first draft of Castilian territorial ambitions in South-east Asia. Hand-drawn on four pieces of parchment measuring 82 × 208 centimetres, the map has no title, no explanatory text, and many of its outlines are sketchy and incomplete. China's coastline is a series of discontinuous lines, the northern outline of the Red Sea is incomplete, and the Nile is not even shown – these regions were of little interest to Ribeiro or his Castilian paymasters. Instead, the map's innovative geography is confined to its eastern and western extremities. Its only inscription is written in a faint hand just inside the North American coastline running from Nova Scotia down to Florida, which reads: '[l]and which was discovered by Esteban Gómez this year of 1525 by order of His Majesty'.[30] On Ribeiro's map, all six of Gómez's new landfalls along the Florida coastline are carefully transcribed.[31] The revised eastern coastline is shown in a sharper but lighter hand than the rest of the map, suggesting that Ribeiro hastily incorporated the results of Gómez's voyage just as the map was being finished in the last few months of 1525.

Ribeiro's innovations did not end with a new outline of the North American coastline. At the bottom right-hand corner of the map, situated right below the Moluccas in the western hemisphere, is a mariner's astrolabe, used for making celestial observations. In the left-hand corner, Ribeiro has drawn a quadrant, used to measure height and declination. Just to the left of the Americas is an enormous circular declination table (the 'circulus solaris'), incorporating a calendar which allowed navigators to calculate the sun's position throughout the year.[32] These make it the earliest known example of a map depicting navigational instruments used at sea, replacing the usual religious or ethnographic icons of earlier world maps.

If this is effectively a sketch map of Castilian overseas imperial policy, why should Ribeiro go to so much trouble to include such carefully drawn scientific instruments? The answer seems to lie in his positioning of the Moluccas. At the map's eastern limits, just above the astrolabe, the 'Provincia de Maluco' is clearly depicted, but it also appears again on the other side of the map, at its most westerly point. In the east the astrolabe flies the flags of both Castile and Portugal, but the Portuguese flag is positioned to the west of the Moluccas, while the Castilian flag is placed to their east. According to the line drawn at Tordesillas, which is shown running down the dead centre of Ribeiro's map and labelled 'Linea de la Partición', the astrolabe's flags show the Moluccas just inside the Castilian half of the globe. As if to emphasize the point, Ribeiro reproduces the islands again on the western side of the map, and positions the rival flags of the two empires to reiterate the Castilian claim. Science, in the shape of Ribeiro's astrolabes, quadrants and declination tables, is appropriated in support of Castile's territorial ambitions: the position of the Moluccas must be correct if the map-maker has recourse to such technically complicated scientific instruments. As a paid servant of Castile, Ribeiro was compiling a comprehensive world map that placed the Moluccas within the Castilian half of the globe, but as a cosmographer committed to the incremental mapping of the known world, he was also carefully incorporating the geographical discoveries made by Gómez and his contemporaries.

In December 1526 Charles V ordered yet another expedition to the Moluccas. But he needed money urgently to sustain an empire that stretched across Europe, Iberia and into the Americas, and which was facing conflict with Turks and Lutherans. Charles had begun to realize that maintaining a claim to the Moluccas was logistically and finan-cially unsustainable, so before the fleet could depart he announced that he was prepared to sell his claim to the Moluccas. It was an unpopular move in Castile. The Cortes, the kingdom's ruling assembly, wanted to bring the spice trade through Castilian ports and was therefore against such a sale, but for Charles there were larger issues at stake. He needed to finance imminent wars with France and England, and to settle his sister Catherine's dowry to King John of Portugal, following their mar-riage in 1525. King John celebrated the marriage by commissioning a series of tapestries entitled 'The Spheres', depicting a terrestrial globe controlled by the king and his new wife. John's sceptre rests on Lisbon, and the globe shows Portuguese flags flying over its possessions through-out Africa and Asia. On the furthest eastern limits of the globe it is possible to see the Moluccas, still flying the Portuguese flag.

Complementing John's marriage to Catherine, Charles married John's sister Isabel in March 1526, in a further attempt to cement the dynastic alliances between the two kingdoms. Notwithstanding the opinions of his new brother-in-law, Charles still insisted on his claim to the Moluccas. He presented the papal ambassador, Baldassare Castiglione, with Diogo Ribeiro's 1525 map of the world, including its obvious placement of the Moluccas within Castilian dominion. It was a fitting gift for Castiglione, better known today as the author of *The Courtier*, one of the Renaissance's greatest manuals on how artfully to make friends and influence people at court. Through their respective use of geography, both emperors were sending out a clear message: they might be closely united in marriage, but they were still divided over their territorial claims to the Moluccas.

Charles knew that Portugal would not give up the islands without major concessions. He had agreed a paltry dowry of 200,000 gold ducats for his sister's marriage to King John (in contrast, King John had paid Charles 900,000 cruzados in cash as Isabella's dowry, the largest in European history). Charles therefore proposed that the 200,000 ducats be forgone in return for granting the Portuguese unlimited access to the Moluccas for six years, after which their ownership would pass to Castile.[33] It was a breathtakingly mercenary offer, compounded by the fact that, as King John prevaricated, Charles offered to sell his claim to the islands to King Henry VIII, even as he contemplated war with his English relative. Robert Thorne, an English merchant living in Seville, wisely counselled Henry to stay well away from such a politically tangled dispute. 'For these coasts and situation of the islands' of the Moluccas, he argued, 'every of the cosmographers and pilots of Portugal and Spain do set after their purpose. The Spaniards more towards the Orient, because they should appear to appertain to the Emperor [Charles V], and the Portuguese more toward the Occident, for that they should fall within their jurisdiction.'[34] Henry sensibly declined an interest in the islands. Charles was left to gamble on John's reluctance to intensify conflict with his brother-in-law over the Moluccas, and he was right. At the beginning of 1529 both sides agreed to conclude a treaty at Saragossa that would finally settle the matter of territorial ownership.

As these machinations continued, Ribeiro set to work redrafting his 1525 map to provide an even more convincing cartographic statement in support of Castile's claim to the Moluccas. In 1527 he completed a second hand-drawn map, based closely on his 1525 map, but slightly larger and finished to a much higher level of detail and artistry. The map's full title, running across its top and bottom, suggests the greater scale of its geographical ambitions: 'Universal chart in which is contained all that until now has been discovered in the world. A cosmographer of His Majesty made it in the year 1527, in Seville.' As well as filling in the gaps left on his 1525 map, Ribeiro adds a series of written legends, mostly describing the function of the scientific instruments, but south-east of the Moluccas is a telling inscription which once again announces the Castilian claim to the islands. Describing '[t]hese islands and province of the Moluccas', the legend explains that they have been positioned 'in this longitude according to the opinion and judgment of Juan Sebastián del Cano, captain of the first ship that came from the Moluccas and the first that circumvented the world to the navigation she made in the years 1520, 1521 and 1522'.[35] Attributing the position of the Moluccas to del Cano's calculations invokes him as a first-hand authority, but it perhaps also betrays Ribeiro's misgivings about locating the islands so far east. Nevertheless, his 1527 world map was clearly intended to provide even more persuasive evidence in support of Castile's claim to the Moluccas.

In April 1529 the Portuguese and Castilian delegations reconvened in the town of Saragossa to renew their negotiations over the Moluccas. After the intense legal and geographical debates that took place at Badajoz–Elvas in 1524, the peremptory discussions were something of an anticlimax. In early 1528, as Charles was about to go to war with France, he dispatched ambassadors to Portugal to propose their neutrality in the coming conflict, in return for a quick settlement of the Moluccas dispute. The terms of the settlement were agreed by ambassadors by early 1529. The final Treaty of Saragossa, ratified by Castile on 23 April 1529, and by the Portuguese eight weeks later, agreed that Charles would give up his claim to the Moluccas in return for substantial financial compensation, and any Castilian found trading in the region could be punished.

Under the treaty's terms the emperor agreed to 'sell from this day and for all time, to the said King of Portugal, for him and all the successors to the crown of his kingdom, all right, action, dominion, ownership, and possession, or quasi-possession, and all rights of navigation, traffic and trade' to the Moluccas. In return, Portugal agreed to pay Castile 350,000 ducats. But Charles also insisted on reserving the right to redeem his claim at any point: he could renew it by returning the cash in full, though this would require appointing new teams to resolve the questions of geographical positioning left unanswered at Badajoz-Elvas. For Charles, this was a clever face-saving reservation, as it was a clause that was unlikely ever to be invoked, but it sustained the fiction of Castile's belief in the validity of its claim.

Both sides decided that a standard map should be created, based not on the accurate measurement of distances, but on the geographical rhetoric produced by the geographers at Badajoz-Elvas. It was a map on which 'a line must be determined from pole to pole, that is to say, from north to south, by a semicircle extending northeast by east nineteen degrees from Molucca, to which number of degrees correspond almost seventeen degrees on the equinoctial, amounting to two hundred and ninety-seven and one-half leagues east of the islands of Molucca'.[36] After six years of negotiations, Portugal and Castile finally agreed on where to place the Moluccas on a world map. A dividing line was drawn right round the globe, taking into account the curvature of the earth. In the western hemisphere the line passed through the islands of 'Las Velas and Santo Thome' in the Cape Verde islands, and crucially continued right round the globe to fall '17 degrees (which equal 297½ degrees) east of the Moluccas', placing the islands firmly within the Portuguese sphere.

The use of maps enshrined in the treaty was unprecedented. Both Castilians and Portuguese acknowledged for the first time the global dimensions of the earth. They also established the map as a legally binding document able to uphold an enduring political settlement. The treaty stipulated that both sides should draw up identical maps enshrining the new location of the Moluccas, and that 'they shall be signed by the said sovereigns and sealed with their seals, so that each one will keep his own chart; and the said line shall remain fixed henceforth at the point and place so designated'. This was more than just the royal seal of approval: it was a way of recognizing that maps were fixed objects, and a means of communication between competing political factions. As documents, they were able to assimilate and reproduce changing information, through which rival states could resolve their differences. The treaty concluded as much in its clause stating that the agreed map 'shall also designate the spot in which the said vassals of the said Emperor and King of Castile shall situate and locate Molucca, which during the time of this contract shall be regarded as situated in such place'.[37] The map thus bound the two empires to agree the location of the Moluccas, at least until they decided to disagree and relocate the islands for whatever diplomatic or political reason arose subsequently.

The official map based on the treaty's terms has not survived. Another map did survive, and was completed just as the treaty received its final ratification: Ribeiro's third and definitive version of his world map, entitled 'Universal chart in which is contained all that has been discovered in the world until now. Diogo Ribeiro, Cosmographer to His Majesty, made it in the year 1529. Which is divided into two parts according to the capitulation which took place between the Catholic Kings of Spain and King John of Portugal at the city of Tordesillas in the year 1494.' The map's basis in Ribeiro's first effort of 1525 is obvious, but its size (85 × 204 centimetres), detailed illustrations and copious inscriptions on expensive vellum testify to its status as a presentation copy designed to convince foreign dignitaries of Castile's claim to the Moluccas. The position of the islands as well as the legend describing del Cano's voyage remain as on the 1527 map. The distance between the west coast of America and the Moluccas is hugely underestimated at just 134 degrees, leaving the islands positioned at 172° 30' W of the Tordesillas line – or 7½° within the Castilian half of the globe.[38] Across the Atlantic and Pacific, ships are depicted plying their trade, but even these apparently innocent decorative flourishes play their part in supporting the Castilian claim. 'I go to the Moluccas,' says one; 'I return from the Moluccas,' says another.[39] Despite the pervasiveness of the Moluccas on the map, many of the previous markers of the diplomatic conflict shown on Ribeiro's earlier maps have gone. The Castilian and Portuguese flags in the eastern and western extremities of the map have disappeared, as has the Tordesillas line, even though the map's title explicitly refers to it.

The map appears to be the final and definitive statement of Castile's seven-year claim to the Moluccas. Charles's decision to relinquish his claim to the islands was unpopular among the Castilian elite. Was Ribeiro's map a last-ditch attempt by those who opposed Charles's strategic surrender of his claim to reassert their authority over the islands? Or did it arrive too late, just as Charles agreed to relinquish his rights to the islands under the terms of the Treaty of Saragossa? Perhaps. But the decorative legends at the bottom of the map suggest another possibility. To the right of the rival Castilian and Portuguese flags, Ribeiro has placed the papal coat of arms. This, and the fact that the map is now held in the Vatican Library in Rome, may indicate that it was created in response to a very specific moment. In the winter of 1529–30 the emperor Charles V travelled to Italy to be crowned as Holy Roman Emperor by Pope Clement VII in Bologna in February 1530.[40] The map appears to have been drawn to intimidate the papal authorities with an image of the world according to the wishes of its emperor. The original Treaty of Tordesillas was ratified by the papacy in 1494; by 1529, the power of Castile and Portugal meant that they paid little heed to the papacy's opinion, unless one of its rulers needed something from them. As Charles V travelled to Italy to receive the crown of Holy Roman

Emperor, he did indeed require papal sanction, even if only for public and ceremonial reasons. Offering the pope a world map adorned with the papal coat of arms may have assuaged fears that the papacy was being sidelined in the momentous global political decisions surrounding the fate of the Moluccas. But it also reminded Pope Clement that it was Charles, not Portugal's King John, who was now the most powerful ruler in Christendom. Just two years earlier, after Clement had decided to transfer his political allegiance to the emperor's great rival, the French king Francis I, Charles had ordered his troops to sack Rome. For diplomatic reasons, the emperor had relinquished his claim to the Moluccas, but Ribeiro's map still reproduced the Castilian delegation's belief in the position of the islands, regardless of the exigencies of diplomacy. Was this the world according to Charles V, for presentation to a humbled pope?

Although Ribeiro's final map was not needed at the negotiating table in Saragossa, it still represents a comprehensive summation of the Castilian case for ownership of the Moluccas, and stands as a remarkable testament to the brilliance of Ribeiro's skilful manipulation of a geographical reality that he probably suspected would eventually disprove its own fine detail. It remained available to the Castilian authorities, should they wish to revive their claim to the islands in future years. The fact that none of Ribeiro's world maps were ever printed, but remained in manuscript form, is a further sign of their political sensitivity. To commit them to print would have fixed the parameters of Castile's claims for the foreseeable future, but if they remained in manuscript they could easily be amended if it became necessary to locate the Moluccas somewhere else to support a future Castilian bid. If Castile had indeed renewed its claim, perhaps Ribeiro's map would have secured even more enduring fame. As it was, Charles V's imperial interests moved elsewhere, and Ribeiro was left to return to his adopted home in Seville, inventing increasingly irrelevant navigational instruments.

Ribeiro died in Seville on 16 August 1533. The contemporary importance of his series of world maps drawn between 1525 and 1529 meant that their innovations, just like those of Waldseemüller, were quickly assimilated by younger mapmakers, who pieced together the mass of travellers' reports and pilots' charts that flooded into Europe from overseas discoveries around a world which first Waldseemüller and then Ribeiro had played such a part in shaping over two decades.

Fig. 18 Detail of Hans Holbein, *The Ambassadors*, 1533.

Traces of Ribeiro's influence endured, and they can still be seen today in one of the Renaissance's most iconic images: Hans Holbein's painting *The Ambassadors*, painted in the same year as the Portuguese cosmographer's death.

Holbein's painting depicts two French diplomats, Jean de Dinteville and Georges de Selve, at Henry VIII's London court on the eve of the English king's momentous decision to marry his mistress Anne Boleyn and sever England's religious ties with the papacy in Rome for ever. The objects placed on the table in the centre of the composition provide a series of moralized allusions to some of the religious and political issues preoccupying the elite of Renaissance Europe. On the bottom shelf is a merchant's arithmetic manual, a broken lute and a Lutheran hymn book, symbols of the commercial and religious discord of the time. In the corner sits a terrestrial globe, just one of the many in circulation since Magellan's circumnavigation of the globe. Looking more closely, it is possible to see the dividing line agreed at Tordesillas in 1494 running down the globe's western hemisphere. We cannot see where this line falls in the eastern hemisphere, because it is tantalizingly obscured in shadow, but we do know that Holbein used a globe attributed to the German geographer and mathematician Schöner, and dated to the late 1520s. The globe itself is lost, but the original gores, the printed segments from which the globe was made, have survived, and are almost identical to the globe shown in Holbein's painting. They trace the route taken by Magellan's 1523 circumnavigation, and clearly show the Moluccas placed in the Castilian half of the globe, in line with Ribeiro's own placement of the islands.

It is a testament to the changes occurring in Europe as a consequence of long-distance travel, imperial rivalry, scientific learning and the religious turmoil of the first half of the sixteenth century that Holbein's painting shares similarities with Ribeiro's maps in placing globes, scientific instruments and mercantile textbooks before religious authority. Traditionally, the depiction of two prominent figures like de Dinteville and de Selve would show them between an object of religious devotion such as an altarpiece or a statue of the Virgin Mary. In Holbein's painting, the central authority of religious belief is replaced by the worldly objects jostling for attention on the table. This is a world in transition, caught between the religious certainties of the past and the political, intellectual and commercial excitement of a rapidly changing present. Religion is quite literally sidelined, its remaining presence that of a silver crucifix barely visible behind a curtain in the top left-hand corner. The global interests of this new world of international diplomacy and imperial rivalry lie elsewhere, on the other side of a newly emerging globe, driven more by imperial and commercial imperatives than religious orthodoxy.[41]

The terrestrial globe was simply too small to be useful in the kind of diplomacy practised by the two French ambassadors, or by the Castilian and Portuguese diplomats who struggled over ownership of the Moluccas throughout the 1520s. What was required to understand this enlarged global world were maps like Ribeiro's, which turned away from the Greek projections of the inhabited world, and instead offered a 360-degree perspective of the entire globe. Unlike a globe, flat maps inevitably contain centres and margins. As the Portuguese and Castile fought over the Moluccas for global pre-eminence Ribeiro provided an object that could be divided according to their particular global interests. This map was flat, but its conception was global.

For most people living in the early sixteenth-century world, like the young boy who bared his buttocks at the Portuguese delegation in Badajoz-Elvas, the dispute over the Moluccas was meaningless; it was a political dispute between two competing empires, with little relevance to most individuals and their everyday lives. Even to those who grasped something of the global implications of the conflict, drawing a line on a map or globe in Seville or Lisbon to represent the partition of the world on the other side of the earth bore little reality to the seaborne activity that continued regardless between Muslim, Christian, Hindu and Chinese pilots and merchants who criss-crossed the commercial worlds of the Indian and Pacific oceans. Both Portugal and Castile's claims to territorial monopolies thousands of kilometres from their imperial centres would prove to be utterly unsustainable. But for the western European empires of first Portugal and Castile, then Holland and England, the act of drawing a line, first on a map, then on a terrestrial globe, and laying claim to places that their putative imperial lords never visited, set a precedent that would be followed through the centuries, and shape so much European colonial policy across the globe over the subsequent 500 years.

# 7

## Toleration

*Gerard Mercator, World Map, 1569*

### Louvain, Belgium, 1544

The arrests began in the February of 1544. During the previous weeks a list of fifty-two names had been drafted in Louvain by Pierre Dufief, the procurer-general of Brabant. Dufief had already established his credentials as a fiercely conservative theologian for his interrogation and execution of the English exile and religious reformer William Tyndale, who had been charged with heresy, condemned, then strangled and burnt at the stake near Brussels in 1536. Forty-three of the names on Dufief's list came from Louvain, the rest from cities and towns – Brussels, Antwerp, Groenendael, Engien – all within a 50 kilometre radius. The list included people from all walks of life – priests, artists and scholars, as well as cobblers, tailors, midwives and widows – all united by the accusation of 'heresy'. Over the next few days Dufief's bailiffs began to round up the accused. Some confessed to denying the existence of Purgatory; others questioned transubstantiation (the belief that the bread and wine of the communion become the body and blood of Christ) and admitted to acts of iconoclasm (destroying images of Christ and his saints). Dufief's interrogation was thorough, and by late spring, although many were released or escaped with banishment and the confiscation of property, a handful had been found guilty of heresy and sentenced: one woman was buried alive, two men were beheaded and one burnt at the stake. Nobody watching their public executions was left in any doubt about the penalty for questioning the religious or political authority of Habsburg rule.[1]

Ever since the Habsburg emperor Charles V had inherited the Low Countries from his Burgundian ancestors in 1519, this fiercely independent patchwork of cities and municipalities had refused to accept what it regarded as the centralization of government and taxation by a foreign power, which ruled through governors-general based in Brussels. Four years before the arrests of 1544, Ghent refused to contribute to the Habsburg war effort against neighbouring France. The subsequent revolt was ruthlessly suppressed by Charles and his sister, Queen Maria of Hungary, governor and regent of the Low Countries. Two years later, anti-Habsburg factions from the eastern region of Gelderland again challenged the authorities, besieging Louvain and forcing Charles to return from Spain and assemble an army to rout his opponents. It was clear to Charles and his sister that the greatest challenge to their authority was not dynastic, but religious. By 1523, Dutch translations of the New Testament based on Martin Luther's writings had been published in Antwerp and Amsterdam, and commentaries on his work published in the same year were banned.[2] The region had a long history of tolerance and plurality in matters of theology and devotional practice, but both Charles and Mary came from a very different Christian tradition. Habsburg experience of the Jewish and Muslim communities in late fifteenth-century Castile helped convince them that any theological deviation from their own particularly orthodox version of Catholicism was a direct challenge to their authority. The arrests and subsequent executions of 1544 represented only a small fraction of an estimated 500 deaths officially sanctioned under Maria's twenty-five-year-long rule, and the estimated 3,000 people who were condemned because of their religious beliefs across Europe between 1520 and 1565.[3]

The lives of many of Dufief's accused are sketchy or non-existent, but records survive of one particular figure, identified on Dufief's list as 'Meester Gheert Schellekens', a resident of Louvain, who was accused of the particularly grave heresy of 'lutherye', or Lutheranism. When Dufief's men came knocking at Schellekens's Louvain home, he was nowhere to be found: he was a fugitive as well as a heretic, and a warrant was issued for his arrest. Within days he was apprehended by the bailiff of Waas, in the nearby town of Rupelmonde, and incarcerated in its castle. Among the many accounts of cruelty, persecution, torture and death during the history of the European Reformation, the arrests and executions of 1544 would be sadly unremarkable were it not for the fact that Schellekens was his wife's maiden name, and 'Meester

Gheert' is better known to history as the mapmaker Gerard Mercator (1512–94).

When pressed to identify a famous cartographer, most people will name Gerard Mercator and the map projection to which he gave his name, his 1569 world map, and which continues to define global mapmaking even today. Variously described as a cosmographer, geographer, philosopher, mathematician, instrument maker and engraver, Mercator was responsible for inventing not only his famed map projection, but also the first collection of maps to use the term 'atlas'. He created one of the first modern maps of Europe, overtook the influence of Ptolemy's *Geography*, and effectively superseded woodcut mapmaking by taking the art of copperplate map engraving to unparalleled heights of beauty and sophistication. We know more about Mercator's life than any of his predecessors because of the increasing professionalization of cosmography and mapmaking. He was one of the first mapmakers to merit his own admiring biography, *Vita Mercatoris* ('Life of Mercator'), published posthumously by his friend Walter Ghim in 1595. His name has become synonymous with his projection, which has been unfairly castigated as the ultimate symbol of Eurocentric imperial domination over the rest of the globe, placing Europe at its centre and diminishing the size of Asia, Africa and the Americas.

To adapt Marx, men make their own geography, but not of their own free will, and not under circumstances they have chosen but under the given and inherited circumstances with which they are directly confronted.[4] The formulation could apply to many of the maps and their makers described in this book, but it applies most directly to the life and works of Gerard Mercator. The age of the Renaissance and Reformation in which Mercator lived is regarded as the great century of individuality, of the rise of biography, the lives of famous men exemplified by Giorgio Vasari's *Lives of the Artists* (1572) and what has become known as 'Renaissance self-fashioning', the ability of individuals artfully to shape their identity by adapting to and exploiting their particular circumstances. Whenever individuals assert themselves, they tend to experience the assaults and limitations of institutions such as Church, State and family; and where they reach for novel and alternative ways of imagining their personal and social existence, these institutions often endeavour to proscribe such alternatives.[5] If the sixteenth century was the great age of selfhood, it was also one of Europe's most intense periods of religious conflict and repression, an era when both Church and State imposed limitations on what people could think and how they should live in pursuit of their own religious, political and imperial goals.

Although the charge of heresy was not explicitly related to Mercator's mapmaking, his career as a cosmographer inevitably asked the kind of questions about creation and the heavens above that brought him into conflict with orthodox religious beliefs of the sixteenth century – both Catholic and Lutheran. Like Martin Waldseemüller, Mercator regarded himself as a cosmographer. He saw his profession as 'a study of the whole universal scheme uniting the heavens of the earth and of the position, motion and order of its parts'.[6] Cosmography was the foundation of all knowledge and 'of the first merit amongst all of the principles and beginnings of natural philosophy'. Mercator defined it as the analysis of 'the disposition, dimensions and organisation of the whole machine of the world', and mapmaking was just one of its elements.

Such an approach to cosmography and geography involved an investigation into the very origins of creation, what Mercator called 'the history of the first and greatest parts of the universe', and 'the first origin of this mechanism [the world] and the genesis of particular parts of it'.[7] This was hugely ambitious – and potentially very dangerous. Neither the Greeks nor later mapmakers like Waldseemüller faced religious injunctions in their quest for the origins of creation through the study of cosmography and mapmaking. But by the mid-sixteenth century, anyone addressing such questions risked incurring the wrath of the righteous on both sides of the religious divide. The problem was that the cosmographer – and by implication his reader – cast his eye across the globe and history, and risked charges of adopting a god-like perspective. The self-belief required to represent divinity was in stark contrast to a reformed religion that stressed humility before Creation. As a result, any cosmographer making a world map in the mid-sixteenth century found it difficult to avoid taking a position on the increasingly contested versions of Christian Creation and some, including Mercator, faced accusations of heresy from religious authorities who were anxious to control anyone who offered a geographical perspective on what the world looked like, and by implication what kind of God created it.

Mercator's career and his mapmaking were indelibly shaped by the Reformation. Following a series of brilliant but ill-advised forays into political and religious mapmaking which possibly contributed to the charge of heresy in 1544, Mercator's 1569 map projection offered navigators a groundbreaking method of sailing across the earth's surface. But seen within the context of the religious conflicts of his time, it also represented an idealistic desire to rise above the persecution and intolerance that touched him and many of those around him, and to establish a harmonious cosmography implicitly critical of the religious discord that threatened to tear Europe apart in the second half of the sixteenth century. Somewhere in the narrow and contested space between social determinism and autonomous free will Mercator managed to transcend the conflicts around him and create one of the most famous maps in the history of cartography, but one which had very different origins from the confident belief in European superiority which it is commonly believed shaped it.

Just like his maps, Mercator was defined by borders and boundaries. Throughout his long life, his travels never took him further than a 200 kilometre radius away from his birthplace of Rupelmonde, a small town on the bank of the River Scheldt in the modern-day Belgian region of East Flanders, where he was born in 1512 and named Gerard Kremer. The region he traversed was (and remains) one of the most densely populated regions in Europe, characterized not only by diversity and artistic creativity, but also by conflict and the competition for scarce resources. His father (a cobbler) and mother both hailed from the German-speaking town of Gangelt in the duchy of Jülich, 100 kilometres west of Cologne, one of the largest and oldest cities in Europe, which lay on the Rhine. To the west of Gangelt lay the Dutch-speaking lands of Flanders, and the continent's commercial hub of Antwerp on the Scheldt. The physical geography of Mercator's early life was shaped by the Rhine–Meuse–Scheldt delta, and the towns, cities and rhythms of life built on the confluence of these three mighty European rivers.

As Mercator grew up, the physical geography of the region was being transformed by the volatile imperatives of human geography. Less than 20 kilometres north of Rupelmonde, Antwerp was growing rich from its traffic in goods from as far away as the New World and Asia. East of the Rhine, Martin Luther was launching his challenge to the papacy, a reformed approach to Christian belief that would quickly spread westwards into the Low Countries. Three years after Luther issued his first public challenge to papal indulgences in Wittenberg, Charles V was elected Holy Roman Emperor by the German princes in the small town of Aachen 600 kilometres to the west, and 100 kilometres from Mercator's home town of Rupelmonde. It was an aggressive statement of imperial intent: Aachen had been the favourite residence of Charlemagne, king of the Franks from 768 and the greatest of all the early Christian post-Roman emperors. By choosing Aachen for his coronation, Charles was signalling his desire to emulate Charlemagne, as well as to extend the geographical limits of the Holy Roman Empire, which traditionally reached as far west as the River Meuse. Charles's coronation not only entitled him emperor of the old western Roman Empire, as well as king of Castile, Aragon and the Low Countries, but also required him to defend the Catholic faith. His religious responsibilities and imperial ambition would place him on a violent collision course with the religious reformers living in the German principalities east of the Rhine.

Mercator's career fell into two halves: the first was shaped by his education and early work in the towns and cities of the Low Countries; the second, following his incarceration in 1544, was spent in Duisburg, a small town in the duchy of Cleves in modern-day western Germany, where he spent the rest of his life, from 1552 until his death in 1594. With the benefit of hindsight, it is possible to see that his career pivoted on the traumatic and near-fatal accusation of heresy. Whatever Mercator might have thought at the time, the ideas and attitudes which brought it down on him can be traced throughout his early years, and its impact is discernible in the maps and geographical books he produced over his four decades in Duisburg.

Mercator received an exemplary humanist education: first at the 'groote school' in 's-Hertogenbosch, one of the best secondary schools in Europe, where the great humanist Desiderius Erasmus (1466/9–1536) studied; then Louvain University, second only to Paris in size and prestige, where he read philosophy. A generation earlier, scholars like Martin Waldseemüller had embraced the new humanist learning offered by universities like Louvain and Freiburg and the challenge of studying classical authors like Aristotle, but by the time Mercator arrived in the 1520s, the excitement had hardened into orthodoxy. For Mercator's philosophical curriculum, this meant a slavish adherence to Aristotle – apart from where the pagan philosopher's doctrines were deemed to contradict established Christian belief.

Although he appears to have followed the usual humanistic fashions, including changing his name to 'Mercator', a Latinized version of the German 'Kremer' ('trader'), the young scholar seems to have left the university with more questions than answers. The humanist study he pursued found it difficult to accommodate either the new reformed theology or the increasingly technical demands of subjects like geography, to which he was drawn as a way of exploring the idea of creation in all its theological, philosophical and practical dimensions. As well as reading authors like Cicero, Quintilian, Martianus Capella, Macrobius and Boethius, he studied Ptolemy and the Roman geographer Pomponius Mela. But it was Aristotle in particular who provided the pious but curious young Mercator with a series of problems: his belief in the eternity of the universe and the everlasting nature of time and matter was at odds with biblical teaching on the creation out of nothing. Louvain's theologians skated over the fine detail of Aristotle's arguments, insisting that his distinction between a mutable earth and fixed heaven corresponded well enough to their Christian equivalents. Within the changing field of geographical study, Aristotle's view of a world separated into *klimata* or parallel zones was proving hard to defend in the wake of Portuguese and Spanish voyages to the New World and South-east Asia. Looking back later in life, Mercator admitted that such apparently irreconcilable differences between the Greek thinkers and Louvain's theologians led him 'to have doubts about the truth of all philosophers'.[8]

Few students were brave enough to challenge Aristotle's authority at Louvain, and Mercator was no exception. A poor student of humble origins, a German native in a Dutch-speaking world with no influential connections, Mercator must have seen a potential career in speculative philosophy as limited. In his 'Life of Mercator', Walter Ghim recalled that 'when it was clear that these studies would not enable him to support a family in the years to come', Mercator 'gave up philosophy for astronomy and mathematics'.[9]

By 1533 Mercator was in Antwerp, studying both disciplines and beginning his association with a group of men who would oversee his transformation from aspiring philosopher into geographer. These were the members of the first generation of Flemish geographers who, following Magellan's voyage, were trying to project the terrestrial world onto globes as well as maps. Three men in particular would offer Mercator different possibilities in the pursuit of his newly chosen profession. The first, Franciscus Monachus, was a Franciscan monk trained at Louvain and living in Mechelen. He designed the earliest known terrestrial globe in the Low Countries (now lost), dedicated to the privy council of Mechelen and accompanied by a surviving pamphlet describing a pro-Habsburg globe (in terms of the Moluccas), in which 'the nonsense of Ptolemy and other early geographers is refuted'.[10] Mercator was also taught geometry and astronomy by Gemma Frisius, the brilliant Louvain-trained mathematician and instrument maker, who was already making a name for himself as a geographer, designing globes and making enormous advances in land surveying. In 1533 he published a treatise on the use of triangulation which used techniques evolved from repeated measurements taken across the flat, featureless landscape of the Low Countries. Frisius also developed new ways of measuring longitude. His terrestrial globes came with pamphlets explaining how to measure longitude at sea by using a timepiece, and although the technology for making such clocks was still rudimentary, he provided the first understanding of how the problem might be successfully resolved. The third figure who had a decisive influence on Mercator in the 1530s was the Louvain-based goldsmith and engraver Gaspar van der Heyden. Both Monachus and Frisius came to van der Heyden's workshop when they needed an artisan to make and engrave their globes, and it was here that Mercator learnt the practical and technical skills involved in making globes, maps and scientific instruments, as well the art of copperplate engraving. A monk, a mathematician and a goldsmith: these three men and the vocations they pursued shaped Mercator's subsequent career. From Monachus he saw that it might be possible to combine a religious life with a scholarly exploration of the boundaries of geography and cosmography; from Frisius he understood the need for mastery of mathematics and geometry in the pursuit of an accurate cosmography; and from van der Heyden he learnt the skills necessary to give physical form to the latest designs in mapmaking, globe construction and instrumentation.

As he struggled to digest such disparate knowledge, Mercator identified one particular skill in which he could excel: copperplate engraving, using the Italian humanist style of italic, 'chancery' hand. Waldseemüller and his generation had relied on the medium of Gothic capitals cut into woodcut blocks when making maps, but, as we have seen, each letter took up a great deal of space, and their upright, square style could look clumsy and easily betray the insertion of different type. Mercator's generation of humanist scholars, by contrast, had gradually adopted the Roman chancery style developed in fifteenth-century Italy, which looked elegant and compact and even had its own mathematical rules. Mercator quickly mastered the style and the skill with which to engrave it onto copper plates. Examples of the italic style used on maps printed in Rome and Venice were beginning to circulate among Antwerp's legions of printers and booksellers, and some printers, seeing the advantages of copperplate engraving, were beginning to experiment with using italic script on their maps. Mercator saw his opportunity to make an impact on his chosen field, and he took it.

The effect on mapmaking was immediate. Copperplate engraving transformed the appearance of maps and globes. Gone were the awkward Gothic lettering and large blank spaces created by the imprint of the woodblock, replaced instead by graceful, intricate letter forms and the artistic rendition of sea and land from the engraver's use of stippling. Engraving also allowed for speedy and virtually invisible corrections and revisions. A copper plate could be rubbed down and re-engraved in just a few hours, something that was physically impossible with woodcuts. Printed maps using the medium suddenly looked completely different; mapmakers now had a method by which to express (and revise) themselves cartographically, and Mercator had placed himself at the forefront of the shift in style. In the space of just four years between 1536 and 1540 Mercator went from an eager pupil of Monachus, Frisius and van der Heyden to one of the most respected geographers in the Low Countries. Four maps published in these crucial years, one in each current field of mapmaking – a terrestrial globe, a religious map, a world map and finally a regional map of Flanders – show him struggling to define his geographical vision even as he refined his distinctive cartographic style.

Having spent just a year in Antwerp, by 1534 Mercator was back in Louvain, and in 1536 he was involved in his first geographical publication, a terrestrial globe. Commissioned by the emperor Charles V, like many globes of its time, this was a collaborative affair, designed by Frisius, printed by van der Heyden, and engraved on copper plates by Mercator in what would become his distinctive elegant italic hand. The finished globe was dedicated to Maximilianus Transylvanus, adviser to the emperor and author of the pro-Habsburg treatise *De Moluccis*

*Insulis.* It is hardly surprising to see that the globe reproduces Ribeiro's political geography in claiming the Moluccas as a Habsburg possession. It also flies the imperial Habsburg eagle over Tunis, captured by Charles from the Ottomans in July 1535, and shows recent Spanish settlements in the New World. Where politics mattered little (in most of Asia and Africa), the globe simply reproduced the traditional Ptolemaic outlines. America was dutifully labelled a Spanish possession but also represented separately from Asia, in line with Waldseemüller's 1507 map, which Mercator probably saw around this time. The globe celebrated for its paymasters the international reach of Habsburg imperial power, but its significance for the scholarly community lay not in its political content but its form. This was the first known terrestrial globe to use copperplate engraving, and the first to use Mercator's italic hand, which established its own geographical conventions, with capitals used for regions, roman script for places, and cursive for descriptive explanations.[11] Nobody had seen anything quite like it before, and this was as much down to Mercator's calligraphy as Frisius's political geography.

For his first independent map, Mercator turned from political geography to religion. In 1538 he published an engraved wall map of the Holy Land which was designed, according to its title, 'for the better understanding of the Bible'.[12] It allowed Mercator to continue his interest in theology, but it also offered him the possibility of urgently needed financial success – no regional maps sold better than those of the Holy Land. Mercator drew on a series of incomplete maps of the Holy Land published by the German humanist Jacob Ziegler five years earlier in Strasbourg, which had provided only a partial historical geography of the region; his beautifully engraved map updated and expanded Ziegler's geography, and added one of the central stories of the Old Testament, the Exodus of the Israelites from Egypt to Canaan.

There was historical precedent for portraying biblical scenes on maps. Medieval *mappaemundi* like the one at Hereford also showed such scenes, including Exodus, and early printed editions of Ptolemy contained maps of the Holy Land too. But Luther's ideas led to a new conception of the place of geography within theology. Before the 1520s, the task of the Christian mapmaker was quite clear: it was to describe the world created by God, and to anticipate the Last Judgement. But one of the many consequences of Luther's challenge to orthodox Christian belief was a different emphasis on the geography of the created

Fig. 19 Lucas Cranach, 'The Position and Borders of the Promised Land', 1520s.

world. Lutheran mapmakers no longer emphasized God as a distant Creator of the world that could be understood only through intercession. Instead they wanted a more personal God whose Divine Providence was present in the here and now of people's lives. As a result, Lutheran statements on geography tended to downplay both Creation and Church history after the peregrinations of the Apostles, choosing instead to show how the world of God functioned. In 1549 Luther's friend Philipp Melanchthon published *Initia doctrinae physicae* ('The Origins of Physics'), in which he wrote:

> This magnificent theatre – the sky, lights, stars, earth – is proof of God the Ruler and Former of the world. Whoever casts his eyes around will recognise in the order of things God the architect who is permanently at work, preserving and protecting everything. In accordance with God's will we may trace His footprints in this world by studying the sciences.[13]

Melanchthon avoided describing God as Creator, calling him instead the 'Former of the world', the divine architect whose hand can be discerned through careful study of the sciences, and geography in particular. God's providential governance of the world can be revealed through empirical scientific study, independent of biblical exegesis. Inadvertently, Melanchthon's arguments would enable later sceptical geographers and mapmakers to question the validity of the Bible's geography.

By the 1530s these reformed beliefs were already affecting maps and their makers, and inspired a completely new genre of maps in Lutheran Bibles.[14] Luther took a more literal approach to geography than Melanchthon, writing that he wanted 'a good geography and more correct map of the Land of Promise' of the Israelites.[15] He tried to obtain maps to illustrate his German translation of the New Testament in 1522, and although this failed, three years later the Zurich printer Christopher Froschauer (who was closely associated with the leader of the Swiss reformed church, Huldrych Zwingli) published an Old Testament based on Luther's translations, illustrated with the first map ever published in a Bible. The subject was the Exodus from Egypt.

In 1526 the Antwerp printer Jacob van Liesvelt reproduced a version of the same map in the first Dutch edition of a Lutheran Bible, which was in turn copied by at least two other local printers prior to the publication of Mercator's map. The map copied in all these Lutheran Bibles was Lucas Cranach's 'The Position and Borders of the Promised Land', a woodblock made in the early 1520s. Like Ziegler, Cranach was a convert to Lutheranism, a close personal friend of Luther, and one of the most prolific and celebrated painters of the German Reformation. Mapping the story of Exodus had particular theological significance for Luther and his followers, because they saw themselves as latter-day Israelites, escaping the corruption and persecution of Rome. Luther interpreted Exodus as representing fidelity to God and the power of personal faith, in contrast to the traditional interpretations (as seen on the Hereford *mappamundi*) of the prefiguration of the resurrection or the importance of baptism.

Lutheran biblical maps concentrated on particular places and stories from the Bible that exemplified the reformed teaching. Maps of Eden, the division of Canaan, the Holy Land in the time of Christ, and the eastern Mediterranean of Paul and the Apostles account for nearly 80 per cent of sixteenth-century Bible maps.[16] In 1549 the English printer Reyner Wolfe published the first New Testament to include maps, telling his readers that 'the knowledge of Cosmography' was essential to 'well read the Bible'. On the map describing St Paul's travels, Wolfe observed that 'by the distance of the miles, thou mayest easily perceive what painful travaile St Paul took in preaching the word of God through the regions of Asia, Africa, and Europe'.[17] Where medieval maps prefigured the end of the world, those of the reformed religion were more interested in tracing the visible signs of God's providence. As Luther continued to stress the importance of the private reading of Scripture over the official doctrine of theological institutions, maps became vital adjuncts to such reading, providing ways of illuminating the Scriptures. They gave the reader a more immediate experience of the literal truth of the biblical events described and provided the faithful with directed readings of the Bible in line with Luther's (or in some cases Calvin's) interpretations.

By the late 1530s maps of the Holy Land portraying the Exodus were exclusively the preserve of Lutheran mapmakers. So why was Mercator, with his close links with the Catholic Habsburgs, drawing so explicitly on not only the geography but also the theology of such maps? The title of his map stated that it was designed for 'the better understanding of both testaments', a typically Lutheran statement. Was this indeed a sign of Mercator's Lutheran sympathies, or simply the enthusiastic naivety of a brilliant young mapmaker caught up in the excitement of a new direction in mapmaking? Flirting with religion on maps was a dangerous business, with potentially fatal consequences. The Spanish scholar Miguel Servetus was condemned repeatedly by both Catholic and Protestant authorities for his 'heretical' publications throughout the 1530s, which included an edition of Ptolemy's *Geography* (1535) that contained a map of the Holy Land on which the Spaniard criticized the fertility of Palestine.[18] Servetus was burned at the stake by the Calvinist authorities of Geneva in 1553.

If Mercator realized the potential dangers of his first independent map, he showed no signs of it. He started work on a second map, using his knowledge of mathematics to design a map of the whole world. Just as Luther's theology had affected mapmaking, so were the seaborne discoveries, already recorded by a number of Portuguese, Spanish and German mapmakers. The surge of interest in representing the globe following Magellan's circumnavigation in 1522 (and discussed in the previous chapter) captured the increasing global awareness of the earth, and gave its rulers a powerful object to hold in their hands as they proclaimed dominion over the whole world. But globe-making only sidestepped the perennial problem of how to project this spherical globe onto a plane surface, which was necessary for accurate navigation right round the globe – a pressing requirement now the Spanish and Portuguese had divided the earth in two. Waldseemüller had tried to do so by returning to Ptolemy's projections, but these methods only covered the *oikoumenē*, the inhabited world, not the earth's whole 360 degrees of longitude and 180 degrees of latitude. Mapmakers like Mercator now faced the challenge of formulating a completely new projection using mathematical rules.

In designing their projections, mapmakers were confronted with three possible options. They could adopt Monachus's solution, of simply doubling the classic circular representation of the *oikoumenē*, to show hemispheres, using straight parallels and curved meridians. They could divide the world into discrete shapes to produce globe gores of the kind designed by Waldseemüller and Frisius. Or they could project the whole globe onto a flat surface using a geometric figure, such as a cylinder, a cone or a rectangle. Each method had its drawbacks. Double hemispheres and globe gores needed to be on an enormous scale to be of any real use. Ptolemy and his predecessor Marinus of Tyre had already struggled with cylindrical and conical projections, and their distortion of size, shape or direction. Initially, Renaissance mapmakers reproduced modified versions of both these projections. But as the new discoveries breached the parameters of the known world and mapmakers like Mercator came into ever closer contact with mathematicians like Frisius, new shapes were proposed to represent the earth: the world became oval, trapezoidal, sinusoidal, even cordiform (heart-shaped).[19] Altogether, at least sixteen methods of projection were in use by the end of the sixteenth century.

Having chosen a shape, the mapmaker was confronted with a further problem. As the margins of the known world were consistently changing, where was its natural centre? Where did the world map begin – and where did it end? One possible answer lay in an even older group of projections, used by the Greek astronomers, and known as azimuthal. An azimuth is an angular measurement within a spherical system, usually (for the Greeks and later mapmakers like Mercator) the cosmos. A common example of an azimuth is to identify the position of a star in relation to the horizon, which acts as the reference plane. If the observer knows where due north lies, the azimuth is the angle between the northern point and the perpendicular projection of the star all the way down to the horizon. From this basic method azimuthal projections can then build up a network of angles based on establishing direction, ensuring that all distances and directions are accurate from a central point, although size and shape are distorted from anywhere else. They came in a dizzying variety: the equidistant projection, which maintained consistent scale and distance between any two points or lines; the orthographic, which enabled a three-dimensional object to be drawn from different directions; the gnomonic, which shows all great circles as straight lines; and the stereographic, which projects the sphere onto an infinite plane from one point on the globe. As many of these names imply, they can be chosen depending on what the mapmaker wanted to highlight – and, by implication, to diminish.

One of the advantages of an azimuthal projection was that it could focus on the equator, the poles, or whatever oblique angle was required by the mapmaker. Projections based on the poles became particularly fashionable, as they provided a new perspective on the recent voyages of discovery, as well as opening up a new area of potential exploration over the North Pole (the search for the North-west and North-east Passages). Placing one or the other pole at the centre of the map also held the distinct advantage of sidestepping the thorny political question

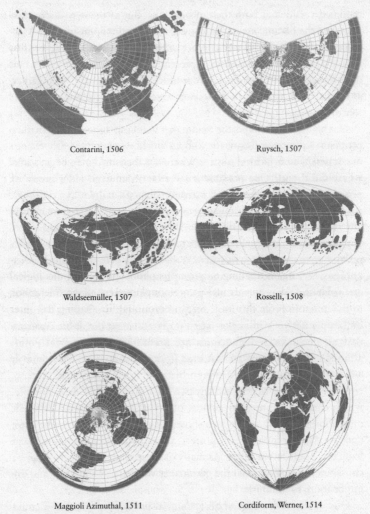

Contarini, 1506    Ruysch, 1507

Waldseemüller, 1507    Rosselli, 1508

Maggioli Azimuthal, 1511    Cordiform, Werner, 1514

Fig. 20 Diagrams of different Renaissance map projections.

of global ownership of the eastern and western hemispheres which had preoccupied mapmakers ever since the 1494 Treaty of Tordesillas.

One of the most extraordinary world maps to develop a double polar projection was the 1531 world map of the French mathematician, astrologer and mapmaker Oronce Finé's. This held the added novel innovation of being shaped like a heart on a revised cordiform projection. On Finé's map the equator runs vertically down the middle of the map, cutting it in half, with the North Pole to the left and the South Pole to the right. The two outermost circular arcs represent the equator, and are tangent to the central meridian that runs horizontally across the centre of the map. It was on Finé's map that Mercator modelled his own – including, however, some changes in line with the latest discoveries. North America (shown by Mercator as 'conquered by Spain') is separated from Asia, but linked to South America, and both are described as 'America' for the very first time. The Malay peninsula shows evidence that Mercator might have seen some of Ribeiro's maps of the region.[20]

None of these geographical innovations take away from the sheer peculiarity of projecting the world onto the shape of a stylized heart. The projection emerged gradually from experiments with Ptolemy's second projection, but in adopting the heart as a defining shape, Mercator was once again treading a dangerous philosophical and theological path. The world as a heart was a commonplace Renaissance metaphor, which played on the idea of the inner emotional life shaping the outer physical world. It would be taken up a century later by John Donne in his poem 'The Good-Morrow', where his lovers 'discover' new worlds of love, in a visual conceit that can only be fully understood in reference to cordiform maps:

> Let sea-discoverers to new worlds have gone,
> Let Maps to other, worlds on worlds have showne,
> Let us possesse one world, each hath one, and is one.
>
> My face in thine eye, thine in mine appeares,
> And true plain hearts do in the faces rest;
> Where can we find two better hemispheres
> Without sharp North, without declining West?[21]

But in the 1530s, the cordiform map projection was associated with controversial religious beliefs. Lutheran theologians like Melanchthon regarded the heart as the seat of human emotions, and as such it was seen as central to the transforming experience of scripture. Appropriating the Catholic symbolism of the heart, Lutheran thinking regarded its representation in books – and maps – as a devotional act of looking into one's heart, or conscience, for signs of grace. While mere mortals could try and interpret their hearts, only God was *kardiognostes* (heart-knower), with the ability to see into the heart without the need for a commentary.[22]

The adoption by cosmographers of the cordiform projection would also become associated with a strand of Stoic philosophy that regarded humanity's pursuit of earthly glory as vain and insignificant when set against the vastness of the larger cosmos. This Stoic version of cosmography drew on Roman writers like Seneca, Cicero, Posidonius and Strabo, and one of its most explicitly geographical expressions was to be found in Macrobius' fifth-century *Commentary on Scipio's Dream* (a text which Mercator would undoubtedly have read whilst studying at Louvain). In Macrobius, Scipio Africanus the Younger is drawn up to heaven in a dream, from where 'the earth appeared so small that I was ashamed of our [Roman] empire which is, so to speak, but a point on its surface'. The lesson in Macrobius' commentary is that 'men of our race occupy only a minute portion of the whole earth, which in comparison with the sky is but a point', revealing 'that no man's reputation can extend over the whole of even that small part' claimed by the Roman Empire.[23] Explaining the geographical power of Stoicism in the Augustan age, Christian Jacob argues such philosophical thinking 'bears witness to the diffusion of the exercise of *kataskopos*, that "view from above" carried across the earthly globe that leads to a relativising of human values and achievements but also to the adoption of an intellectual perspective, the spiritual gaze that discloses the beauty and order of the world beyond the shimmering of appearances and the limitations of human knowledge'.[24] By the early sixteenth century, as the world expanded but conflict and intolerance only intensified with religious turmoil and the pursuit of imperial power and glory, cosmographers like Finé, Abraham Ortelius and Mercator developed a Stoic contemplation of the harmonious relations between the individual and the cosmos in response to the bigotry and prejudice that seemed to be engulfing 'that small part' of the world called Europe.

To make a heart-shaped map in the first half of the sixteenth century was a clear statement of religious dissent. It invited its viewer to look to their conscience, and to see it within the wider context of a Stoic universe. But such flirtations with 'pagan' philosophy were not always welcomed by Catholic or Protestant authorities. Oronce Finé was so involved in the study of occult philosophy that he was briefly arrested in 1523; indeed, virtually every sixteenth-century mapmaker who adopted the cordiform projection harboured hermetic and reformed sympathies.[25] Mercator dedicated his world map to his friend Joannes Drosius, a cleric who, six years later, would be accused of heresy alongside him. Mercator chose a method of projection which, on mathematical, philosophical and theological grounds, could be interpreted as at least unorthodox, and at worst heretical.

It was probably just as well that such a derivative and relatively unusual map was not a great success. Mercator never used it again, and did not even mention it in any of his subsequent publications and correspondence, and was probably anxious to distance himself from it as the work of a still relatively inexperienced mapmaker. Ghim's 'Life of Mercator' passed over the 1538 world map in silence, recording instead that Mercator turned to the other growth area in early sixteenth-century geography, regional mapmaking. 'Responding with enthusiasm to the urgent request of a number of merchants, he planned, undertook and, in a short space of time, completed a map of Flanders'.[26] This map, completed in 1540, would prove to be one of Mercator's most popular early maps, reprinted fifteen times over the next sixty years.

The map was commissioned by a group of Flemish businessmen who wanted Mercator to replace a map of the region which appeared to challenge Habsburg rule. Pierre van der Beke's map of Flanders, published in Ghent in 1538, seemed to side with the city's rebellion against Queen Maria of Hungary's attempt to raise funds for the Habsburg war effort by flagrantly refusing to endorse Habsburg sovereignty over the region. The map was lined with references to Ghent's civic authorities, noble families and feudal rights, and represented an early appeal to a Flemish 'patrie', or Fatherland, in opposition to Habsburg rule.[27] By 1539, as Ghent descended into rebellion and Charles V mobilized his army to march on the city, mercantile factions, horrified at the consequences, decided the least they could do was commission a map which took the opposite approach to van der Beke's. Mercator's map was finished so quickly that one of its decorative frames was left blank, but otherwise it excised every potentially patriotic reference contained in van der Beke, and made the region's Habsburg allegiances as explicit as possible, culminating in a loyal dedication to the emperor who was bearing down on the city even as the map was nearing completion.[28]

Alas, it had no discernible impact. Charles entered Ghent with an army of 3,000 German mercenaries in February 1540, beheaded the rebellion's ringleaders, stripped the guilds of their commercial privileges, and tore down the old abbey and city gates. The emperor was far more successful than Mercator and his map in leaving his mark on the civic spaces of Flemish cities like Ghent.[29]

Nevertheless, judging from its numerous reprints, Mercator's map of Flanders was a commercial success, and brought him to the notice of Charles V again, thanks to the political support of his old university friend Antoine Perrenot, recently appointed bishop of Arras, whose father, Nicholas Perrenot de Granvelle, was the emperor's first councillor. With their backing, Mercator began work on a series of globes and scientific instruments, including a terrestrial globe, completed in 1541, dedicated to Granvelle, which updated his previous collaborative effort with Frisius and van der Heyden. Everything seemed to be going well for Mercator, still in his thirties and a highly respected geographer with a growing reputation as an instrument maker. Then came the winter of 1544, and the accusation of heresy.

The evidence of the time, as well as Mercator's subsequent religious writings, suggests that his beliefs were far more complicated than simply 'Lutheran'. From the late fifteenth century a more inward, private version of religious belief began to characterize the educated classes in the cities of northern Europe. Diarmaid MacCulloch has argued that such people associated 'the more demonstrative, physical side of religion with rusticity and lack of education, and treat[ed] such religion with condescension or even distaste, seeing rituals and relics as less important than what texts can tell the believer seeking salvation'. Such believers became known as 'spirituals', and were characterized by 'a conviction that religion or contact with the divine was something from within the individual: God's spirit made direct contact with the human spirit'.[30] If these spirituals were understandably sceptical about the rituals of Catholicism, they also eschewed the increasingly prescriptive teachings of Luther and certainly those of Calvin. In 1576 Mercator wrote to his son-in-law on the controversial subject of transubstantiation – a belief that the bread and wine of communion are the body and blood of Christ, which Lutheranism regarded as more of a symbolic union between Christ and his believers. For Mercator, 'this mystery is greater than people can understand. Moreover, it is not reckoned under the articles of the faith which are necessary for salvation ... Therefore, let anybody be of this opinion: so long as he is pious and does not utter any other heresy against the word of God, he should according to my conviction not be condemned. And I feel that one should not break community with such a man'.[31] Such arguments lead us to believe that, from his Catholic rural background to his exposure to the learned environment of Louvain and thinkers like Frisius and Erasmus, Mercator should be considered a 'spiritual', who grasped the necessity of reform, but still expressed a pre-Reformation conviction that an individual's religion was a private matter. His religious beliefs influenced everything he published (including his maps), but they were not defined by a public profession of faith. In the early 1520s, such beliefs might have gone unnoticed, but by 1544 they were easily interpreted as heretical.

At a time when the Catholic Habsburg authorities were increasingly scrutinizing people's religion, it seemed almost inevitable that Mercator's unorthodox beliefs would finally catch up with him. The wider circumstances which led to his arrest were sparked by a conflict between the two patrons who would shape his career – the emperor Charles V and Wilhelm, duke of Jülich-Cleves-Berg. Upon his accession to the duchy in 1539, Wilhelm inherited the duchy of Guelders on the northeastern borders of the Low Countries, which lay outside the emperor's inherited dominions, despite his ambition to unify the region under Habsburg rule. Having allied himself with German Lutheran principalities and France, Wilhelm marched into the Low Countries in the summer of 1542, and by July his forces were besieging Louvain, Mercator's adopted home town. Once again Charles was forced to return from Spain at the head of a massive army. As French opposition evaporated, Charles attacked the duchy of Jülich, and Wilhelm quickly capitulated. In September 1543 he signed a peace in which he kept his Rhineland territories on the condition they remained Catholic, and relinquished his claim to Guelders, giving Charles effective control over the seventeen provinces that would eventually constitute the Netherlands.[32]

The relief for the besieged citizens of Louvain was only temporary. Shaken by events, Charles's sister Maria began rounding up those suspected of holding reformed religious sympathies. Within months Mercator was under arrest. The substance of the accusation of heresy remains obscure, although the surviving documents refer to 'suspicious letters' sent to Minorite friars in Mechelen (possibly Monachus). Perhaps the letters discussed theology, or geography, or both. Without any first-hand accounts of Mercator's stated religious beliefs we shall probably never know if there were any real grounds for the accusations, but they left Mercator languishing in the castle at Rupelmonde for nearly eight months. Fortunately, both his local priest and the authorities at Louvain University petitioned for his release by late summer. As the executions of the condemned began, Mercator was suddenly freed and all charges against him were dropped.

He returned to Louvain, where he found the atmosphere more threatening than ever. The taint of imprisonment still hung over him, and was reinforced by the news of the execution in November 1545 of the printer Jacob van Liesvelt, found guilty of publishing heretical works. As the wave of persecution increased during the following months and years, it became obvious that, despite their intellectual and cosmopolitan attractions, cities like Antwerp and Louvain were no longer safe for spiritual thinkers interested in the foundational questions of cosmography.

It was clearly time to leave, but Mercator still needed to make a living. Over the next six years there were no maps, just a handful of dutiful but uninspiring mathematical instruments dedicated to the emperor Charles V (accidentally destroyed in one of the early clashes between the emperor's Catholic armies and the Schmalkaldic League of Lutheran Princes in 1548). Mercator began to retreat into a Stoic contemplation of the stars. In the spring of 1551, ten years since his last cartographic publication, Mercator published a celestial globe to sit alongside his earlier terrestrial globe. It would be the last thing he made in Louvain. Less than a year later, as war and rebellion threatened to engulf the region yet again, he finally left for good and headed back towards the Rhine.

Mercator probably never grasped the irony that the man whose actions in 1543 indirectly led to his imprisonment in 1544 was responsible for providing a refuge in 1552. Following his humiliation at the hands of Charles V, Duke Wilhelm of Jülich-Cleves-Berg had returned to his ducal lands to recover his pride and invest in building, learning and education. He designed Italianate palaces at his residences in Jülich and Düsseldorf, and in Duisburg, 30 kilometres north of Düsseldorf, Wilhelm planned to build a new university. In 1551 he invited Mercator there, and although details of his offer remain obscure, it seems he wanted Mercator to take the chair of cosmography.[33] For Wilhelm, the attraction of luring one of Europe's leading cosmographers to his new centre of learning was obvious; for Mercator, the opportunity of securing an academic position and the chance to escape the oppressive atmosphere of Louvain was too good to miss. In 1552 he set off on the 200 kilometre journey to Duisburg, via his parents' home town of Gangelt almost halfway along the route. Compared to Antwerp and even Louvain, Duisburg was a small, insignificant town, but it enjoyed the tolerant rule of a duke who resisted the demands for theological conformity of either Rome or, increasingly, of Geneva, embracing instead an Erasmian pursuit of a 'middle way' that regarded faith as a strictly private matter.

Secure in the protection of a benign patron, Mercator resumed his mapmaking. In 1554 he published an enormous, fifteen-sheet wall map of contemporary Europe based on the latest surveying methods, which finally turned its back on the Ptolemaic understanding of European geography by reducing the continent's dimensions by nine degrees from the Greek geographer's overestimation. It proved to be his most successful map to date, selling 208 copies in 1566 alone, and was lauded by Walter Ghim as drawing 'more praise from scholars everywhere than any similar geographical work which has ever been brought out'.[34] It was followed in 1564 by another popular map, this time of the British Isles, published in the same year that Mercator was named as Wilhelm's official cosmographer.[35]

Confident in his new home and free from financial or theological worries, Mercator was finally able to pursue the career for which his theological interests and academic training had prepared him. In the mid-1540s he began to plan an extraordinarily ambitious cosmography 'of the whole universal scheme uniting the heavens and the earth and of the position, motion, and order of its parts'.[36] It would involve the study of creation, heaven, earth and what he called 'the history of the first and greatest parts of the universe': in other words, a chronology of the universe from its creation. The plan would revolve around a world map, but, unlike Mercator's earlier derivative cordiform world map, this one, by a completely different method, would ensure its distinctiveness. But before he could embark on it, he needed to complete his proposed chronology of the world.

Ever since antiquity, geography and chronology had been regarded as the two eyes of history, and both were now undergoing radical reassessments in the light of the recent voyages of discovery. The encounter with the New World alone required new cosmographies to understand the changing terrestrial space of the known world; its inhabitants and their histories posed equally difficult questions for Christian chronology. Why were such people not mentioned in the Bible? How should their history be assessed within Christian Creation – especially where it threatened to predate it? In the sixteenth century, cosmography and chronology were central to providing answers to some of the most contentious questions of the time.

Both subjects appealed to brilliant, unorthodox and in some cases dissident thinkers. To many, the cosmographer seemed to adopt a divine perspective from which to gaze upon the earth, while also looking up and speculating on the structure and origins of the universe. But as Mercator knew, it also risked accusations of pride, or hubris – or heresy. Nor was chronology immune from such charges. The study of the arrangement of historical events in time, and their assignation to an agreed dateline, had fascinated scholars since classical times, but by the sixteenth century the practical and moral value of establishing such a line had become a preoccupation for scholars.[37] 'What chaos would there be in our present life', asked Mercator's contemporary, the astrologer Erasmus Rheingold in 1549, 'if the sequence of years were unknown?'[38] Without accurate chronology, how could you correctly celebrate Easter? And without temporal accuracy, how could you prepare for the predicted end of the world? At a more practical level, from the later fifteenth century people demanded increasingly accurate measurement of both clock and calendrical time. The development of mechanical escapement clocks introduced a new sense of time that drove people to work and to prayer, and these new technologies were supplemented by the publication of ever more complex chronologies, calendars and almanacs.

By the mid-sixteenth century, people also 'turned to chronology in the hope of finding an order that the chaos of the present denied'.[39] But with such hopes and fears came suspicions. Mercator's near contemporaries, the Catholic Jean Bodin (1530–96) and the Huguenot Joseph Scaliger (1540–1609) both wrote vast, learned chronologies drawing on classical sources which appeared to contradict the biblical account of the Creation. In private, Scaliger worried over everything from Jesus's genealogy to the date of the crucifixion, and concluded that chronology was not necessarily defined by religion. Both chronologers and cosmographers inevitably came to the attention of Catholic and Protestant authorities. As well as the accusations levelled against cosmographers like Finé, Bodin was accused of heresy, and Scaliger fled religious persecution in France; and many of their works ended up on the papacy's Index of Prohibited Books.

By reviving his career as a cosmographer, and beginning with chronology, Mercator was trying to find a new way of answering the questions about creation and the origins of the cosmos that had occupied him since he had been a young student in Louvain. It was a more recondite route, but perhaps the secrets of chronology could disclose the past and even more importantly the future, putting the current apocalyptical times into greater perspective; and, as many including Mercator believed, chronology could disclose an imminent eschatology. After the publication of the *Chronologia* he would write to a friend that 'I remain convinced that the war which is now being waged, is the one of the Hosts of the Lord, which is mentioned at the end of the seventeenth chapter of the Revelation of St. John; wherein the Lamb and the Elect will prevail, and the Church will flower as never before'.[40] Whether this represented an attack on the excesses of reformed religious assaults on Rome is unclear, but it shows that Mercator certainly believed that the end of the world was imminent, and that chronology might reveal its exact date.

In 1569 Mercator's *Chronologia* was published in Cologne. It drew on a vast assortment of Babylonian, Hebrew, Greek and Roman sources in an attempt to provide a coherent history of the world in accordance with the Scriptures.[41] His solution to the problem of a chronology that could acknowledge all these sources and their divergent temporalities was to plot a table across which it was possible to compare each Christian date with Greek, Hebrew, Egyptian and Roman calendars. Readers could therefore navigate their way across time, slicing through a particular moment to compare it with other moments in world history. On page 147, for example, they could locate Christ's Crucifixion as the fourth year of the 202nd Greek Olympiad, 780 in the Egyptian calendar, year 53 since the third destruction of the temple of Jerusalem in the Hebrew calendar, 785 in the Roman calendar, and 4,000 years since

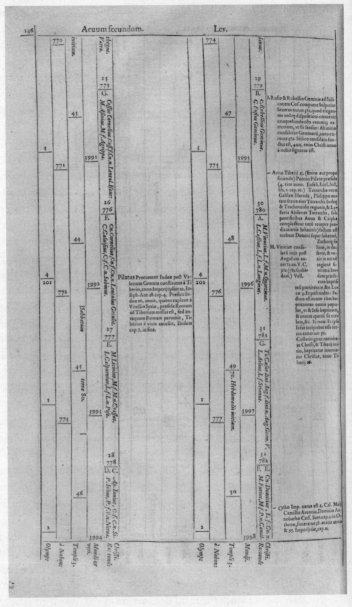

Fig. 21 Gerard Mercator, pages from *Chronologia*, 1569.

778

782

*Ahenobarbus.*
*Iul.Sri.Ionian.*

31
781
D.

*Ser.Sulpitius C. f Ser.n. Galba.*
*L.Cornelius.L.f.P.n.Sulla.*

51

779

3999

*Sabbatum*

*terra 81.*

52

34
784
C.

Vicesimo anno Tiberij L2. Vitellius & Fabius Priscus Coss. fuerunt.Dion lib.58.pag.811.a.

*Paullus Fabius Paulli f.Q.n.P.n.Persicus.*
*L.Vitellius.P.f.Q.n.*

— CHRISTVS Iesus Dominus noster mortem in cruce subijt 2.die Aprilis, feria 6. Luna 15. anno 4,Olymp. 202.Græca suppu tationis, Vide 3.cap.

— Pentecoste & Spiritus sancti missio. Act.2.

4
205
780

4000

3

782

— Nero natus 18. Cal. Ianuarij, ante 9. menses quam Tiberius excessit. Suet.cap.2. Obijt æt.ætatis anno. Suet.cap.57. Aurel.Victor. aut legendum est Ante 9. menses, aut post 9. menses, vt sequenti anni die hoc natus sit.

— CALIGVLA incipit imperare die obitus Tiberij, 17.Cal.Apri lis. Cn.Acerronio Proculo & C.Porrio Nigro Coss. Regna uit ann.3.menses 10. dies 8. occisus 9.Cal.Febr. Suet. Agebat annum ætatis 25. quinqs adhuc ad eum explendum mensibus, & diebus 4. insigens cum imperium assumeret. Dion lib.59. pagina 650,a. Caius cum 3.annis, mensibus 9. diebus 28.ea quæ retulimus registet, recipia competit se non esse Deum, Idem pag.649,a. Agrippa filius Aristobuli, paucis diebus post mortem Tiberij coronatus est rex Iudææ à C.Cæsare, anno autem Cai 2,Hierosolymam redijt. Iosep. Ant.18.cap.15.pag.437.h.Regnavit, annis. Euseb. Regnavit 4 ann sub Caio Cæs cum Philippi Tetrarchia tribut, quartum vero cum Herodis, tres ann reliquos sub Claudij complevit imperio. Iosep. Ant.19.cap.7. in fine.

Philippus Tetrarcha Iudex, filius Herodis Magni, frater Herodis Tetrarche, Galilææ obijt anno 22. Tiberij. Iosep.Ant.18.cap.9. in principio. Succedet ei maxilli Agrippa.

Marcellus, procurator Iudeæ, Pilati loco constituitur à Vitellio Syriæ præside, Iosep.Ant.18,cap.7. in fine. Vitellij præsidis mentionem quoq; sub his Coss. facit Tacit.lib.5.pag.118,h.

*niuli.*

36
787
F.
*Cn.Acerronius Proculus.*
*C.Pontius Nigrinus.*

55

4003

4007.b.398.

37
788
E.
*C.Iulius Germa.Cæf.Aug.Germ. II.*
*L.Apronius,L.f.L.n.Cæsianus.*

783

32
782
B.

*C.Sestius Galus.*
*M.Servilius.M.f Ros.n Nonianus.*

Pauli conversio.
5.Cal. Aprilis M.Servilio, C. Sestio Coss. fuius coruit celebris. exquit à populo Romano curatum est. Plin.lib.10.cap.43.

53

781

4001

56

4
204

784

4004

Cycli decemnovalis initium iuxta Dionysij Exigui abbatis Romani rationem. Deduximus autem hos Dionysij cyclos , ab anno 532, quo primus ab eo institutus est, vsque ad proxima passioni Christi tempora, vt cum inferta passim Astronomicis observationibus facilius coferri, & quantum primis hisce temporibus à veritate aberrent , deprehendi possint. Putat autem Dionysius primo anno cycli sui Lunam decimamquartam, Nonis Aprilis, vt testatur Beda, vnde si quis anno Domini 133.inchoando intervallum ad secundum Lunæ eclipsin à Ptolomæo, anno 135. observatam colligat, & fyzygias luminarium ei convenientes aptet, inveniet Lunam decimamquartam dicto 133. anno, diebus circiter 12. prius ex Ptolomæi calculo deprehendi, qui ex Dionysiaci cycli continuatione posita intelligatur.

35
786
A. G.
*Q.Plautius Plautianus.*
*Sex.Pepuslus.Q.f.Galit.*

54

781

4002

38
789
D.
*M.Aquilius E.Iulianus.*
*P.Nonius,M.f.Asprenas Ros.consu.*

57

2

1

2

*Olymp.*

*à Nabon?*

*Mundi.*

*Templi 3.*

*Christi. Rec.condi*

4003

*Olymp.*

*à Nabon?*

*Templi 3.*

*Christi Rec.cond.*

4005

N ij

Creation.[42] The problem that Mercator (along with other chronologers) faced was how to arrive at these calculations based on divergent estimates of the passage of time between the Creation and the coming of the Messiah. The Greek text of the Old Testament claimed 5,200 years separated the two events, while the Hebrew text claimed it was 4,000 years. Like many other chronologers, Mercator endorsed the Hebrew version, with slight revisions based on his reading of classic authors like Ptolemy.[43]

Compared to subsequent chronologers like Scaliger, Mercator's theological chronology was quite traditional, although because it included references to reformed religious events and individuals it was soon placed on the Index of Prohibited Books. But it was his method of organizing his material that was so significant. In aligning simultaneous historical temporal events on the same pages, Mercator was attempting to establish chronology, and to reconcile apparently incompatible historical data, in the same way that the cartographer was trying to square the spherical globe and project it upon a flat surface.

Mercator's *Chronologia* represented part of his wider cosmographic ideal, uniting the study of chronology and geography to rise above contingent, earth-bound behaviour. Its inspirations were Plato, Ptolemy and the Stoic philosophy of Cicero's 'Scipio's Dream', adopting a transcendent, cosmic gaze, rising up to look down on the world from above, indifferent to its petty terrestrial conflicts.[44] This was the immediate context for the creation of the world map on Mercator's famous projection: just as the *Chronologia* invited its readers to navigate across time, his world map would offer a spatial navigation across the globe, which also needed the guiding hand of the cosmographer to transform it into a flat surface. Rather than celebrating the virtues of European civilization by placing it at the centre of his work, the map was part of a cosmography that aimed to transcend the theological persecution and division of sixteenth-century Europe. Instead of exhibiting a confident Eurocentrism, Mercator's world map would provide an oblique rejection of such values and a search for a larger picture of harmony across universal space and time.

The *Chronologia* was not a great success. Mercator had little if no reputation as a chronologer, and the book's traditional interpretation of dates and events (notwithstanding its unusual layout) meant that it received little popular or critical attention; indeed, depite the fact that

Mercator spent more than a decade writing it, the *Chronologia* is usually overlooked when compared to his geographical achievements, and
in particular the map he was about to publish.

Several months after the printing of his chronology, Mercator released
the next instalment of his cosmography: a world map, published in
Duisburg, entitled *Nova et aucta orbis terrae descriptio ad usum navi-
gantium emendata accommodata*, or 'A New and Enlarged Description
of the Earth with Corrections for Use in Navigation'. Mercator's 1569
projection may be the most influential map in the history of geography,
but it was also one of the most peculiar. Nothing prepared Mercator's
contemporaries for such a strange object: not its scale, its appearance,
or its claims 'for use in navigation'. As a cosmographer interested in
mapping the heavens onto the earth, Mercator had shown little or no
previous interest in the practical applications of maps for the pursuit of
accurate navigation; indeed, his only previous attempt at a world map
using the cordiform projection in 1538 reflected a fascination with the
theology of the heart more than navigating across the terrestrial globe.

This world map was enormous. Engraved on eighteen sheets, it was
intended to hang on a wall, and when assembled it measured more than
2 metres in length, and was nearly 1.3 metres high, similar in size to
Waldseemüller's 1507 world map. But even more surprising was its
strange layout. On first inspection it looks more like a work in progress
than a triumphant moment in global cartography. Large areas of the
map are given over to elaborately decorated cartouches containing
extensive legends and complicated diagrams. North America, which on
Waldseemüller's map looked like a modest wedge of cheese, was transformed by Mercator into 'India Nova', a sprawling behemoth, with its
northern land mass covering more space than Europe and Asia put
together. South America, with its inexplicable south-western bulge, bore
little resemblance to its presentation by Ribeiro and other mapmakers
as an elongated pendulum. Europe covered twice its true area, Africa
appeared reduced in size compared with contemporary maps, and
South-east Asia was unrecognizable to those brought up on the Ptolemaic overestimations of its shape and size.

Even more peculiar is Mercator's depiction of the polar regions,
which were shown running the full width of the map's top and bottom,
making no apparent concession at all to the earth's sphericity. Mystified
viewers could consult the legend in the map's bottom left-hand corner,

which calmly informed them that Mercator based his conception of the northern polar regions on a mythical voyage undertaken by a fourteenth-century Oxfordshire monk called Nicolas of Lynn, who used his 'magical arts' to sail all the way to the North Pole. Mercator concluded that the polar region was composed of a circular land mass, 'the ocean breaking through by nineteen passages between these isles forms four arms of the sea by which, without cease, it is carried northward there being absorbed into the bowels of the Earth'. On one of the land masses Mercator wrote, 'Here live pygmies whose length in all is four feet, as are also those who are called Screlingers in Greenland'.[45]

In its fine detail the map looks more obviously poised between an older cosmographical tradition and a newer mathematical understanding of geography – much like Mercator's religious beliefs. Mercator's delineation of Asia is drawn from Marco Polo's travels, but the map's legends also record in some detail the recent political manoeuvrings surrounding the voyages of da Gama, Columbus and Magellan. There are long written digressions on the existence of the fabled Christian ruler Prester John, alongside very precise revisions of the Ptolemaic geography of the Nile, Ganges and the location of the 'Golden Chersonese'. But across Africa and Asia Mercator also reproduces Pliny's 'Samogeds, that is the people who devour each other', 'Perosite, with narrow mouths, who live on the odour of roast flesh', and 'men who unearth the gold of ants'.

Mercator's map shows the study of cosmography stretched to its furthest limits. In an attempt to combine the synoptic desire of cosmography with the mathematical rigour of the new techniques of surveying and navigation, the map looked backwards to classical and medieval authorities as much as it looked forwards to embrace a new conception of geography. But the great discovery Mercator made in his years of studying chronology alongside geography was a method of plotting a spherical earth on a plane surface, a mathematical projection that would transform mapmaking and signal the beginning of the end of cosmography.

In his address to the reader, contained within the enormous legend conveniently obscuring most of North America, Mercator explained that 'in making this representation of the world we had three preoccupations'. They were 'to show which are the parts of the universe which were known to the ancients', so that 'the limitations of ancient geography be not unknown and that the honour which is due to past centuries be given to them'. The ancients, and in particular Ptolemy, were being politely given their due and quietly shown the door. Secondly, Mercator aimed 'to represent the positions and dimensions of the lands, as well as the distances of places, as much as in conformity with very truth as it is possible so to do'. But finally, and most importantly, his intention was

> to spread on a plane the surface of the sphere in such a way that the positions of places shall correspond on all sides with each other both in so far as true direction and distance are concerned and as concerns correct longitudes and latitudes; then, that the forms of the parts be retained, so far as is possible, such as they appear on the sphere.

Mercator's two aims here sound like basic common sense. Most people now would assume that a world map ensures that geographical features on a map have the same shape as on a globe, and that directions and distances are accurately represented. But Mercator knew from thirty years of making globes that it is not possible to retain both features on a plane surface. For the mid-sixteenth-century mapmaker, the problem was compounded by the fact that the representation of large areas was predominantly the domain of the cosmographer, who sought to show continents and seas from an imaginary point located above the earth, while direction and distance was of almost exclusive interest to the sea pilot, navigating across open water, with little or no interest in the shape of land masses.

Prior to the sixteenth century, none of this really mattered. Cosmography pursued its classical ideals, projecting geometrical principles onto the surface of a vaguely defined world. At the other extreme, the portolan sailing charts used in the Mediterranean required extremely basic methods of navigational projection, as they covered such a tiny fraction of the earth's surface. As a result, they developed geometrical networks of criss-crossing straight lines to sail from one location to another. These were known as 'rhumb lines' – from the Portuguese *rumbo* ('course' or 'direction'), or the Greek *rhombus* ('parallelogram'). In reality, rhumb lines were curved due to the sphericity of the earth's surface. If they were extended over great distances, the distortion would lead to a pilot sailing way off course, but across the relatively short distances of the Mediterranean, such discrepancies were of little serious consequence.

Once the Portuguese began sailing longer distances down the coast of Africa and across the Atlantic, one of the many problems they faced was how to draw maps with straight rhumb lines that took into account the earth's curvature.

Technically, a rhumb line is what later mathematicians would call a loxodrome (taken from the Greek *loxos*, or 'oblique', and *dromos,* or 'course').[46] As its root suggests, a loxodrome was a diagonal line of constant direction that intersects all meridians at the same angle. Rhumb lines were not the only method of navigating across the earth's surface. A navigator could use the traditional portolan-style straight line sailing method (and many navigators, fearful of change, continued on this path for decades), but beyond the Mediterranean it left sailors so far adrift that it soon became unsustainable. The other method was great circle sailing. A great circle is, as its name suggests, the largest circle that can be drawn around the globe, with its plane running through the earth's centre. The equator and the meridians are all great circles. The advantage of great circle sailing was that great circles always represent the shortest route between any two points on the earth's surface. But the likelihood of charting a route from one location to another that involved sailing exactly along the equator or a parallel was not only highly unlikely, but also technically very difficult, as the bearing of the curved arc is constantly changing, requiring pilots to repeatedly adjust their direction.

Rhumb lines represented a *via media*, or 'middle way'. They were the most likely directions sailed by navigators, particularly once diagonal east–west routes via the Cape of Good Hope and Magellan's Straits became vital for sixteenth-century European seaborne trade (the kind of routes traced by the ships depicted on Mercator's map). But another complicated feature of any rhumb line drawn across the earth's surface was not only that it curved, but that, if followed indefinitely, it traced a spiral that ends up infinitely circling one or other of the poles, because of the gradual convergence of the meridians. For mathematicians, a loxodrome's spiral is a beguiling geometrical feature, but for navigators, turning it into a straight line was a frustrating exercise. Mercator confronted the problem as early as 1541, when he traced a series of rhumb lines across the surface of his terrestrial globe. Portuguese cosmographers had already described the loxodrome in the 1530s when trying to explain why pilots navigating across the Atlantic found themselves

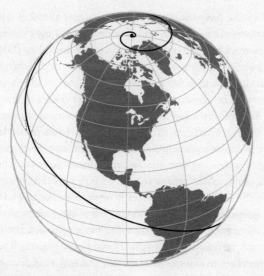

Fig. 22 Model of a spiral loxodrome.

gradually sailing off course. Unfortunately the Portuguese could offer no solution as to how to flatten out a loxodrome accurately onto a plane surface.

In his address to the reader, Mercator proposed the ingenious solution to the problem that lay at the heart of his new projection, which was the curvature of the meridians. 'Indeed,' he wrote, 'the forms of the meridians, used till now by geographers, on account of their curvature and convergence to each other, are not utilisable for navigation', because, as he went on, 'at the extremities they distort the forms and positions of regions so much, on account of the oblique incidence of the meridians to the parallels, that these cannot be recognised nor can the relation of distances be maintained'. Mercator then concluded famously, 'It is for these reasons that we have progressively increased the degrees of latitude towards each pole in proportion to the lengthening of the parallels with reference to the equator'. How did he reach this conclusion, and how did it work?

Mercator's projection is based on a cylindrical perception of the earth. Later interpreters used the analogy of the earth as a balloon to explain the method. Put the balloon inside a cylinder with the same diameter as the balloon at the equator. If the balloon is inflated, its curved surface will be pressed and flattened against the cylinder's walls. The curved meridians are 'straightened' as they meet the cylinder, as are the parallels. One consequence of such stretching and flattening is that the North and South poles can never touch the cylinder's walls, and effectively stretch onwards into infinity. If the cylinder is then unrolled with the imprint of the balloon's meridians and parallels, the resulting rectangle approximates to the Mercator projection. This description offers one plausible explanation as to how Mercator developed it. Having spent decades on the mathematical and practical creation of terrestrial globes, Mercator was able to conceptualize how to adapt accurately the surface of such globes to represent a flat map. Take a segment, or gore of a globe, which looks rather like a vertical slice of an apple or an orange. Redraw it on a flat piece of paper, but retain the width of each meridian at the equator from top to bottom. Then stretch out the parallels to compensate for the gradual straightening of the meridians (as with the balloon analogy), and you are left with a thin rectangle. If the same method is applied to each globe gore, you have a series of rectangles which, when stitched together, make a flat map.[47]

The result still caused distortion of land masses at the northern and southern extremities, but if Mercator could accurately calculate how far apart to space his parallels he could achieve something unique: what cartographers call 'conformality', defined as the maintenance of accurate angular relations at any point on a map. Notwithstanding the distortion of land mass, navigators could plot a straight line across the map's surface and, if they maintained a consistent angle of bearing, would still arrive at their projected destination. For Mercator, this meant straightening the meridians and calculating how far apart his parallels should lie if they were to retain a straight line of bearing. So, for example, at the equator the distance between any two meridians is twice the same distance along the parallel running 60° N, due to the convergence of the meridians. On his map Mercator therefore widened the parallel at 60° N to twice its actual length, ensuring that an oblique angle running through it would be straightened.[48] All the other parallels were subject to the same calculation, and were lengthened accordingly.

Mercator produced what cartographers today term the first cylindrical conformal projection – it treated the globe like a cylinder, and maintained accurate angles across its surface. Sixteenth-century pilots, of course, could not care less about its title; Mercator's method simply allowed them to 'straighten' the meridians, which no longer curved inwards towards the poles, but instead ran perpendicular to the parallels. They could now plot a rhumb line using Mercator's projection, but rather than following a spiral and running off course as on earlier charts, a straight rhumb line now retained its accuracy in navigating from one location to another. It was a relatively simple but ingenious solution to the problem of projecting the entire earth onto a plane surface which had preoccupied mapmakers since the time of Ptolemy. Mercator appeared to have finally squared the geographical circle. It was a decisive breakthrough which would change maps for ever, and immortalize Mercator.

However, even a cursory glance at the shape of Mercator's world reveals some obvious problems inherent in the projection. As the balloon analogy shows, the meridians are never allowed to converge on a single point, so the polar regions stretch into infinity, always already beyond the map's rectangular frame. This is just one of the reasons that Mercator requires his small inset map to explain the geography of the North Pole. The mathematical stretching that takes places at the poles also affects the relative size of land masses in higher altitudes, which is why in the southern hemisphere Antarctica dwarves every other continent, and Greenland looks the same size as South America, when in terms of surface area it is actually only an eighth as large. In contrast, Europe appears twice the size of South America, when it actually covers just half the surface area. The stretching of the parallels north to south also means that Mercator's projection distorts distances between locations over long seaborne journeys – though at the time it seemed more important to ensure pilots arrived at their location rather than how long it took them, particularly in an age before the advent of steam.

There still remained one crucial problem. Mercator could not offer a reproducible mathematical formula for his projection which would allow mapmakers and pilots to copy his methods. Neither the logarithms nor the integral calculus required to reproduce trigonometric tables delineating the projection's parallels and meridians were available to him. This made his empirical achievement all the more remarkable (and an enduring mystery), but it still meant pilots found it difficult to use the projection. Writing in 1581, the Elizabethan mathematician William Borough said of Mercator's explanation of his method, 'by augmenting his degrees of latitude towards the Poles, the same is more fit for such to behold as study in cosmography by reading authors upon the land, than to be used in navigation at sea'.[49] Although Mercator had effectively solved a centuries-old problem of mapmaking for navigators, he appeared surprisingly indifferent to providing an explanation that would ensure its immediate fame and longevity; without further mathematical explanation it remained the preserve of scholarly cosmographers.

Nothing seemed to prepare Mercator for resolving the age-old conundrum of plotting the entire surface of the globe onto a flat map and retaining conformality for navigation. His greatest work up to this point in his career remained the 1541 terrestrial globe. On its spherical surface, Mercator had no difficulties projecting the curvature of the earth, but a method of transferring such an image onto a flat map had eluded him for more than three decades. There is, however, an intriguing possibility: the years spent working on connecting events through time in the *Chronologia* enabled him to imagine a new way of connecting places in terrestrial space on a flat map. Throughout the 1560s, as he laboured on his chronology, he also compiled the data and invented the projection that would culminate in the 1569 world map. Both publications came out within months of each other. Perhaps, in the same way that Mercator allowed believers to read laterally across time and navigate the different religious temporalities in the *Chronologia*, his new projection allowed navigators to travel across the space of God's earth, using a rhumb line to pursue a 'middle way' between inaccurate straight-line and impractical great-circle sailing, 'accurately' connecting places in space, just as the *Chronologia* 'correctly' located different events in time.[50]

Unlike any of his earlier geographical efforts, Mercator's map is noticeably bereft of imperial patronage, religious affiliations or political boundaries. There are no imperial eagles and few far-flung territories claimed on behalf of European rulers expressing their global dominion. It offered a more accurate method of navigating across the earth, but it also offered its Christian readers a vision of spiritual peace and concord that drew on the Stoic principles of Cicero and Macrobius. In the rarely read dedication to the map, Mercator honours his patron Duke Wilhelm, but he takes it as an opportunity to describe the world's people and countries within a cosmic image of harmony that evokes the classical gods of antiquity, but which subsumes them under the sway of a Christian God who is unconcerned with wars, famine and religious conflict:

> Happy countries, happy kingdoms in which Justice, noble progeny of Jupiter, reigns eternal and where Astraea, having regrasped her sceptre, associates herself with divine goodness, raising her eyes straight to the heavens, governs all in accordance with the will of the Supreme monarch and devotes herself to the submission of unfortunate mortals to His sole empery, seeking happiness . . . and though Impiousness, the enemy of virtue, causing Acheron to riot, raises some gloomy disorder, no terror is felt: this allgood Father who, residing on the crest of the world, orders all things by the nod of His head, will never desert His works or His kingdom. When the citizen is in this wise governed he fears no ambush, he has no dread of horrible wars and mournful famine, all pretences are swept away from the unworthy backbiting of sycophants . . . dishonesty, despised, lies prone, virtuous deeds everywhere call forth friendship and mutual treaties bind men solicitous of serving their King and their God.

A contemplation of the world on his map enables Mercator's readers to understand that, as long as they are 'governed' by a belief in God, regardless of their religious belief, then riot, conflict and the destructive pursuit of earthly glory can be seen as transitory, and insignificant when viewed from a cosmographical perspective.

Such an interpretation of the projection might be appropriately 'oblique', rather like a rhumb line. But then again, Mercator had himself withdrawn into a world of coded critique and arcane symbolism following the traumatic events of the 1540s. Today, both supporters and detractors of Mercator's projection tend to judge it as a disinterested mathematical innovation, and believe its wider theological and cosmographical contexts, as well as Mercator's own life, to be relatively incidental. But Mercator's career shows that in the mid-sixteenth century it was impossible to separate science from history, history from geography, geography from cosmography, and cosmography from theology. For Mercator, everything was connected, but also ultimately subsumed under a spiritual authority, a divine architect who oversaw everything including the projection of the world He created.

Within his lifetime, Mercator's projection was a qualified failure. Sales were slow, and many like Borough complained that Mercator's inability

.to explain his methods made them virtually useless for practical use for seaborne navigation. It took an Englishman, Edward Wright, in a series of mathematical tables in his book *Certaine Errors in Navigation* (1599), to provide the calculations required to translate the projection for the use of pilots, who slowly began to adopt the method during the course of the seventeenth century.

Mercator himself seemed indifferent to his achievement, and spent the final three decades of his life continuing work on his cosmographic project, of which the *Chronologia* and his world map represented just two elements. In 1578 he published an edition of Ptolemy's *Geography*, which lovingly reproduced the Greek geographer's maps as historical curiosities, an important but now redundant conception of the earth as it was understood by the Hellenic world. The edition effectively ended the classical geographer's influence on contemporary mapmaking. From now on, mapmakers trying to map the world would chart their own path rather than revising and updating Ptolemy.

Mercator continued to write theological works with a direct bearing on his cosmography, including a study of the Gospels, *Evangelicae historiae*, published in 1592. Finally, just a year after his death in 1594, the culmination of his cosmography was finally published. The *Atlas sive cosmographicae meditationes de fabrica mundi et fabricate figura* (subsequently published in English as 'Atlas, or Cosmographic Meditations on the Fabric of the World and the Figure of the Fabrick'd') was the first modern atlas to use the title, with 107 new maps of parts of the world, although it omitted to use the 1569 projection on its world map, another sign of Mercator's indifference to its scientific innovations. Instead he chose a double hemispherical stereographic projection to depict the world. In the *Atlas*, Mercator offers an illuminating reflection on the place of his earlier world map within his cosmography. He tells his readers that he has used geographical information from his earlier map of Europe and the world map of 1569, and implores them to turn to cosmography, 'the light of all history, both ecclesiastical and political, and the idle onlooker will learn more from it than the wayfarer from his lengthy, irksome, and expensive labours (who "often changes skies but not his mind")'.[51] Mercator takes a quote from the Roman poet Horace's *Epistles*, which was subsequently used by the Roman Stoic philosopher Seneca, to emphasize the true value of cosmography: to meditate on spiritual conscience rather than terrestrial orientation. Developing this

Stoic perspective, Mercator invited his reader to 'think diligently of the glory of your dwelling-place, which is only temporarily granted to you, along with the poet George Buchanan, who thus compares it to the celestial realm in order to draw forth your souls, immersed in terrestrial and transitory affairs, and show the way to higher and eternal things'.[52] Buchanan (1506–82) was an internationally renowned Scottish historian and humanist scholar, a Lutheran sympathizer, tutor to Mary, queen of Scots and her son, the future King James I of England, and a well-known Stoic. It is typical of Mercator that he quotes a poem by Buchanan rather than choosing his own words to summarize his Stoic approach to mapping the earth and the heavens:

> May you perceive how small a portion of the universe it is
> That we carve out with magnificent words into proud realms:
> We divide with the sword, and purchase with spilled blood,
> And lead triumphs on account of a little clod of earth.
> That strength, seen separately by itself,
> Is great indeed, but if you compare it with heaven's starry roof, it is as
> A dot or the seed from which the old Gargettian [Epicurus]
>     created innumerable worlds.

In a final explicit evocation of Macrobius' 'Scipio's Dream', Buchanan concludes that, because humanity is confined to such a small part of the universe, the pursuit of worldly glory is foolish:

> How tiny the part of the universe is where glory raises its head,
> Wrath rages, fear sickens, grief burns, want
> Compels wealth with the sword, and ambushes with flame and
>     with poison;
> And human affairs boil with tremulous uproar![53]

Speaking through Horace, Seneca and their neo-Stoic follower Buchanan, Mercator recommended that an individual simultaneously retreat from and transcend the religious and political discord of his generation, and instead pursue spiritual shelter by an acceptance of a larger cosmic harmony. Only cosmography could provide a suitable perspective from which to view the theological conflict of the Reformation, and to offer a way of turning away from its intolerance to embrace a more inclusive perspective of divine harmony.

By the end of the sixteenth century, much of Mercator's innovative work was either being more efficiently marketed by lesser geographers, or was starting to look intellectually outdated. His *Chronologia* quickly became obsolete with the publication in 1583 of Scaliger's more comprehensive *De emendatione temporum* ('Study on the Improvement of Time'). His younger disciple Abraham Ortelius had already published an atlas of the world in Antwerp in 1570, although it did not use the term, choosing instead the title *Theatrum orbis terrarum* ('Theatre of the World'). As a member of the dissenting religious group, the Family of Love, with its connections to the Protestant Anabaptist sect, Ortelius was freer than Mercator to use his *Theatrum* to embrace an explicitly Stoic attitude towards the description of the world (just six years before the forces of Philip II of Spain brutally sacked Antwerp in November 1576, killing an estimated 7,000 people). The descriptive cartouches on Ortelius's world map offer a more explicit version of Mercator's cosmographical philosophy of peace, concord and indifference to worldly glory. They included quotes from Seneca, asking 'Is this that pinpoint which is divided by sword and fire among so many nations? How ridiculous are the boundaries of mortals', and Cicero's rhetorical question, 'what can seem of moment in human occurrences to a man who keeps all eternity before his eyes and knows the vastness of the universe?'[54]

Mercator's *Atlas* still deserves its description as the first modern atlas (it was far more innovative than Ortelius's limited but shrewdly packaged publication), and it established the layout and running order for most subsequent atlases. It sold well and the name ultimately stuck, but Mercator had missed his moment. The earth was moving under his feet (quite literally, according to Copernicus's new theories of a heliocentric universe) and his cosmography represented the discipline's zenith. It finally displaced Ptolemy's *Geography*, but whereas the Greek's influence lasted for more than a millennium, Mercator's cosmographical publications barely reached into the following century before, like Ptolemy, they became another historical curiosity. Instead, it was his 1569 world map that would endure.

The pace of political, intellectual, theological and geographical change was simply too fast for any one scholarly individual to explain, and led to what has been called the crisis of cosmography, in which the polarized religious atmosphere would no longer tolerate the pride of the cosmographer's god-like perspective. The sheer complexity of representing the natural world meant that no single figure could any longer provide a convincingly synthetic and comprehensive vision of everything. Collections of travellers' reports and voyages compiled by more modest intellects such as Giovanni Battista Ramusio, Richard Hakluyt and Théodore de Bry began to supplant the singular perspective of cosmographers like Mercator. Subsequent geographers like Jodocus Hondius and Willem Blaeu in the Low Countries and the Cassinis in France turned into dynasties, drawing on generations to work on globes and atlases using state finance and employing vast teams of scholars, surveyors and printers. Cosmography fragmented into a series of discrete practices, and its theological and moral power gave way to that of mathematics and mechanics.[55]

If this fragmentation was seen by some as progress, it also reduced mapmaking's ability to transcend worldly conflict and intolerant attitudes in favour of a larger understanding of secular and sacred space. As David Harvey has ruefully pointed out, 'the Renaissance tradition of geography as everything understood in terms of space, of *Cosmos*, got squeezed out'. As cosmography withered away, geography 'was forced to buckle down, administer empire, map and plan land uses and territorial rights, and gather and analyse useful data for purposes of business and state administration'.[56] But though Mercator's cosmography quickly became an irrelevance, his map projection, inspired by cosmographical concerns, became central to this new geography. Its mathematical principles were appropriated for measuring nation states and Europe's growing colonial possessions. The projection was adopted by the English Ordnance Survey, the British Navy's Admiralty charts, and in a suitably cosmographical twist, by the NASA space agency to map various parts of the solar system. The great cosmographer would undoubtedly have approved.

Mercator made his own geography, but not of his own free will. His 1569 map of the world on his now famous projection was determined by a very particular concatenation of forces, which allowed him to imagine cosmography as a scholarly discipline from which a more tolerant, harmonious vision of the individual's place in the cosmos could be imagined. Ultimately, such a vision was unsustainable, and it hastened the decline of cosmography. But the 1569 projection would endure, shaped by the religious intolerance of European civilization, rather than its inherent superiority over the rest of God's earth.

# 8

# Money

*Joan Blaeu,* Atlas maior, *1662*

## Amsterdam, 1655

On 29 July 1655 the new Amsterdam Town Hall was officially opened with a banquet attended by the city's councillors and dignitaries. Designed by the Dutch architect Jacob van Campen, and taking over seven years to complete, the building was the largest architectural project ever undertaken by the Dutch Republic in the seventeenth century. Van Campen's aim was to produce a building that would rival the Roman Forum, and announce to the world the emergence of Amsterdam as the new centre of political and commercial power in early modern Europe. Addressing the banquet, the renowned scholar and diplomat Constantijn Huygens recited a poem commissioned for the occasion, in which he extolled the city's councillors as 'the founders of the eighth wonder of the world'.[1]

The building's greatest wonder, as well as its most interesting innovation, lay at its heart, the vast Burgerzaal, or People's Hall. At 46 metres long and 19 metres wide, with a height of 28 metres, the People's Hall was the largest unsupported civic space then in existence. Unlike the great Renaissance royal palaces of the fifteenth and sixteenth centuries, the People's Hall was open to everyone. It also departed from earlier monumental built spaces for another reason. Rather than adorning its walls with tapestries or paintings, the principal decoration of the People's Hall was on its polished marble floor in the form of three flat, hemispherical globes.

As visitors walked into the hall, the first image showed the western terrestrial hemisphere, the second the northern hemisphere of the heavens, and the third the northern terrestrial hemisphere. Carefully inlaid into the marble floor, rather than hung on walls, enclosed in books or locked away by their owners, like so many earlier maps, the images in the People's Hall were displayed for all to see. Amsterdam's citizens, so many of whom had personal or indirect experience of long-distance sea-borne travel, were now given the novel sensation of walking across the earth. The world, it seemed, had come to Amsterdam. Such was their confidence, the Dutch Republic's burghers did not even feel the need to place their city in the middle of their marble hemispheres: for them, Amsterdam was the centre of the world.

The three hemispheres were inlaid in the Hall's floor by the Dutch artist Michiel Comans, but they are reproductions of a world map printed seven years earlier and made by arguably the greatest, certainly most influential, Dutch mapmaker in the history of cartography: Joan Blaeu (1598–1673). Printed on twenty-one sheets, more than 2 metres in length and nearly 3 metres high, Blaeu's vast copperplate engraved world map, depicting the twin terrestrial hemispheres, was noticeably different from Mercator's 1569 world map with its odd projection and speculative western and southern continents. Unlike Mercator, Blaeu was able to draw on his role as an institutional mapmaker: from 1638 he was the official cartographer to the Dutch East India Company, the *Vereenigde Oostindische Compagnie* (otherwise known as the VOC), which allowed him unrivalled access to records of more than fifty years of Dutch commercial voyages to the west and east of Europe, as well as to the latest pilots' maps and charts tracing the route to the Indies and beyond. This enabled him accurately to depict the tip of South America and New Zealand ('Zeelandia Nova'). It was also the first world map to show both the west coast of Australia – labelled 'Hollandia Nova detecta 1644' – and Tasmania, named after Abel Janszoon Tasman, the first European to reach the island and claim formal possession of it in December 1642.[2]

But Blaeu's map was also created to celebrate a specific political event. It was dedicated to Don Casparo de Bracamonte y Guzman, count of Penaranda, the leading Spanish representative at the diplomatic negotiations which culminated in the Peace of Westphalia that ended both the Thirty Years War (1618–48) and the even longer-running Eighty Years War (also known as the Dutch War of Independence) between Spain and the territories that would eventually comprise the United Provinces. The peace agreement divided the northern (predominantly Protestant) republican provinces of today's Netherlands from the southern (traditionally Spanish-dominated) regions of modern-day Belgium, granting the United Provinces independence as well as the right to freedom of religious expression for its Calvinist majority. The new Republic became the hub of the commercial world, with the VOC and its headquarters in Amsterdam at its centre. Blaeu's map was a shrewdly pitched celebration of political independence, and a prefiguration of the Dutch domination of seaborne trade that would quickly follow the treaty's ratification.[3]

Blaeu's map of 1648 is probably the first to be reproduced in this book that is immediately recognizable as a modern map of the world. Even though it remains sketchy on the topography of the Pacific, and the mapping of Australia's coastline is incomplete, it is more familiar to us than Ribeiro's seemingly unfinished map, or Mercator's projection. The Blaeu map's familiarity is partly based on its incremental accumulation of geographical data, which by the mid-seventeenth century had produced a reasonable consensus among European mapmakers about what the world looked like. But if we look more closely at its six inset images, the map appears to be celebrating more than just a new era of peace in Europe and the standardization of a particular image of the world. In the top left- and right-hand corners, Blaeu has represented the northern and southern celestial hemispheres. Between these two images, just below the Latin word 'terrarum' in the map's title, is another inset diagram. This depicts the solar system according to the heliocentric theories of Nicolaus Copernicus, which show the earth revolving around the sun, overturning centuries of first Greek then Christian belief in a geocentric universe. Although Copernicus's groundbreaking book *On the Revolutions of the Spheres* had been first printed in 1543, just over a century earlier, Blaeu was the first mapmaker to incorporate his revolutionary heliocentric theory into a map of the world. As if to emphasize the point, an inset diagram at the bottom of the map shows a map of the world as it looked in 1490 in the middle, with a diagram portraying the Ptolemaic cosmos on the left, contrasted with the great Danish astronomer Tycho Brahe's diagram of a 'geo-heliocentric' cosmos (first published in 1588) on the right.

In reproducing Blaeu's 1648 world map on the floor of the People's Hall, the city's councillors were self-consciously creating a whole new world picture, one that effectively signalled the end of the European

Renaissance. They were paying not only for a new kind of map, but for a new philosophy of the world, in which the earth, and by implication humanity, was no longer to be found at the centre of the universe. It was also a world in which the scholarly pursuit of geography and mapmaking were now fully institutionalized within the apparatus of the state and its commercial organizations – which in the Dutch Republic meant the VOC.

The VOC transformed the practice of trade and the involvement of the public in funding commercial activities. Managed by a board of seventeen directors known as the *Heeren XVII*, the company was divided into six chambers across the seventeen Provinces. As a joint-stock company, the VOC offered any Dutch citizen the opportunity to invest and claim a share of its profits. This proved to be very attractive: in 1602 the Amsterdam chamber attracted more than 1,000 initial subscribers from a population of just 50,000. With an average dividend of over 20 per cent on an investor's initial investment, and a growth in public subscriptions from 6.4 million guilders at its inception, to more than 40 million by 1660, the VOC's methods revolutionized European commercial practice, valorizing risk and encouraging the monopolization of trade in a way never seen before.[4]

One consequence of these changing methods of financing long-distance trade was a transformation of the role of maps. The Portuguese and Spanish empires had established their commercial importance as route-finding devices, and tried to standardize them through the creation of organizations like the Casa de la Contratación. But these initiatives, like all overseas activities, were controlled by the crown. The maps they produced were invariably hand-drawn in a futile attempt to limit their circulation, and because the Iberian peninsula did not have the extensive printing industry that emerged in northern Europe from the late fifteenth century. Although the Dutch commercial companies that were founded in the 1590s lacked the money and manpower commanded by their Spanish and Portuguese rivals, they were able to draw on an established body of printers, engravers and scholars experienced in collating the latest geographical information on maps, charts, globes and atlases. Mapmakers like Waldseemüller, Mercator and Ortelius had already turned mapmaking into a profitable business, selling authoritative and beautiful maps on the open market to anyone that could afford them. The Dutch commercial companies saw the opportunity to capitalize on this development by employing mapmakers to create manuscript charts and printed maps providing the safest, quickest and most profitable routes from one commercial location to another. It also made sense to bring together teams of mapmakers to standardize information and encourage commercial collaboration and competition.

As a result, by the early 1590s a variety of Dutch mapmakers were competing to provide commercial companies with maps to assist in developing overseas trade. In 1592 the States General, the body of the elected legislative delegates from the Republic's provinces, granted the mapmaker Cornelis Claesz. (c. 1551–1609) a twelve-year privilege to sell a variety of charts and wall maps that could be bought for anything from 1 guilder for a map of Europe, to 8 guilders for a bound collection of maps of the East and West Indies. In 1602 the mapmaker Augustijn Robaert began supplying charts to the VOC, sometimes charging up to 75 guilders for each one with their comprehensive depiction of these newly discovered regions.[5] Maps were becoming a relatively profitable trade, and their makers were being gradually institutionalized by the companies that needed them. With money to be made, a new generation of talented mapmakers emerged, sometimes collaborating but also competing with each other for patronage of the new commercial companies as well as merchants and pilots working independently of organizations like the VOC. Petrus Plancius, Cornelisz Doetsz, Adriaen Veen, Johan Baptista Vrient and Jodocus Hondius the Elder all sold maps, charts, atlases and globes to the VOC as well as to private individuals according to their particular needs. Maps were now being reproduced, bought and sold for specific commercial purposes.[6] The Portuguese had introduced the scientific craft of modern mapmaking, but it was the Dutch who turned it into an industry.

On the new Dutch maps, far-flung territories no longer simply faded away on the margins, nor were the world's edges fearful, mythical places full of monstrous people to be avoided wherever possible. Instead, on maps like Petrus Plancius's map of the Moluccas (1592), the world's borders and margins were clearly defined and identified as places for financial exploitation, with its regions labelled according to markets and raw materials, and its inhabitants often identified according to their commercial interests. Every corner of the earth was being mapped and assessed for its commercial possibilities. A new world was being defined by new ways of making money.

The world map that expressed the concerns of the time was not, like Blaeu's 1648 publication, laid on a floor or put on a wall. Instead it was to be found in a book or, more precisely, an atlas. The 1648 map was just one of many Blaeu made in preparation for his greatest cartographic publication, one of the largest books produced in the seventeenth century. This was the *Atlas maior sive cosmographia Blaviana*, or 'Large Atlas or Blaeu's Cosmography', published in 1662, and which has been described as 'the greatest and finest atlas ever published'.[7] In its sheer size and scale it surpassed all other atlases then in circulation, including the efforts of his great predecessors Ortelius and Mercator. It was a truly baroque creation. The first edition alone ran to eleven volumes containing 3,368 pages written in Latin, with 21 frontispieces and a staggering 594 maps, giving a total of 4,608 pages across the eleven volumes. Subsequent editions were published in French, Dutch, Spanish and German throughout the 1660s, adding even more maps and text. The *Atlas* was not necessarily the most up-to-date geographical survey of the world, but it was certainly the most comprehensive, and established the format of the atlas as the primary vehicle for disseminating standardized geographical information about the shape and scale of the world and its regions. It finally achieved what mapmakers had tried but failed to do for decades since the first printed editions of Ptolemy's *Geography* were published in the late fifteenth century: it bound the world in a book (or in this case, many books), and it would never be equalled.

The *Atlas* was partly the product of the emergence of a Dutch Calvinist culture that celebrated the pursuit and acquisition of material wealth while also fearing the shame of its possession and consumption – what Simon Schama has famously described as an 'embarrassment of riches'.[8] It was also shaped by a specifically Dutch visual tradition that Svetlana Alpers has called 'the art of describing' – the impulse to observe, record and define individuals, objects and places as real, without the kind of moral or symbolic associations which shaped Italian Renaissance art.[9] But the detail of the *Atlas*'s creation, and the part played by the Blaeu dynasty throughout the first half of the seventeenth century in establishing it as Europe's premier geographical atlas, reveals other facets of a story characterized by religious conflict, intellectual rivalry, commercial innovation and financial investment in a new scientific conception of the earth's place in the wider cosmos. The result was a change in the perception of the role of geography and the status of the mapmaker within Dutch culture and society, first developed by figures like Claesz. and Plancius, and cemented by the Blaeus. As mapmakers were increasingly institutionalized, they were granted unprecedented political influence and wealth. Nowhere was this more obvious than in the case of the Blaeu dynasty.

Joan Blaeu was part of a line of mapmakers that spanned three generations, beginning with his father, Willem Janszoon (1572–1638), and ending with Joan's son Joan II (1650–1712). At the heart of the dynasty stands Joan Blaeu, who collaborated with his father to build up the family business, before its gradual decline in the hands of his three sons Willem (1635–1701), Pieter (1637–1706) and Joan II. In 1703 Blaeu control of Dutch mapmaking came to an end when the VOC stopped using the family's name on its maps.[10]

The origins of the *Atlas maior* lay in the remarkable career of Joan's father, Willem. Born Willem Janszoon in Alkmaar or Uitgeest, around 40 kilometres north of Amsterdam, Willem adopted the surname 'Blaeu' from his grandfather's nickname, 'blaeuwe Willem' ('blue William'), although he only began to sign maps with his adopted surname from 1621.[11] Born into a prosperous but undistinguished merchant family, Willem began his working life as clerk to a local herring dealer. But his ambition and aptitude for mathematics soon led him to leave the business and by 1596 he was studying under Tycho Brahe on Hven (an island situated between Denmark and Sweden). Brahe was one of the most innovative and admired astronomers of his time, who had established a research institute and astronomical observatory on Hven in 1576, from where he conducted some of the most accurate observations of the planets of the time. His work led him to produce a modified geocentric model of the solar system, which he rather immodestly called the Tychonic system. Poised between Ptolemy's geocentric theories and Copernicus's heliocentric beliefs, Tycho posited a compromise whereby the earth still remains at the centre of the universe, with the moon and sun orbiting it, while the other planets circle the sun.

Although Blaeu spent only a few months on the island, he appears to have assisted Brahe in his astronomical observations, learning basic skills in celestial cosmography and mapmaking.[12] As well as developing the practical skills that would sustain him for the rest of his life, Blaeu also inherited from Brahe scepticism towards Ptolemy's geocentric universe. Over the next few years he gradually embraced the new heliocentric model developed by Brahe's most famous student, Johannes Kepler. By 1599 Blaeu was back in the Netherlands, where he made one of his first scientific objects, a celestial globe based on Brahe's star catalogue. Strangely overlooked by historians of science, Blaeu's globe is the first known non-Ptolemaic representation of the heavens.

It was an ambitious beginning for a young man starting out in a scientific world which prized empirical research and practical outcomes over more speculative approaches to the natural sciences. The struggle for independence from Spain had led many craftsmen, merchants, printers, artists and religious dissidents to leave the southern Spanish-controlled provinces, especially following the sack of Antwerp in 1585, and move north to cities like Amsterdam. The result was a sudden influx of new religious, philosophical and scientific ideas. From the 1580s the Flemish mathematician and engineer Simon Stevin (1548–1620) worked in Leiden as a military engineer to the army of Prince Maurice of Orange in its struggle against the Spanish, as well as writing a series of innovative works in Dutch on mathematics, geometry and engineering. Stevin pioneered the use of decimal fractions in coinage and weights, and was the first scientist to understand tides according to the attraction of the moon. A variety of other books on compound interest, trigonometry, algebraic equations, hydrostatics, fortifications and navigation were all aimed at specific practical applications – 'at which', Stevin wrote, 'theory should always aim'.[13] In astronomy the Dutch reformed minister Philips Lansbergen (1561–1632), who moved north after the sack of Antwerp, settled in Middleburg and began working on a set of astronomical tables and observations on the motion of the earth. His works, which supported Copernicus's heliocentric theories, soon became bestsellers, and were used subsequently by Kepler and Galileo in their astronomical writings. Another reformed minister, Petrus Plancius (1552–1622), yet another who fled north and settled in Amsterdam, not only worked alongside commercial mapmakers like Cornelis Claesz., but also pioneered astronomical observations in an attempt to determine longitude. Plancius invested heavily in the VOC, advised them on emerging overseas markets, named new constellations, and adopted Mercator's projection in a series of regional and world maps that championed Dutch commercial interests.

The overriding interest in the practical (and particularly commercial) impact for science of these men was not lost on Blaeu, who understood that he could not sustain a living from simply endorsing the new scientific ideas of Brahe and Kepler. By 1605 he was in Amsterdam, the obvious destination of a young man interested in science and business. Blaeu soon became one of more than 250 booksellers and printers working in the city, which was now overtaking Venice as the centre of the European book trade. The capital drew on the Republic's relative tolerance in matters of politics, religion and science to publish and sell books by figures like Stevin and Plancius on a wide variety of topics and printed in a bewildering number of languages, from Latin and Dutch to German, French, Spanish, English, Russian, Yiddish and even Armenian.[14]

Blaeu opened his own printing business in Amsterdam, publishing poetry as well as practical seamen's guides, including his bestselling *Light of Navigation* (1608), which again drew on Brahe's astronomical observations to assist in more accurate seaborne navigation. But he also grasped the commercial potential for exploiting the growing market for a new kind of maps, and over the next three decades his business flourished. He employed copper engravers to make his maps and once his son Joan was old enough he increasingly delegated their editing to him. Willem only published maps for which there was an established demand. The most popular subjects were the world, Europe, the four continents, the Dutch Republic, Amsterdam, Spain, Italy and France. Despite his grasp of mathematical cartography learned from Brahe, and his obvious sympathy towards the new science, Willem was first and foremost an entrepreneur. Although he published approximately 200 maps, he signed himself as the actual creator of fewer than 20.

Blaeu realized that if he was to establish himself as a cartographic printer of any significance, he needed to produce high-quality world maps that would excel those of competitors like Plancius, Claesz., Doetsz and Robaert. In 1604 he launched plans to publish no fewer than three distinct world maps, each on a different projection. Employing engravers to copy and amend maps still in circulation, he began by publishing a world map on a simple cylindrical projection, followed by one using a stereographic projection, and finally, in 1606–7, a beautifully engraved world map on four sheets using the Mercator projection. The map, which has since been lost and only survives in a poor photographic reproduction, is one of the most important world maps of seventeenth-century Dutch cartography. As well as acknowledging the influence of Plancius by using Mercator's projection, it provides an encyclopedic portrayal of the political, economic and ethnographic preoccupations of the Dutch Republic at the beginning of the seventeenth century.

The map's representation of the world only takes up half of its printed surface. Across the top of the map ten of the most powerful emperors of the time are shown on horseback (including the Turkish, Persian, Russian and Chinese emperors); in the left- and right-hand borders are twenty-eight topographical views of the world's major towns and cities, from Mexico in the west to Aden and Goa in the east. Next to these, running along the bottom of the map are thirty illustrations of local inhabitants of the regions depicted, including Congolese, Brazilians, Indonesians and Chinese, portrayed in what Blaeu imagined to be their national dress. Framing the world to the left, right and below is a Latin description of the earth, with ten further engravings depicting various scenes and figures from history.[15]

The map's title, *NOVA ORBIS TERRARUM GEOGRAPHICA ac Hydrogr. Tabula, Ex Optimis in hoc opere auctorib' desumpta auct. Gul. Ianssonio*, or 'New World Map by Willem Janszoon based on data borrowed from the best makers in this field', suggests how Blaeu composed his map, a point on which he expanded in one of its many legends. 'I thought it appropriate', writes Blaeu, 'to copy the best sea charts available from Portuguese, Spaniards and from our compatriots, and included all discoveries made hitherto. For decorative purposes and pleasure I filled the borders with pictures of the ten most powerful sovereigns ruling the world in our times, the principal towns and the large variety of costumes of the different peoples.' Blaeu carefully describes the application of Mercator's projection, conceding that it 'did not enable me to represent the north and south part of the globe as a plane'. The result is a vast and largely speculative southern continent, the result of using Mercator's projection, but also a response to the still uncharted territories of Antarctica and Australasia. To the left and right elaborately engraved cartouches explain the mathematical projection, while across the bottom lines in verse comment on the scene above, where Europe sits in majesty receiving gifts from her subject peoples:

> To whom do the Mexicans and Peruvians offer gold necklaces and shining silver jewels? To whom does the armadillo bring skins, sugar cane and

Fig. 23 Willem Blaeu, world map drawn on the Mercator projection, 1606–7.

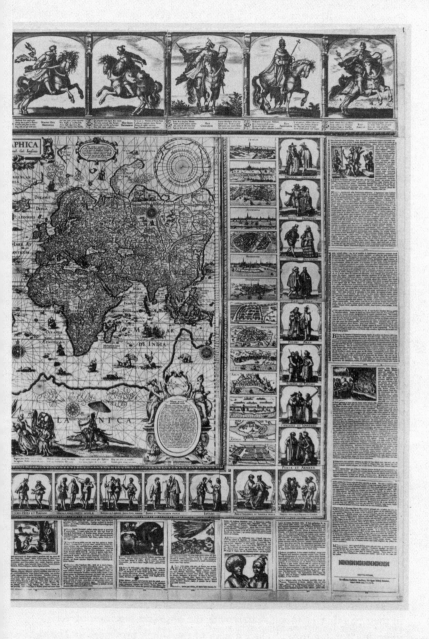

spices? To Europe, enthroned on high, the supreme ruler with the world at her feet: most powerful on land and at sea through war and enterprise, she owns a wealth of all goods. O Queen, it is to you that the fortunate Indian brings gold and spices, while the Arabs bring balsamic resin; the Russian sends furs and his eastern neighbour embellishes your dress with silk. Finally, Africa offers you costly spices and fragrant balsam and also enriches you with shining white ivory, to which the dark coloured people of Guinea adds a great weight of gold.[16]

Blaeu's map, depicting the global imperial landscape, the world's great commercial cities and its range of people, reflected the new mercantile imperatives of the Dutch Republic. Its coverage of the known world assessed everywhere and everyone for their commercial potential, from Europe as a personification of trade, to Africa and the Mexicans offering up their wares to enrich it as the globe's pre-eminent continent.

A measure of Blaeu's success can be gauged by how far later maps, charts and globes were reproduced in the paintings of Dutch interiors and still lifes by a whole range of seventeenth-century Dutch painters. Of these, none was more fascinated by maps than Johannes Vermeer. At least nine of his surviving pictures painstakingly depict wall maps, sea charts and globes in the kind of exquisite detail that has led one critic to write about the painter's 'mania for maps'.[17] Vermeer's painting The Geographer, dated around 1688, shows a young man absorbed in the act of mapping, with the paraphernalia of his trade scattered all around him. On the cabinet behind him sits a globe, and on the wall hangs a sea chart which is identifiable as Willem Blaeu's 1605 'Sea Chart of Europe'. In one of his earliest paintings, The Soldier and a Laughing Girl, dated around 1657, Vermeer depicts a map of Holland and West Friesland (oriented with west at the top), which hangs on the wall behind the domestic scene of a woman and a soldier; it is as visually arresting as the painting's main subjects. As well as painting this map, Vermeer used a variety of other maps by Dutch mapmakers, including those of the seventeen provinces by Huyck Allart (fl. c. 1650–75) and Nicolaus Visscher (1618–79), and maps of Europe by Jodocus Hondius the Elder (1563–1612). Other artists shared Vermeer's interest in maps – Nicolaes Maes (1634–93) and Jacob Ochtervelt (1634–1682) both featured maps in their paintings, although rarely with the obsessive precision of Vermeer. In choosing to reproduce a map of the Dutch provinces in The

*Soldier and a Laughing Girl*, Vermeer followed his artistic contemporaries in displaying popular pride at the political and geographical unity of the recently independent Republic.

So precise is Vermeer's rendering of this particular map, even down to its title, that it is easily identified as the creation of a well-known contemporary Dutch mapmaker, Balthasar Florisz. van Berckenrode. In 1620 the States General granted Berckenrode a privilege to publish this map, which he then sold for 12 guilders a copy. In the seventeenth century, print privileges, which prevented the copying of particular texts or images for a specified period of time, represented the closest equivalent to modern-day copyright. Infringement of privileges was punishable by a substantial fine, and as such sanctions were enforced by the States General, which meant that they effectively represented a political endorsement of a printed work's content.[18] Granting a privilege did not automatically ensure commercial success: despite its patriotic appearance Berckenrode's map was, according to written accounts, not terribly popular, and no copies of his 1620 edition are known to have survived. Perhaps as a result of its disappointing sales, Berckenrode sold the copper plates and publication privileges to the map in 1621 to Willem Blaeu, who seems to have had more success with it: he persuaded Berckenrode to remap its northern regions with greater accuracy, and it became increasingly popular throughout the 1620s.[19] Blaeu proceeded to reproduce it until 1629, when the privilege expired, and it is an edition of this map with Blaeu's name on it that Vermeer reproduces in his painting. Although Blaeu had no involvement in the map's design or engraving, he effectively turned it into a Blaeu map by signing it, and this is probably how Vermeer understood it when he painted it in the late 1650s (and on at least two other occasions over the next fifteen years). This was not the first or the last time that Blaeu and his sons would appropriate maps for their own commercial advantage, but it is a telling example of the ways in which the family business prospered.

Towards the end of the second decade of the seventeenth century Blaeu had established himself as one of Amsterdam's leading printers and mapmakers. His success was partly due to his unique talents as an engraver, scientist and businessman, a combination that most of his rivals lacked, and which enabled him to produce distinctively beautiful and precisely engraved maps, but he was also fortunate to emerge at a particularly crucial moment in the young Republic's history. Slightly younger than rivals like Claesz. and Plancius, he was also in a position to take advantage of the commercial opportunities provided by the twelve-year truce agreed between Spain and the Republic in 1609, which briefly allowed the Republic to pursue international trade unfettered by Spanish military and political opposition. But the decision to sign the truce had been extremely contentious, and caused a disastrous split between the Stadholder (the effective head of state) of the United Provinces, Prince Maurice of Orange, who opposed it, and the Land's Advocate of Holland, Johan van Oldenbarnevelt, who supported it. The settlement initially brought commercial prosperity, but it divided the provinces into two opposing camps. Their differences were intensified by a complex theological division between Calvinists (broadly supported by Prince Maurice and many of the directors of the VOC), and their opponents, the Arminians or 'Remonstrants' (supported by Oldenbarnevelt), who took their name from a petition known as the Remonstrance, which attempted to enshrine their theological differences from Calvinism. As tensions mounted and both sides took up arms, Maurice marched on Utrecht in July 1618. Oldenbarnevelt was arrested and, after being tried by a court led by the director of the VOC, Reynier Pauw, a staunch Calvinist and Contra-Remonstrant, he was beheaded in The Hague in May 1619.

Blaeu suddenly found himself on the wrong side in the dispute. Born into the Mennonite movement, an offshoot of the sixteenth-century Anabaptists, with their strong tradition of personal spiritual responsibility and pacificism, his sympathies were decidedly libertarian, and many of his friends were Remonstrants or 'Gomarists' (named after the Dutch theologian Franciscus Gomarus, 1563–1641). Just as the Contra-Remonstrants were putting Oldenbarnevelt on trial for his life, the VOC was trying to limit the circulation of maps relating to Dutch overseas commercial navigation by appointing an official cartographer responsible for drawing up and correcting the company's logbooks, charts and maps. Blaeu was the obvious candidate, but his political and religious persuasions meant that his appointment by the predominantly Contra-Remonstrant VOC was out of the question. The directors instead appointed one of his protégés, Hessel Gerritsz, who was regarded as a politically safer choice than his mentor.[20]

Blaeu continued to build his business throughout the 1620s, by now with the help of his son Joan. At the end of the decade he began to broaden his cartographic range even further. Having risen to prominence by producing single-sheet maps as well as globes, composite wall maps and travel writing, he now diversified into atlases, making an acquisition that sparked one of the bitterest rivalries in seventeenth-century mapmaking, and which ultimately led to the creation of Joan Blaeu's *Atlas maior*. In 1629 Blaeu acquired around forty copperplate maps from the estate of the recently deceased Jodocus Hondius the Younger. Hondius was himself part of a mapmaking dynasty started by his father, one of the early suppliers of maps to the VOC. In 1604 Jodocus Hondius the Elder spent what he described as 'a considerable sum' buying the copper plates for Mercator's *Atlas* from the cartographer's surviving relatives at an auction in Leiden. It was a publishing coup for Hondius, and within two years he published a revised and updated version of the *Atlas* in Amsterdam. It boasted 143 maps, including 36 new ones, some made by Hondius, but most acquired from other mapmakers, and a dedication to the States General of the United Provinces. Although he destroyed the design and integrity of Mercator's original *Atlas* by trading on the great cartographer's name (and what he had produced) Hondius achieved immediate financial success. The new atlas was so popular that in just six years prior to his death in 1612, he issued seven editions in Latin, French and German.[21] He even authorized an engraving in the *Atlas*'s opening pages showing himself sitting opposite Mercator, both happily working on a pair of globes, even though Mercator had by this time been dead for nearly twenty years. What is today known as the *Mercator-Hondius Atlas* was far from comprehensive in its geographical scope, and its additional maps varied in quality. But it became the leading atlas of its day by virtue of its appropriation of Mercator's imprimatur, and because its only competitor, Ortelius's *Theatrum orbis terrarum* (1570), which was no longer being updated, appeared terribly old-fashioned; it was also too costly for potential rivals to compete with Hondius's *Atlas* by making nearly 150 new maps from scratch.

When Hondius died in 1612, the business was taken over by his widow, Coletta van den Keere, and their two sons Jodocus Hondius the Younger and Henricus Hondius. Sometime around 1620 the brothers fell out and went their separate ways. Jodocus began to prepare maps for a new atlas, while Henricus went into business with his brother-in-law, the publisher Johannes Janssonius.[22] Before he was able to publish his new atlas, Jodocus died suddenly in 1629, aged just 36. Blaeu now saw his opportunity. Though the Hondius atlas had come to dominate the market, the family squabble prevented subsequent editions from incorporating any new maps, and it was effectively stagnating. As the family tussled over the estate, Blaeu seized the chance to acquire Henricus's new maps and launch his own rival work.

How Blaeu managed to acquire the maps is unknown, but it is clear how he used them. His first atlas, entitled *Atlantis Appendix* – literally an atlas that supplemented the work of Mercator and Hondius – published in 1630, contained sixty maps, mostly of Europe and with virtually no regional coverage of Africa and Asia. Of these sixty maps, no fewer than thirty-seven came from Hondius, whose name was simply deleted and replaced with Blaeu's imprint. It was an audacious move, compounded by Blaeu's cheeky refusal to even acknowledge Hondius's maps in his preface to the reader. 'I admit', wrote Blaeu in acknowledging the precedence of the works of Ortelius and Mercator, 'that herein figure some maps, which have already been published either in the *Theatrum*, or in the *Atlas*, or in both, but we give these maps in another form and with another appearance, and made, augmented and supplemented with greater diligence, care, and accuracy, so that, with the rest, they may be called almost new.' Blaeu grandly concluded with almost comical disingenuousness that his maps 'have been composed with diligence, truthfulness, and correct judgement'.[23]

Blaeu's actions were in part motivated by a longer history of commercial conflict with Janssonius. As early as 1608 he had addressed a plea to the States of Holland and West Friesland, demanding security against the loss of income caused by pirated editions of his maps, a thinly veiled attack on Janssonius for the striking resemblance between his 1611 world map and Blaeu's 1605 map.[24] In 1620 Janssonius struck again, printing copies of Blaeu's *Light of Navigation* with plates designed by Pieter van der Keere, Jodocus Hondius the Elder's brother-in-law. As Blaeu's privilege to print his book had run out, his only way to defend himself against Janssonius's flagrant piracy was to publish a new pilot's guide, at great expense.[25] Until 1629, it must have seemed to Blaeu that Janssonius, now helped by Henricus Hondius, was commercially unassailable. To have apparently triumphed now over his adversary with the publication of the *Appendix* must have brought Blaeu a degree of personal satisfaction, even though it also magnified the professional rivalry between the two families that would last for more than thirty years.[26]

Just like the *Mercator-Hondius Atlas*, Blaeu's *Atlas Appendix* was an uneven affair in its geographical coverage and printed quality. Nevertheless, it was an immediate success, as wealthy members of the public were eager to buy and inspect a new atlas different from those produced by Hondius. Henricus Hondius and Johannes Janssonius were understandably appalled that their control of the market was now being challenged by an atlas primarily composed of maps made by their dead relative. They quickly responded, later in 1630, by publishing an appendix to their atlas, followed in 1633 by a newly enlarged French edition of the *Mercator-Hondius Atlas*, in which they directly attacked Blaeu's *Atlas Appendix* as 'a hotch-potch of old maps', which also copied maps from Jodocus the Younger's atlas.[27]

Hondius and Janssonius's criticism of Blaeu's hastily printed atlas was totally justified – although it was a charge that could equally be levelled at their own atlas too. The competition made both sides realize that atlases composed of a bricolage of old maps and hastily commissioned or pirated new ones were unsustainable. A completely new atlas was required that included up-to-date maps incorporating recent discoveries, including some on the VOC's manuscript charts of South-east Asia. But such a venture required massive capital investment (in skilled workmanship, labour hours and the sheer volume of printed text involved), as well as access to the latest navigational information. In the latter half of the 1620s, the changing political and commercial climate meant that Blaeu held the advantage over his rivals: the power of the Contra-Remonstrant political faction gradually diminished, and Blaeu's Remonstrant allies found new favour within both the city's civic authorities and the VOC. They included his close friend Laurens Reael, one of the city's most powerful and influential figures, related by marriage to Arminius, a former Governor-General of the East Indies, and a director of the VOC.[28]

For Blaeu, this shift in power reached a climax in 1632 when the post of official cartographer to the VOC fell vacant on the death of Hessel Gerritsz. Whereas Blaeu's appointment had been almost unthinkable in 1619, by 1632 the position was his for the taking, and when the VOC's directors (including Reael) visited him in December 1632 to offer him the post, he accepted immediately. He was formally appointed on 3 January 1633. His contract stipulated that he was responsible for keeping a record of the logbooks of VOC pilots travelling to South-east Asia, correcting and updating the company's sea charts and maps, appointing 'trustworthy' individuals to make the maps, maintaining absolute secrecy, and providing a biannual report to the directors on this and the rest of his cartographic endeavours. In return he received a yearly salary of 300 guilders, a modest salary in line with those of comparable state officials, but one which could be supplemented by individual payments from the VOC for each chart and map he made.[29] It put Blaeu right at the heart of the Republic's political and commercial policies, and gave him a position of unprecedented power and influence within the Dutch mapmaking profession.

Even as he was being appointed, Blaeu was working on yet another attempt to corner the market, his *Novus Atlas*, which (it was promised in a pre-publication notice) would be 'entirely renewed with new engravings and new detailed descriptions'. Published in 1634, it was the first Blaeu atlas to name the involvement of Willem's son Joan, even though he had been helping his father since at least 1631. Unfortunately the *Novus Atlas* did not live up to its publicity. Although it included 161 maps, over half had been published before, 9 were published in an incomplete state, and 5 had not even been intended for inclusion![30] Blaeu's duties as VOC cartographer and his desire to rush out an atlas before the competition probably led to the mistakes.

His appointment as the VOC's cartographer nevertheless gave Blaeu the confidence to expand the scope of his atlases, for which the tools were to hand. At his death in 1632, Gerritsz's estate had included six copperplate engravings of India, China, Japan, Persia and Turkey, all commercially sensitive regions in which the VOC were busy trading and mapping. The VOC's privileges meant that they were in effect the Company's possessions, but Blaeu, probably with the help of Reael, who was one of Gerritsz's executors, managed to acquire the plates for his own use. In 1635 Blaeu published an even bigger atlas, this time in two volumes, which included 207 maps, 50 of them new, and which made even grander claims to comprehensiveness. 'It is our intention', wrote Blaeu in the preface, 'to describe the whole world, that is the heavens and the earth, in other volumes such as these two, of which two about the earth will shortly follow.'[31] The atlas reproduced a Gerritsz map of India and South-east Asia which simply added decorative cartouches at the top and left-hand corner, and in the right-hand corner a scene of putti playing with navigational instruments and plotting their way across a terrestrial globe with a pair of compasses. The cartouche on the left reveals that the map was dedicated to none other than Laurens Reael.

Such manoeuvres clearly show Blaeu's pragmatic pursuit of domination of the market in atlases, but his motivations were not always straightforward. In 1636, following Galileo Galilei's condemnation by the Catholic Inquisition for his heretical heliocentric beliefs, a group of Dutch scholars hatched a plan to offer the Italian astronomer asylum in the Dutch Republic. The plan was floated by the great jurist, diplomat (and Remonstrant sympathizer) Hugo Grotius – whose books were published by Blaeu – and was enthusiastically supported by Laurens Reael and Willem Blaeu. Beyond their intellectual belief in a heliocentric universe, all three men also had vested commercial interests in offering such an invitation. Grotius, having already written on the subject of navigation, was hoping to lure Galileo to Amsterdam so that he would offer the VOC a new method of determining longitude which, if successful, would give the Dutch complete domination of international navigation.[32] Blaeu's somewhat nonconformist intellectual beliefs coincided with his eye for a novel commercial opportunity: Galileo represented a new way of looking at the world, but it was also one that Blaeu might have calculated would give him a decisive edge in cartographic publishing in the 1630s. Ultimately, the plans to invite Galileo came to nothing, as the astronomer pleaded that ill health (and undoubtedly the terms of his house arrest by the Inquisition) prevented him from making what would have been a sensational defection to Europe's leading Calvinist republic.

The scheme's failure made little difference to Blaeu, who went from strength to strength. In 1637 he expanded the family business by moving the printing works to a new building on the Bloemgracht in the Jordaan district in the west of the city, home to the dyeing and painting industries. With its print foundry and nine letterpresses, six of which were dedicated to mapmaking, the new building was the largest printing house in Europe. Willem, alas, had only had a year in which to enjoy his pre-eminence as Europe's greatest printer. In 1638 he died, bequeathing the family business to his sons Joan and Cornelis (c. 1610–42).

Willem's death marked the end of the first phase of the Blaeu dynasty's rise to almost complete domination of printing and mapmaking in the Dutch Republic. He had carved out a career placing him at the forefront of printing and mapmaking in Amsterdam. Willem's world maps and navigational guides superseded those of previous geographers, and his published atlases challenged those of Ortelius and Mercator. He led the way in putting cartography at the heart of the state's political and commercial policies, culminating in his VOC work, and publishing maps and books describing a heliocentric world in which the earth was no longer positioned at the centre of the universe. But for Joan and Cornelis, the exigencies of the business of publishing, the competition with Hondius and Janssonius and the ongoing demands of the VOC's commissions meant that they needed to consolidate their father's achievements before their competitors moved in.

Following their father's death, Joan and Cornelis's business was given a boost by the news that Henricus Hondius had inexplicably withdrawn from making atlases with his brother-in-law, leaving Janssonius to continue on his own. The Blaeus' position was further reinforced in November 1638, when Joan was confirmed in his father's job as official cartographer to the VOC. Under Willem's tenure, the post had expanded in line with the increasing volume of trade between Amsterdam and the VOC's Indonesian headquarters in Batavia (modern-day Jakarta); by the time Joan was appointed, the Dutch Republic's merchant fleet had grown to around 2,000 vessels, eclipsing every other European maritime sea-power. With a capacity of approximately 450,000 tons, and employing about 30,000 merchant sailors, the VOC received investors' subscriptions of an estimated 40 to 60 million guilders each year; at the same time, its profits continued to grow and its markets expanded into spices, pepper, textiles, precious metals and luxury items like ivory, porcelain, tea and coffee. Throughout the 1640s it was dispatching more than 100,000 tons of shipping eastwards each year, and by the end of the century it had sent an estimated 1,755 vessels and over 973,000 people to Asia (of whom 170,000 lost their lives en route).[33]

All these ships needed maps and charts to navigate from Texel to Batavia. The skipper and the chief and junior pilots were each given a complete set of at least nine charts, and the third watch a more limited set. All of them were made by Blaeu and his assistants. The first chart showed the route from Texel to the Cape of Good Hope; the second depicted the Indian Ocean from the east coast of Africa to the Sunda Straits separating Java and Sumatra; the following three showed the Indonesian archipelago on a larger scale, followed by charts of Sumatra, the Straits, Java and finally Batavia (including Bantam on the Indonesian island of Java). Each set was accompanied by globes, manuals, logbooks, blank sheets and even a tin cylinder for storing the charts. In an attempt to restrict their circulation, the VOC ordered that any charts not returned at the end of a voyage would have to be paid for.

Blaeu's role as official cartographer to the VOC connected him with everyone from the third watch on board a VOC East Indiaman all the way up to the directors of the company and its policy-making decisions. The master and mates of each VOC vessel were required to show the company's cartographer their logs, journals and any topographical sketches they made en route to the East, and Blaeu had to check and approve every log before depositing it in the VOC's East India House on Oude Hoogstraat. Blaeu then drew sea charts, known as 'leggers', a template for the subsequent finished maps, based on what he read. These charts were simple in outline, and on the same scale as was used in the final maps. They incorporated new material whenever appropriate, and formed the basis of the standard set of charts used by all VOC pilots. Up to four assistants were then employed to execute handdrawn charts on parchment – drawn by hand rather than printed to try to prevent their details being easily circulated on the open market, and on parchment because of its durability on the long sea voyage. Making charts in this way also allowed for a quick and ingenious method of updating the original charts. These would be revised by pricking out new coastlines or islands with a needle and then placed on top of a blank sheet of parchment and dusted with soot. Once removed, the specks of soot left on the new sheet of parchment through the needle pricks could then be carefully joined up by Blaeu's assistants to form a new and more accurate representation of coastlines.[34]

The costs involved were considerable: each new map that Blaeu made cost the company between 5 and 9 guilders (the price of a small painting), so outfitting a ship with a complete set of new charts would cost at least 228 guilders. Blaeu's costs were probably no more than 2 guilders for each chart, giving him an enormous profit margin of at least 160 per cent. These figures are of course tentative, as it is impossible from the small number of charts that survive to estimate how many were returned and reused, nor how often Blaeu was required to update each chart. But there seems no doubt that his post was extremely lucrative. In 1668 Blaeu invoiced the company for a staggering 21,135 guilders – an astonishing figure considering that his own annual salary was 500 guilders, similar to that of a master carpenter (and the average cost of a house in Amsterdam). It probably included bills for charts, but also larger, luxury items such as globes and handpainted maps for presentation to foreign dignitaries. In 1644 Blaeu was paid 5,000 guilders for a gigantic handpainted globe presented to the king of Makassar (in modern-day Indonesia), and other records show payments of anything from a few hundred guilders to tens of thousands for globes, atlases and decorative maps.[35] In contrast, Blaeu's assistants appear to have been poorly paid by their employer. One of them, Dionysus Paulusz, drew a map of the Indian Ocean for which Blaeu charged the VOC's directors 100 guilders, even though Paulusz complained that he received little more for it than 'a sip of water'.[36]

Blaeu's appointment reflected the peculiar balancing act between official exclusivity and private entrepreneurship that characterized the edicts of the VOC. Although insisting that his sea charts were the exclusive property of the company, and that the methods of their creation must remain a secret, the directors gave Blaeu remarkable autonomy in how he exploited his new-found cartographic knowledge in his other printing projects. This knowledge even allowed him to block the company's proposed reforms of their navigational practice. Throughout the 1650s and 1660s, the directors suggested the printing of a standardized navigational manual, and although Blaeu was involved in the discussion, he persistently prevaricated. It was simply not in his interest to support such an initiative, especially as he began work on the *Atlas maior*.[37]

The VOC post therefore brought Blaeu more than just very considerable financial profit. It gave him unparalleled access to the latest cartographic information for his charts, and the ability to influence (and if necessary block) new initiatives. It also brought him enormous cultural and civic influence. Over the next three decades he took on a series of public positions: he served on the city council, including a stint as alderman, captain of the civic guard and commissioner of fortifications.[38]

Blaeu also expanded the activities of his printing house on

the Bloemgracht, publishing religious works by Catholics as well as Remonstrants and Socinians (a liberal sect that rejected the idea of the Trinity, and who were despised by Calvinists as much as if not more than the Catholics), despite the objections of Amsterdam's civic authorities. Blaeu was so confident of his political position that in 1642 he even survived a raid on his press by the *schout*, the city's legal prosecutor, for his publication of a Socinian tract. The *schout* ordered that the books be burnt and the Blaeu brothers fined 200 guilders, but Blaeu's influence quickly led the city's burgomasters to quash the verdict (although too late to save the books from the fire). As ever, Blaeu turned the controversy to his advantage, publishing a subsequent edition in Dutch advertising its scandalous nature as a book that was 'publicly executed and burnt by fire'.[39] The apparently liberal disposition of the printing practices Blaeu inherited from his father continued to define his publishing decisions, but they were inevitably influenced by commercial considerations. He also used his wealth to invest in the cultivation of the Virgin Islands, undertaking to supply African slaves to work on its plantations.[40] Taken together with Paulusz's claims regarding Blaeu's meanness as an employer, his activities as a slave-trader show that he inherited both his father's libertarian beliefs and his ruthless entrepreneurial streak.

Blaeu's enduring ambition as a printer was to dominate the trade in atlases once and for all, but despite his appointment as the VOC's cartographer and the privileged information it gave him Blaeu still faced relentless competition from Johannes Janssonius. Free of their father and business partner respectively, the two men were now locked in a fierce competition to produce the finest atlas on the market. Each redoubled his efforts, printing ever larger and more ambitious atlases throughout the 1640s and 1650s, jettisoning references to earlier mapmakers like Mercator, and even using the same title, *Novus Atlas*, to emphasize the modernity of their products. Blaeu concentrated on simply adding further volumes to the initial structure of the atlas inherited from his father. In 1640 he issued a new atlas in three volumes, introducing new maps of Italy and Greece. In 1645 he published a fourth volume of England and Wales, dedicated to King Charles I, just as England's civil war began to turn in favour of the king's republican adversaries. There was a brief hiatus in Blaeu's atlas production in the late 1640s – partly due to a series of publications produced in response to the signing of the 1648 Treaty of Westphalia, including his twenty-one-sheet map of the world in two hemispheres that would be used as the basis for the floor of the Burgerzaal. In 1654 he added a further volume on Scotland and Ireland, and in 1655 a sixth volume of seventeen new maps of China, which drew on his extensive contacts within the VOC's operations in the Far East. One volume of the atlas sold for between 25 and 36 guilders, and the complete edition of all six volumes cost 216 guilders.

But Janssonius continued to match Blaeu volume for volume, even claiming that subsequent editions of his atlas would provide a comprehensive description of the whole world, including the heavens and the earth, surpassing not only Blaeu's efforts, but also the great cosmographical treatises of the sixteenth century. By 1646 he had also published four new volumes, adding a fifth sea atlas in 1650, and in 1658 completing a sixth volume consisting of 450 maps – even bigger than Blaeu's six-part atlas containing 403 maps.

By 1658, the two publishers had fought each other to a stalemate. If anything, despite Blaeu's obvious advantages in terms of printing resources and access to VOC material, Janssonius's atlas was more balanced and comprehensive. But by this time Blaeu, already in his late fifties, had taken a momentous decision. He decided to embark on a publishing project that was designed to eclipse Janssonius once and for all: a comprehensive description of the earth, the seas and the heavens. He proposed calling the venture *Atlas maior sive cosmographia Blaviana, qua solum, salum, coelum, accuratissime describuntur*, or 'Large Atlas or Blaeu's Cosmography, in which the Land, the Sea and the Heavens are Very Accurately Described'. Blaeu envisaged a three-stage publishing project starting with the earth, proceeding to the seas and finally to the heavens. Janssonius had already promised such an atlas, but lacked the resources to publish a truly definitive edition. Blaeu now channelled all his formidable resources into what would be his last and greatest publishing achievement.

In 1662, as work on the first part of the project neared its conclusion, Blaeu announced he was giving up the bookselling branch of his business empire to concentrate on printing the atlas, and held a public sale of all the stock in his bookshop to raise revenue for its imminent completion. When it was published later that year, it became clear why Blaeu needed all the capital he could muster. The first edition of the *Atlas*

MONEY

*maior*, published in Latin, was simply enormous. Nothing like it had ever been printed before, and its 11 volumes, 4,608 pages and 594 maps dwarfed all Blaeu's previous atlases, as well as those of Janssonius. But Blaeu's plan to dominate the European atlas market meant that he embarked on not one but five editions of the *Atlas maior* simultaneously. The first was in Latin – a prerequisite for the learned elite – the rest in more popular and profitable vernacular languages. The second, published in 1663, in 12 volumes including 597 maps, was in French, to supply Blaeu's largest market. The third was in Dutch, for his home audience, published in 1664 in 9 volumes containing 600 maps. The fourth was in Spanish, the language of what was still regarded as the continent's great overseas empire. The fifth and rarest edition was published in German in 1658. Blaeu began work on this atlas first, but delayed it to ensure that the more important Latin and French editions came out first. It was published in an abbreviated format in 1659, although its complete version ran to 10 volumes and 545 maps. Each edition varied according to the regions it portrayed and the printing formats used, but in most cases they duplicated the same text and maps as a gesture towards the standardization required of an atlas.[41]

The statistics involved in the creation of these atlases over a period of nearly six years from 1659 to 1665 are astonishing. It is estimated that the print runs of all five editions amounted to 1,550 copies, with the Latin edition the largest at 650 copies. But this apparently modest figure represented a phenomenal cumulative total of 5,440,000 pages of text, and 950,000 copperplate impressions. The time and manpower involved in putting all of this together was extraordinary. Setting the initial 14,000 pages of printed text across the five original editions, based on a calculation of eight hours to compose one page, involved five typesetters working for 100,000 hours. This represented a team of compositors working full-time for 2,000 days, or six years. In contrast, printing the 1,830,000 sheets of text involved was a relatively quick process. Assuming that at full capacity Blaeu's nine presses were able to print fifty sheets per hour, the printed text for all four editions could in theory have been completed in just over ten months. Printing the engraved copperplate maps was another matter, not least because they needed to be executed on the reverse of sheets which had already been printed; probably only ten impressions of one copperplate could be printed per hour. Based on 950,000 copperplate impressions across all four editions,

Blaeu's six presses would need to have been in full-time operation for nearly 1,600 days, or four and a half years. Many of the maps were also handcoloured, which gave the buyer the satisfying illusion of purchasing a custom-made object, although Blaeu put this out to piece-workers at 3 stuivers a map, making it difficult to assess the time involved. Then the careful binding of just one multi-volume atlas could take at least a day. And all this (as well as other printing jobs completed during the same period) was undertaken by a workforce of no more than eighty employed by Blaeu in his Bloemgracht workshop.[42]

Such a massive and potentially risky capital investment was reflected in the sale price of the *Atlas maior*'s different editions. Each one cost substantially more than Blaeu's previous atlases, most of which sold for just over 200 guilders. A handcoloured Latin atlas cost 430 guilders (although uncoloured it was only 330 guilders), while the larger French atlas cost 450 guilders in colour, 350 guilders uncoloured. These prices made the atlas not only the costliest ever sold, but also the most expensive book of its day; 450 guilders was a decent annual salary for a seventeenth-century craftsman and the equivalent in today's currency to roughly £20,000. The *Atlas maior* was clearly not a publication aimed at modest working people: its buyers were either those associated with its creation, or people who might be able to assist the Dutch in their political and commercial expansion: politicians, diplomats, merchants and financiers.

After so much effort and expectation, it is remarkable that the *Atlas maior* was so unadventurous. Not only its layout but also its maps suggested that Blaeu had little appetite for reform or innovation. Previous atlases by both Blaeu and Janssonius suffered from a simply cumulative approach where quantity triumphed over quality, and in which huge areas of the globe were covered in minute detail while others were almost completely neglected, and there was little coherence to the running order of the maps. The *Atlas maior* made no effort to rectify these deficiencies, nor did it offer a substantially new body of maps reflecting contemporary geographical knowledge. For example, the first volume offered a world map, followed by maps of the Arctic regions, Europe, Norway, Denmark and Schleswig. Of its 22 maps, 14 were new, but some of the rest were more than thirty years old. The third volume focused exclusively on Germany, in 97 maps, but only 29 were reproduced for the first time. The fourth volume shows the Netherlands, with

63 maps, 30 of which were technically new, but most were actually old maps printed in a Blaeu atlas for the first time. Blaeu even opened the volume by reproducing a map of the Seventeen Provinces that was first published by his father in 1608! The fifth volume, dedicated to England, contained 59 maps; all but 18 were simply copied from John Speed's *Theatre of the Empire of Great Britain* (1611). The atlas only left Europe in volume 9, which showed Spain and Africa, while volume 10 consisted of just 27 maps of Asia; all but one of them had been previously published, and none showed much evidence of the VOC's extensive exploration of the region.[43]

Thus, in the case of the *Atlas maior*, the medium of print did not help cartographic innovation, but hindered it. The maps were beautifully reproduced, their typography still regarded today by connoisseurs of copperplate engravings as in a class of its own. But with the money invested in such a vast undertaking, Blaeu faced a problem: did he risk introducing new and unfamiliar maps of places that might alienate his conservative (and necessarily wealthy) buyers, or did he gamble on the popular appetite for innovation – for which the immediate prehistory of the sales of printed maps offered little evidence? Throughout his career Blaeu proved reluctant to introduce innovative knowledge gleaned from the VOC records into his printed maps, preferring instead to keep this for his paid work on the company's manuscript charts. In this respect, the *Atlas maior* was no different: it was simply, as its title implied, 'grander' in terms of its size than any preceding atlas.

Nowhere is this more evident than in the atlas's very first map: *Nova et accuratissima totius terrarum orbis tabula*, or 'New and Very Accurate Map of the Whole World'. Unlike so many of the atlas's other maps, this one was relatively new. Until now, Blaeu's atlases had reproduced a version of his father's 1606–7 world map drawn on Mercator's projection. This new world map abandons the Mercator projection: instead, it returns to Mercator's convention, established in his 1595 *Atlas*, of representing the earth as twin hemispheres, depicted on a stereographic equatorial projection that also showed close affinities with Blaeu's earlier 1648 world map on which the marble maps in the Burgerzaal were based. The stereographic projection imagines a transparent earth marked with lines of latitude and longitude, sitting on a piece of paper where, in Blaeu's example, the equator touches the flat surface. If light is cast through the earth, the shadows cast on the paper show curved meridians and parallels converging on a straight line representing the equator. The method was not new (even Ptolemy wrote about it), but during the Renaissance it was mainly used by astronomers for making star charts, or globe-makers like Blaeu who were primarily interested in representing the curvature of the earth's surface. Nevertheless, Blaeu knew full well that the VOC were beginning to appreciate the superiority of Mercator's projection, especially for navigation. If anything, the choice of the stereographic projection over his father's preferred use of the Mercator projection in his new world map catered to established public tastes for twin hemispherical projections, evident in the Burgerzaal and the 1648 world map, but which stretched back as far as the 1520s following the completion of Magellan's first circumnavigation of the globe.

The map's intentions were not simply to adopt the most saleable projection available. Its geography adds little to the 1648 map. In the eastern hemisphere Australia, labelled 'Hollandia Nova', remains incomplete, with the tentative suggestion that it might be joined to New Guinea. In the western hemisphere the north-west coast of North America is similarly left incomplete, with the erroneous portrayal of California as an island. What has changed in the 1662 map is its elaborate border decorations. At the bottom there are four allegorical personifications of the seasons, with spring on the left, winter on the far right, and autumn and summer in the middle. Above the twin hemispheres is an even more elaborate allegorical scene: to the left, above the western hemisphere, stands Ptolemy, holding a pair of dividers in one hand and an armillary sphere in the other. Opposite him, in the top right-hand corner of the eastern hemisphere, is Copernicus, placing a set of dividers onto the surface of a terrestrial sphere. Between them are ranged personifications of the five known planets, depicted according to their classical gods. From the left, next to Ptolemy, sits Jupiter with thunderbolt and eagle, followed by Venus with her Cupid, Apollo (or the sun), Mercury with his caduceus, Mars in his armour, and finally, just above Copernicus, Saturn, identified by the six-pointed star that adorns his flag. Below Apollo, the moon peeps out, represented by a foreshortened head and shoulders between the two hemispheres.

This is an image of the world situated within the wider cosmos, with the earth at its centre, apparently unfolding its twin hemispheres. Or is it? Evoking Ptolemy's geocentric beliefs side by side with Copernicus's heliocentric theories suggests that the map is trying to have its cosmology both ways. The map actually follows an oblique Copernicanism in showing the planets in order of their proximity to the sun. Mercury is slightly closer to Apollo with his sceptre, next comes Venus, followed by the moon and the earth. Mars then precedes Jupiter and finally comes Saturn, in exactly the order promoted by Copernicus's followers.

In the 'Introduction to Geography' that prefaces his atlas, Blaeu acknowledges that 'cosmographers are of two discrepant opinions concerning the centre of the world and the movement of the celestial bodies. Some place the earth at the centre of the universe and believe it to be motionless, saying that the sun with the planets and fixed stars revolve around it. Others place the sun at the centre of the world. There, they believe, it is at rest; the earth and the other planets revolve around it.' In what could act as a direct commentary on his world map, Blaeu goes on to explain the cosmography of the Copernicans. 'According to them,' he writes,

> Mercury rides its course from west to east in the first sphere, the one nearest the sun, in eighty days while Venus in the second sphere takes nine months. They also affirm that the earth – which they take to be one of the luminaries or a planet like the others, and which they place in the third sphere with the moon (which moves around the earth, as if in an epicycle, in twenty-seven days and eight hours) – completes in one natural year its revolution of the sun. In this way, they say, the seasons of the year are differentiated: Spring, Summer, Autumn and Winter.[44]

Blaeu goes on to describe Mars, Jupiter and Saturn in their respective positions, just as on his world map.

In a wonderfully disingenuous argument, Blaeu goes on to insist that 'it is not our intention here to specify which of these opinions is consistent with truth and best befits the natural order of the world'. He leaves such questions 'to those versed in the science of celestial matters', airily adding that there is 'no noticeable difference' between the geocentric and heliocentric theories, before concluding that 'since the hypothesis of a fixed earth seems generally more probable and is, besides, easier to understand, this introduction will adhere to it'.[45] So speaks the entrepreneur, not the scientist, the publisher rather than the geographer.

Nevertheless, in its placement of the planets, the 1662 *Atlas maior*'s world map is the first evocation of a heliocentric solar system that dislodges the earth from the centre of the universe. This, aside from the sheer scale of its production, is the atlas's historical achievement, but the commercial exigencies of its publication meant that, unlike his father, Joan Blaeu diluted its radical science. If anything, despite the Dutch Republic's sympathy towards scientific challenges to the geocentric orthodoxy, the 1662 world map takes a step back from the 1648 map, with its insets of the Copernican and Tychonic challenges to the prevailing Ptolemaic model. The 1662 map itself offers an image of the earth within a heliocentric solar system, but wrapped in so many classical mantles and diminished by Blaeu's preceding comments that most historians have failed to realize its significance.[46] It seems that Blaeu was simply unsure if his support for the new scientific theory was good for business or not; in the end, his convoluted account of Copernicus's theory created a magnificent but generally overlooked cartographic image of a heliocentric world.

Judged on its own commercial terms, Blaeu's decision seems to have been an astute one, because the *Atlas maior* was a remarkable success. Copies were bought by wealthy merchants, financiers and political figures in Amsterdam and throughout Europe. Blaeu also dedicated the various editions to some of the continent's most politically influential figures, dispatching it with customized colouring, binding and armorial stamps. Many of these copies were held in specially designed and elaborately carved cabinets of walnut or mahogany, adding to its status as an object that was more than just a book or a series of maps. The Latin edition was dedicated to Emperor Leopold I of Austria, and the French edition to King Louis XIV, who received his with Blaeu's accompanying gloss on the importance of his subject. 'Geography', wrote Blaeu (paraphrasing Ortelius), 'is the eye and the light of history', and he held out before the king the prospect that 'maps enable us to contemplate at home and right before our eyes things that are farthest away'.[47] He also dispatched volumes to influential dignitaries, and took payment from the Dutch authorities for customized copies to be offered as exotic gifts to foreign rulers, including a velvet-bound Latin atlas sent to the Ottoman sultan by the States General in 1668 in an attempt to cement the political and commercial alliance between the two states. It found such favour that it was copied and translated into Turkish in 1685.[48]

Publication of Blaeu's *Atlas maior* also marked the end of the fifty-year rivalry with Janssonius, but not because of any inherent geographical superiority. In July 1664 Janssonius died. In the same five years as those of Blaeu's editions Janssonius had succeeded in publishing Dutch, Latin, and German editions of his own *Atlas maior*. The Dutch edition, published in nine separate volumes between 1658 and 1662, followed a similar running order to Blaeu's, and contained 495 maps. The eleven-volume German edition of 1658 featured no fewer than 547 maps. Janssonius may have lacked Blaeu's publishing resources and his political connections, but virtually up to the day he died he continued to match his great rival in the publication of vast atlases.[49] If he had lived longer, the story of Blaeu's ultimate command of Dutch mapmaking might have been very different.

Blaeu was so successful that in 1667 he extended his printing empire to new premises on Gravenstraat. But his triumph was short-lived. In February 1672 a fire swept through the new building, destroying much of Blaeu's stock and presses. The official account into the fire reported that, as well as the loss of books, 'the large printing works with everything in it was damaged to such an extent that even the copper-plates stacked in the far corners melted like lead in the flames', and put Blaeu's losses at a staggering 382,000 guilders.[50] If Blaeu held any hope of completing the atlas with the two promised sections on the sea and the heavens, it was now gone for ever. Even worse was to follow. In July 1672, as the Dutch Republic faced imminent war with France, the States General offered William of Orange the title of Stadholder. The shift in power led to the dismissal of anti-Orangist members of the Amsterdam Council, including Blaeu. His publishing house literally in ruins and his political influence over, Blaeu quickly went into decline, and on 28 December 1673 he died, aged 75.

Blaeu's death signalled the demise of the family business. His sons carried it on, but they lacked either their father or grandfather's brilliance and drive. The market for maps had also changed, and the political climate discouraged large capital investments in multi-volume atlases. It was just too risky. The deaths of Janssonius and Blaeu also meant that the commercial rivalry which had driven so many atlas publications between 1630 and 1665 was gone. There was neither supply of nor demand for new atlases. Between 1674 and 1694 the copper plates used to print the *Atlas maior* and which had survived the fire were sold off and dispersed in a series of sales and auctions.[51] In 1696 the family business was finally wound up, and in 1703 the VOC used its printer's mark for the last time, ending its long and successful association with the family that had proved so central to its mapmaking.

The history of Blaeu's atlas was probably also not what its creator had envisaged. Having failed to complete his project, but tailoring so many copies of its first part for individual recipients, Blaeu inadvertently sparked an entirely new approach to atlas consumption: what became known as 'composite atlases'. Late seventeenth-century buyers began to copy Blaeu by supplementing their copies of his atlas with new maps and drawings. The Amsterdam lawyer Laurens van der Hem (1621–78) bought a Latin copy of the *Atlas maior*, and used it as the basis for amassing an extraordinary forty-six volumes with 3,000 maps, charts, topographical drawings and portraits, which he carefully organized and bound professionally as if it were a gigantic extension of Blaeu's original work. Van der Hem's atlas was so impressive that the grand duke of Tuscany offered to buy it for 30,000 guilders – quite a return on his original investment of 430 guilders, even though he had expanded it enormously.[52] Others similarly customized their Blaeu atlases according to their personal tastes in navigation, cosmography, even orientalism. Rather like Blaeu's own atlases, these customized examples were endlessly extensible and potentially infinite: only the collector's death signalled their completion.

Ironically, because Blaeu concentrated primarily on the marketing of his atlas for a commercial audience at the expense of geographical or astronomical innovation, subsequent atlas publication moved away from the publisher/geographer as the organizer of the text, and instead put the decision of what to include in the hands of the purchaser. Italian printers began to publish maps in standard formats, which customers could then buy and assemble into their own atlases. Subsequently known by map-dealers as IATO atlases (Italian, assembled to order, first made in the sixteenth century), they can be more accurately called Italian composite atlases, as it was the collector and not necessarily the publisher who selected the maps. The emergence of these composite atlases was a symptom of the dilemma experienced by mapmakers and printers at the end of the seventeenth century: the sheer amount of geographical data they possessed had never been greater, and the print technology at their disposal had reached such a level of speed and precision that it could reproduce such information in the finest detail, but no one was clear

MONEY

how it should all be organized and presented. When could geographical knowledge be regarded as complete? And how could such projects make money? Surely this was an endless task, best left to individuals to make their own decisions about the geography they required.

With its beautiful typography, elaborate decoration, exquisite colouring and sumptuous binding, Blaeu's *Atlas maior* was unparalleled in seventeenth-century printing. It was the product of a Dutch Republic that, following its violent struggle to break free of the Spanish Empire, created a global marketplace that preferred the accumulation of wealth over the acquisition of territory. Blaeu produced an atlas that was ultimately driven by the same imperatives. For him, it was not even necessary to place Amsterdam at the centre of such a world; Dutch financial power was increasingly pervasive but it was also invisible, seeping into every corner of the globe. In the seventeenth century as today, financial markets make little acknowledgement of political boundaries and centres when it comes to the accumulation of riches.

In fact, the *Atlas*'s success hindered rather than helped geographical developments towards the end of the seventeenth century. It represented the end of the classically inspired tradition of acquiring universal geographical knowledge that had driven mapmakers since Ptolemy. The sheer scale of Blaeu's publication could not compensate for its inability to offer any new geographical methods for creating an image of the world, as it catered for a map-buying public who were more interested in the decorative value of their maps and atlases than in their scientific innovations or geographical accuracy. It offered no new method of seeing the world in terms of scale or projection, though it did subtly present a world no longer positioned at the centre of the universe. But for Blaeu, the heliocentric theory was only as good as its sales figures. The *Atlas maior* was a truly baroque creation, which decisively broke with its Renaissance lineage. Where earlier mapmakers like Mercator sought to produce a singular scientific vision of the place of the world in the cosmos, Blaeu simply accumulated ever more material on the world's diversity, driven by the market rather than by a desire to establish a particular understanding of the world. Bereft of a defining intellectual principle, the *Atlas maior* grew and grew, a flawed and unfinished masterpiece, driven by money as much as knowledge.

# 9

# Nation

*The Cassini Family, Map of France, 1793*

*Paris, France, 1793*

On 5 October 1793, the National Convention of Republican France issued a 'Decree establishing the French Era'. The edict introduced a new calendar designed to mark the official proclamation of the French Republic just over a year earlier on 22 September 1792. It was part of a series of reforms designed to sweep away every vestige of the recently toppled *ancien régime*, from its methods of absolutist rule to the way it marked the passage of calendrical time. According to the National Convention, the date was now officially 14 Vendémiaire II, or Year II of the Revolution (the first month of autumn, named after *vendange*, the grape harvest). Just weeks before the calendar's inception, the Convention received a report from one of its more radical deputies, the actor, dramatist and poet, Fabre d'Églantine. Having already voted in favour of King Louis XVI's execution and featured prominently on the committee entrusted to create the new calendar, d'Églantine now turned to maps. He drew the Convention's attention to 'the general map of France, called the map of the Academy', which he complained 'had been produced in very great part at the expense of the government; that it had then fallen into the hands of a private individual who treated it as his own property; that the public could only have use of it by paying an exorbitant price, and that they even refused to send maps to the generals who asked for them'.[1]

The Convention agreed with d'Églantine, and ordered that the plates and sheets related to the map be confiscated and transferred to the military office of the Dépôt de la Guerre. The decision was greeted with triumph by the Dépôt's director-general, General Étienne-Nicolas de

Calon. 'By this act,' he announced, 'the Convention snatched back from the greed of a company of speculators a national achievement, the fruit of forty years of work by engineers, which had all the more to be completely available to the Government as its loss or abandonment would reduce its resources and increase those of the enemy.'[2]

D'Églantine's attack and Calon's glee were aimed at confiscating the map of France and bringing down Jean-Dominique Cassini (1748–1845). Jean-Dominique had the misfortune of being the last of four generations of the distinguished Cassini mapmaking dynasty, and putative owner of the map of France, a vast project which was tantalizingly near to completion when the National Convention sequestered it. For the staunchly royalist Jean-Dominique, the nationalization was a political catastrophe and a personal tragedy. 'They took it away from me,' he lamented in his memoirs, 'before it was entirely finished and before I had added the final touches to it. This no other author has suffered before me. Is there a painter who has seen his painting seized before having put the final touches to it?'[3]

The struggle for ownership was over what the revolutionaries referred to as 'the general map of France', and what, to the obvious annoyance of d'Églantine and Calon, Cassini and his associates proprietorially called the *Carte de Cassini*. It was the first systematic attempt to survey then map an entire country according to the science of triangulation and geodesy, or the measurement of the size and shape of the earth's surface. On its projected completion, the *Carte de Cassini* would consist of 182 separate sheets, all on a uniform scale of 1:86,400, which once joined together formed a map of the entire country that was almost 12 metres (40 feet) high by 11 metres (38 feet) wide. This was the first modern map of a nation, using innovative scientific surveying methods to comprehensively represent a single European country; but by 1793 the question was: who owned it? The new revolutionary nation it represented, or the royalists who spent four generations making it?

The map's origins stretched back to the early 1660s, beginning with Jean-Dominique's great-grandfather Giovanni Domenico Cassini (1625–1712), or Cassini I.[4] Giovanni was de facto the first director of the Paris Observatory founded in 1667 by the King Louis XIV. For more than a hundred years Giovanni's heirs – his son Jacques Cassini, or Cassini II (1677–1756); his grandson César-François Cassini de Thury, Cassini III (1714–84), and finally his great-grandson and namesake

Jean-Dominique (Cassini IV) – worked consecutively on a series of nationwide surveys operating on the strict scientific principles of verifiable measurement and quantification. Despite the practical, financial and political vicissitudes of the project, and the different directions pursued by each generation of Cassinis, their method of unifying geodesy with surveying would affect all subsequent Western mapmaking. Their principles still define most modern scientific maps, from world atlases to the Ordnance Survey and online geospatial applications, all of which are still based on the methods of triangulation and geodetic measurements that were first proposed and practised by the Cassinis. What began life as a survey of a royal kingdom would provide the template on which all modern nation states would be mapped for the next 200 years.

The proclamation of 1793 was the first ever state nationalization of a private mapmaking project. The intimate relationship between each generation of Cassinis and the French royal family who part-funded the project made it an obvious political target for the revolutionaries, but individuals like d'Églantine and Calon had also grasped the greater value of appropriating the Cassini survey for their own particular political agenda, and despite their royal associations, the maps printed from the surveys would eventually become a symbol of the new 'French Era', a blueprint for fashioning a conception of France as a modern, republican nation state. Everyone saw the military value of the maps. At a time when the fledgling republic faced imminent invasion from hostile neighbouring kingdoms, the Cassinis' detailed maps of each region of France and its borders would prove vital in defending the new regime. But the National Convention had already sought to rationalize the country's administration by reforming its bewildering assemblage of ecclesiastical provinces, *parlements*, chambers and dioceses into eighty-three *départements*, and the nationalized Cassini maps would also play a central role in enabling the state to define and administer these regions.[5]

It would also have a deeper, more intangible impact. In the hands of the Republic, the Cassini surveys would foster the belief that this was a map of the nation, for the nation. It would enable the French public, invoked by d'Églantine in his demand for the survey's nationalization, to 'see' their nation, and identify with it in one of the first cartographic manifestations of national consciousness. The surveys both responded

to and drew on the emergence of what thinkers like Charles de Secon-
dat, baron de Montesquieu (1689–1755) and Jean-Jacques Rousseau
(1712–78) began defining as 'the general spirit of nations' throughout
the eighteenth century.[6] The Bourbon monarchs had encouraged the
survey as a way of celebrating their rule, centred on Paris. Under the
Republic it would be seen as defining every inch (or metre, following the
Convention's adoption of the metric system in April 1795) of territory
mapped as French, binding people and land together in allegiance, not
to a monarch, but to an impersonal imagined national community
called France.[7] Political rhetoric would now claim that the physical territory of the nation and the sovereignty of the state were one and the
same thing, an idea that would be exported across Europe and ultimately the rest of the world.

The Cassini surveys were not primarily interested in producing maps
of the world, although their endeavours utilized geodesy and the accurate measurement of the earth's shape and size. By implication their
ambition was to map France and then extend the established principles
of surveying and mapmaking throughout nation states right round the
globe. But their contribution to the history of mapmaking has also been
neglected beside that of the story of the British Ordnance Survey, to
which it gave rise. Although the Ordnance Survey has become widely
known, it was the Cassinis who first established the enduring principles
of Western cartography, and who were responsible for the perception
and function of maps within the administration of modern nation states.

Mid-seventeenth-century France was an unlikely place to transform the
future of mapmaking. Spanish and Portuguese mapmakers had dominated the field throughout most of the sixteenth century, and the shift to
the Low Countries in the early seventeenth century had largely bypassed
France, which had little involvement in the seaborne discoveries or
joint-stock company initiatives which flowered to its south and north.
Its monarchy had been ruled since the late sixteenth century by the
Bourbon dynasty, which came to power through a series of protracted
internecine wars over religion. In response both to these internal threats,
and to the powerful regional independence of the kingdom's provinces,
the Bourbon monarchs established one of the most centralized political
states in Europe. This centralizing tendency and the regionalism which
resisted it clearly needed management, and one obvious method was to

map the realm outwards from its political centre. Other European monarchies would come to similar conclusions: the Holy Roman Emperor Joseph I (1678–1711) commissioned large-scale survey maps of Hungary, Moravia and Bohemia in the first decades of the seventeenth century, and throughout the 1770s the Comte de Ferraris (1726–1814) produced a *Carte de Cabinet* based on detailed surveys of Austrian possessions in the Low Countries. But in France the inherent difficulty of the task was compounded by the kingdom's sheer size. At around 600,000 square kilometres, France was the largest country in Europe. Over half its total boundary length of more than 6,000 kilometres was made up of land borders, many of which were shared with rival dynasties; it became obvious to the monarchy's ministers that an effective mapmaking strategy was required not only to administer the interior, but also to defend the kingdom from invasion.

More than any other country in early modern Europe, France was preoccupied with drawing consistent and enduring political boundaries on its maps and atlases. In Abraham Ortelius's *Theatrum orbis terrarum*, 45 per cent of the maps contained irregularly marked political boundaries, but by the time of the publication of Nicolas Sanson's atlas *Les Cartes générales de toutes les provinces de France* in 1658–9, 98 per cent of his maps applied a new, systematic method for representing political boundaries using standard colours and dotted outlines that distinguished *parlements* or judicial regions from more traditional ecclesiastical divisions.[8] Sanson (1600–67) was the *géographe du roi*, the Bourbon monarchy's official geographer as it began to consolidate its authority over its provinces. He was understandably interested in drawing dividing lines between countries and their subdivisions, whether his maps depicted France and its regions or the various kingdoms of Africa.

The genesis of the Cassini map of France lay not in the measurement of the earth and its man-made divisions, but the observation of the stars. In December 1666, the young King Louis XIV (1638–1715) formed the Académie des Sciences at the instigation of his controller-general of finance, Jean-Baptiste Colbert. The first meetings involved a select group of twenty-two astronomers and mathematicians including Jean Picard (1620–82) and the Dutchman Christiaan Huygens (1629–95). The foundation of the Académie also included plans for a scientific observatory, and the following year work started on a site in Faubourg Saint-Jacques, south of central Paris. By 1672 the Paris Observatory was operational.

The founding members of the Académie were joined by Giovanni Domenico Cassini (Cassini I), who would become the observatory's first unofficial director. Cassini I was a brilliant Italian astronomer, internationally famed for his research in Bologna and Rome. His work on the movements of Jupiter's moons expanded on Galileo's research, and also offered a way of determining the age-old problem of longitude. Astronomers and geographers understood that longitude is a measure of distance corresponding to differences in time. The problem was how to record such differences accurately. Cassini understood that if the time of a celestial phenomenon such as the eclipse of one of Jupiter's moons could be recorded simultaneously in two places, the results could lay the foundation for determining degrees of longitude. At an astronomical level, these calculations could help to determine the exact circumference of the earth; at a geographical level, they could provide statesmen like Colbert with the information they needed to map an entire country comprehensively.

Colbert's plans for a scientific academy grew out of a new understanding of the role that science could play in the management of the state. In England and Holland, empirical observation and experimentation were challenging the classical certainties of natural scientific enquiry. Francis Bacon's *New Atlantis* (1627) envisaged an academy of experimental scientists that prefigured the creation of the Royal Society (founded in 1662). Colbert's interests in science were more pragmatic. He wanted to sponsor scientific research projects that would directly benefit his attempt to build a French state apparatus that would be the envy of Europe.[9] For Colbert, absolute information would inform and strengthen political absolutism.

One of the Académie's secretaries, Bernard le Bovier de Fontenelle, later wrote that Colbert

> supported scholarship, and did so not only because of his natural inclination, but for sound political reasons. He knew that the sciences and arts alone suffice to make a reign glorious; that they spread the language of a nation perhaps even more than do conquests; that they give the reign a control over knowledge and industry which is just as prestigious and useful; that they attract to the country a multitude of foreigners who enrich it by their talents.[10]

Eager to distract Louis from going to war, and receptive to the lobbying of France's leading astronomers, Colbert enthusiastically endorsed the creation of the Académie, paying 6,000 livres for the site on which the observatory would be built, and more than 700,000 livres to complete it. Colbert even granted its members annual pensions of up to 3,000 livres, as was given to Cassini upon his arrival in 1668.[11] The pensions acknowledged a change in the social status of experimental scientists, who were now being incorporated into the apparatus of state power at the highest level.

Like Ptolemy's Alexandria and al-Idrīsī's Palermo, Colbert's Paris Observatory became a centre of calculation, a place where diverse information could be gathered, processed and disseminated to a wider audience in the interests of the state authorities,[12] but it would operate on a scale and with a level of precision of which Ptolemy and al-Idrīsī could only dream. Initially, the appearance of a series of comets, as well as eclipses of the sun and moon earlier in 1666, ensured that the astronomers dominated the new observatory. But Colbert's ambitions required that the Académie's remit extended beyond astronomy, and that this would be a very different place for organizing scientific knowledge from those that had gone before it like Alexandria, Palermo or the Casa de la Contratación in Seville.

As Fontenelle observed, Colbert's interest in supporting the Académie derived from his programme for bureaucratic management of the royal state. Even before the Académie's foundation, Colbert was anxious to commission a large-scale, up-to-date map of the entire kingdom to assess its resources. He requested that provincial officials submit every available map of the regions, to assess 'whether they are inclined to war or agriculture, to commerce or manufacture – and also of the state of roads and waterways, the rivers in particular, and of possible improvements to them'.[13] These would then be collated and corrected by Nicolas Sanson. In theory it was a great plan, but the responses revealed the formidable political and logistical problems that would have to be overcome to complete such a project. Only eight provinces bothered to respond to Colbert's request. The rest remained silent, either lacking the cartographic resources or being anxious that the results might lead to higher taxation. Despite his interest in marking political boundary lines, Sanson was more at home making handcoloured maps of the ancient world, and was unsurprisingly intimidated by the physical scale of the project. In a memorandum written in 1665, he acknowledged the need for two related projects: the creation of one general map of France, and another of regional maps showing their administrative divisions. These regional maps would show Colbert every feature, 'including the smallest hamlets and assarts [land cleared for cultivation], even châteaux, farms and single private houses that stand alone and away from the parishes'. Considering the size of France and the variety of terrain, this would be a physically daunting, technically challenging and very expensive operation. If traditional surveying methods were used – pacing the land with measuring rods, consulting the locals and deferring to ancient statutes – then 'the task would never be ended were one to employ all the surveyors and geometers in the world'.[14] There had to be another way, so Colbert asked his new Académie to develop a new method of surveying large swathes of territory; he was so impatient that its members even discussed the matter during their first meeting in December 1666.

Their recommendations proposed a novel fusion of astronomy and geography. The expertise used in making scientific instruments for mapping the heavens would be applied to the tools used in topographical surveying, and Cassini's astronomical observations could be applied in the determination of longitude. Funding was provided to refine established scientific instruments, including quadrants, used to measure the altitude of celestial bodies, and their navigational equivalent, the sextant, as well as the alidade, used in surveying to determine direction and orientation. The Académie decided to apply their new principles and instruments in a series of 'observations'. First Paris then the entire country would be surveyed and mapped using the latest scientific innovations. The Académie's methods brought together two strands of scientific measurement. Cassini provided astronomical observations that offered the most accurate calculation of longitude. Abbé Jean Picard, a French priest, astronomer, surveyor and founding member of the Académie, provided topographical precision based on practical surveying techniques. When these two were combined they would offer a powerful method for undertaking a countrywide survey of France.

Picard was already well known for his adaption of measuring instruments to enable much greater precision in the observation of celestial phenomenon and topographical surveying. His primary interest was in solving a scientific conundrum which was at least as old as Eratosthenes: how to calculate the earth's diameter accurately. Whereas Cassini was interested in calculating longitude from east to west, Picard was concerned with measuring an arc of the meridian from north to south. Such an arc (or line) could be drawn due north–south anywhere on the planet, tracing an imaginary arc from pole to pole right around the earth's circumference. Such an arc could ascertain the latitude of any particular place, as well as the diameter and circumference of the earth.

Picard's method of surveying involved two kinds of measurement: the first, a celestial measurement to establish the surveyor's latitude; the second, a series of angular terrestrial measurements which made possible accurate triangulation. A new micrometer (a gauge used to measure the angular size of a celestial object) allowed Picard to calculate planetary dimensions more accurately, while his telescopic quadrant replaced the usual method of using pinholes for sighting, enabling unprecedented accuracy in measuring both celestial altitudes and terrestrial angles. Armed with these new instruments, he was now ready to conduct the first modern geodetic survey of the earth's surface and in 1669 he set out to measure a meridian line between Malvoisine, south of Paris, and Sordon near Amiens, which he calculated were on the same meridian. He used 4 metre long wooden rods in calculating the distance of just over 100 kilometres, and his results were impressive. Picard computed that one degree of latitude was 57,060 *toises*. One *toise* was 6 French feet, or just under 2 metres, giving a final estimate of 111 kilometres (69.1 miles). Using these figures as a multiplier he also calculated that the earth's diameter was 6,538,594 *toises*, or 12,554 kilometres (7,801 miles). The actual figure is today calculated at 12,713 kilometres (7,899 miles).

The implications of Picard's survey for astronomy were sensational. His calculation of the earth's size verified Isaac Newton's hypothesis of universal gravitation, encouraging the Englishman to finally publish his arguments in the *Philosophiae Naturalis Principia Mathematica* (1687).[15] The practical impact of Picard's methods on mapmaking was also substantial. To establish his meridional arc, Picard had measured a base line along which it was also now possible to 'triangulate' distances and directions. Knowing the exact length between two points on his base line, Picard could identify a third point in the landscape and, using trigonometrical tables, precisely calculate its distance. The result looked like a triangulated snake moving across the base line. The method was used in Picard's *Mesure de la terre* (1671) and in the Académie's first 'observation', the 'Carte particulière des environs de Paris', completed by Picard in the late 1660s and first published in 1678. On its scale 1

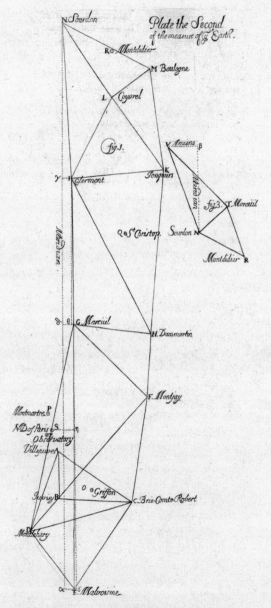

Fig. 24 Diagram of triangles, from Jean Picard, *Le Mesure de la terre*, 1671.

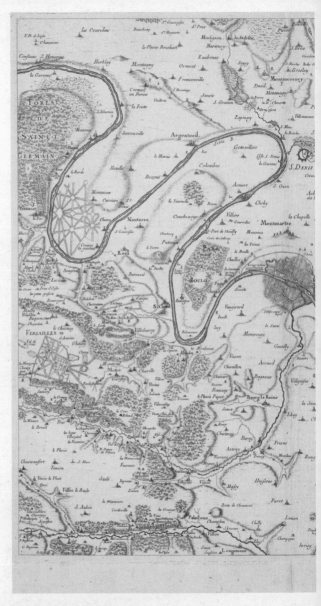

Fig. 25 'Carte particulière des environs de Paris', 1678.

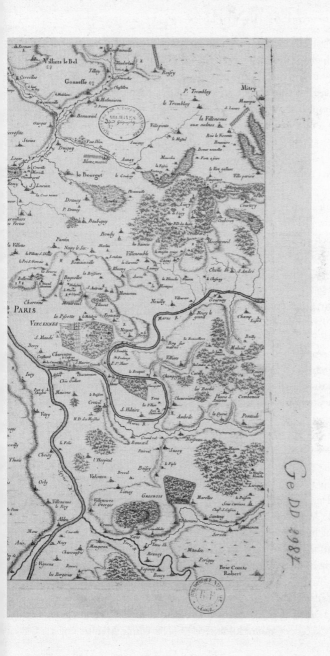

*ligne* (the smallest unit of pre-revolutionary measurement, one-twelfth of an imperial inch, or approximately 2.2 millimetres), represented 100 *toises* on the ground, giving a scale of 1 : 86,400. This would become the standard scale for all subsequent regional maps produced by the Cassinis. Both examples show that at this stage the Académie's main aim was to provide a new geometrical framework for the subsequent mapping of the country. Distances were measured according to the mathematics of triangulation, allowing the correct location of places to be plotted across empty space. The result resembles a chain of abstract geometry rather than a depiction of a chaotic, thriving country.

The Académie's next 'observation' gave a much better indication of the political power of the new methods. Picard was again chosen to lead the project, which would involve mapping the entire coastline of France. Having established the principles of a meridian line from which a triangulation survey could map the interior, Picard agreed with Cassini that an outline of the whole country would require a different method. This time Cassini's observations of the eclipses of Jupiter's moons would be used to calculate longitude. In 1679 Picard went back into the field. With the help of Philippe de la Hire, another member of the Académie, Picard spent the next three years calculating positions along the coastline. Previous maps of France had calculated positions according to a prime meridian running through the Canary Islands, a hangover from the sixteenth-century methods of calculating longitude inherited from the Greeks. But longitudinal distance between the meridian through the Canary Islands and any meridian in France had yet to be established. Picard now based observations on a prime meridian through Paris. He gradually moved down the coast and over to the Mediterranean, taking measurements in Brittany (1679), La Rochelle (1680) and Provence (1682).[16]

The finished map, entitled the *Carte de France corrigée*, was finally presented to the Académie in February 1684. The Academicians, not to mention the king himself, were shocked. As if to emphasize the modernity of their calculations, Picard and La Hire plotted their new coastline in bold over the top of the traditional outline estimated by Sanson. The new map showed the meridian of Paris for the very first time, but it also dramatically reduced the size of France from Sanson's calculation of over 31,000 square leagues (150,000 square kilometres) to just over 25,000 square leagues (120,000 square kilometres).[17] The whole Atlantic coast

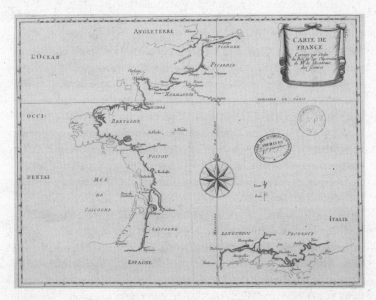

Fig. 26 Jean Picard and Philippe de la Hire, *Carte de France corrigée*, 1693 edition.

was shifted eastwards, while the Mediterranean coast retreated north-wards. The map showed that strategically important naval ports like Cherbourg and Brest had been plotted several kilometres out to sea on Sanson's earlier map. Fontenelle captured the mixture of scientific excite-ment and political concern created by the map's unveiling. 'They effected a very substantial correction to the coast of Gascony,' he recalled, 'making it straight where before it had been curved, and bringing it closer in; so that the King [Louis XIV] had occasion to say, jokingly, that their journey had brought him nothing but loss. It was a loss that enriched geography, and rendered navigation more certain and safe.'[18] The message was daunt-ing but clear: the traditional map of France had to be torn up and calculated again by a new kind of geometrical measurement.

By the mid-1680s, everything was in place for a comprehensive sur-vey of the entire country. The combination of Cassini's astronomical observations and Picard's methods of triangulation had established a general geodetic framework from which a detailed survey of the country's interior could now be undertaken. But Colbert's demand for geographical information from the regions had still not been satisfied, and as far as the Academicians were concerned the main purpose of their work up to the 1680s was still the larger measurement of the earth's size and shape. Even as the surveyors completed their work, Louis's armies were on the march, invading part of the Spanish Nether-lands and sparking war (1683–4). Together with the deaths of first Picard (1682) and then Colbert (1683), Louis's military expenditures meant that lack of money put paid to any immediate support for extend-ing the work Cassini and Picard had begun. In 1701 Louis's dynastic ambitions embroiled him in yet another European conflict, this time over hereditary rights to the vacant Spanish throne. Horrified at the spectre of Spain and France united under the Bourbon monarchy, Eng-land, Holland and Portugal launched a long and bloody war against them that stretched across Europe, North America and even the Carib-bean. By the time the twelve-year War of the Spanish Succession had come to a bitter and inconclusive end in 1713, Louis's territorial ambi-tions remained unfulfilled, and his treasury was massively depleted. With Cassini I's death in 1712, there was little political appetite or intel-lectual leadership for ambitious surveying and mapmaking projects.

Work continued intermittently on extending the measurement of the Paris meridian across the length of the country from north to south, but this was regarded as a geodetic project designed to answer the question consuming late seventeenth-century scientists: what was the earth's defini-tive size and shape? Isaac Newton's theory of gravity assumed that the earth could not be a perfect sphere because its force seemed to vary between the equator and the poles. Newton concluded that the earth was not a perfect sphere but an oblate spheroid, slightly bulging at the equator and flattened at the poles. Cassini I and his son Jacques (Cassini II) were unconvinced, and followed the theories of René Descartes (1596–1650). Revered across Europe as the great philosopher of the mind, Descartes was also renowned as a 'geometer', or applied mathematician, who put forward the argument that the earth was a prolate ellipsoid, bulging at the poles but flatter at the equator, like an egg. His theory was widely accepted by the Académie, and the resolution of the controversy soon became a matter of national pride on both sides of the English Channel.[19]

Neither group had much empirical evidence to support its claims. Newton's supporters pointed to unverified reports that the effect of gravity on pendulum measurements increased towards the poles. Jacques Cassini attempted to assert his authority as his father's successor as head of the Paris Observatory in 1712 by endorsing the Cartesian position. Delivering a paper to the Académie in 1718, Cassini II argued that the surveys supervised by his father and Picard in the 1680s revealed that degrees of latitude shortened towards the North Pole, confirming Descartes's prolate ellipsoid.[20] In a reversal of national stereotypes, English speculative theory was pitted against French empirical observation. In an attempt to resolve the dispute, the Academicians lobbied the new king, Louis XV, and his naval minister, to support scientific expeditions along the equator and near the poles to measure their respective degrees of latitude. As well as offering to resolve a scientific debate in France's favour, the Academicians also pointed to the commercial and colonial benefits of such adventures. Louis agreed, providing financial backing to two expeditions, 'not only for the progress of the sciences, but also for commerce, in making navigation more exact and easier'.[21] The precise astronomical observations and surveying practices developed by Cassini and Picard would now be tested in distant parts of the globe to solve one of science's great foundational questions. The Académie's original surveying mission had suddenly become international, in pursuit of the resolution of a dispute that overshadowed its previous preoccupations with the borders and regions of France.

In 1735 the first expedition set off to the Spanish colony of equatorial Peru, followed by the second the following year to Lapland in the Arctic Circle. Only the comparative measurement of the length of a degree at the equator and in the Arctic Circle could resolve the controversy, because if the earth was oblate (as Newton claimed) the length would increase, but if it was prolate (as Descartes claimed) it would decrease. Both teams intended to reproduce Cassini's surveying methods of determining latitude through astronomical observations and measuring distance by triangulation. The Peruvian mission was beset with disasters, from earthquakes and volcanic eruptions to civil wars, and took eight years to return. The Lapland venture was more successful, and in August 1737 its leader, Pierre-Louis Moreau de Maupertuis, was back in Paris.[22] Maupertuis reported his findings three months later to the Académie, as well as to Louis and his ministers. Cassini II's horror was undisguised: Maupertuis's estimates of the degree of latitude confirmed Newton's belief that the earth bulged very slightly at the

Fig. 27 Pierre-Louis Moreau de Maupertuis, 'A Map of the country where the arc of the meridian was measured', *The Figure of the Earth*, 1738.

equator. Picard's measurements in 1669 had strengthened Newton's thesis concerning universal gravity, and the Cassini family's methods now also provided irrefutable empirical evidence, against themselves, of Newton's theory that the earth was an oblate spheroid. The French Newtonians were triumphant. They included none other than Voltaire, who wrote to congratulate Maupertuis, mischievously addressing him as 'My dear flattener of worlds and the Cassinis'.[23]

The Peruvian expedition returned in 1744 and also confirmed Newton's theory. Despite the blow to the Académie's prestige, the controversy over the earth's shape proved that the Cassini surveying method could be exported and practised anywhere in the world. The fact that it disproved the Cassinis' own belief in Descartes's shape of the earth only strengthened the growing realization that this was a scientific method that could present a verifiable, disinterested representation of the world, regardless of faith and ideology. A further consequence of the debate over the earth's shape now presented itself. Initially Cassini I and Picard had undertaken their first surveys based on the assumption of a perfectly spherical earth. Now that Newton's theory had been verified, all their calculations needed to be revised.

The appointment in 1730 of Philibert Orry (1689–1747) as Louis XV's controller-general renewed Colbert's original interest in a countrywide survey 'for the good of the State and the convenience of the public'.[24] Orry had little interest in what he considered esoteric debates over the earth's shape: he was more concerned that the Department of Public Works (the Ponts et Chaussées) lacked accurate maps to develop France's transport network, and in 1733 he ordered Cassini II to resume triangulation of the whole country. Unlike Colbert, Orry wanted to establish the state's control over the recruitment and training of engineers and surveyors (or 'geometers'). Louis XIV and Colbert had patronized a group of savants chosen for their family connections and individual brilliance. Orry, by contrast, understood that the state needed to establish scientific colleges to recruit and educate students in the requisite skills of surveying and mapmaking. He wanted standardized maps to provide the navy with accurate charts and allow the army to build its fortifications and fix the kingdom's boundaries. He would later issue a proclamation calling for the survey to trace 'road plans following a uniform type in all the kingdom's généralités'.[25] The aims and even the language associated with the survey were beginning to change. The role of the state, the public interest and the importance of standardization were now replacing royal patronage, elite scientific speculation and astronomy in supporting its completion. But until a new generation of trained geometers emerged, Orry had no other choice but to turn to Cassini II to complete the survey.

Cassini had a very different agenda. Having married into a family of the *noblesse de robe* in 1711, he saw astronomy as a much loftier pursuit than geography, and was concerned to protect the reputation of his father and the family's scientific lineage; perversely, he regarded the opportunity to restart the survey primarily as an opportunity to counter the Newtonians and conclusively prove Descartes's theory of the earth's shape once and for all. Work began again on the painstaking process of measuring base lines and triangulating distances in 1733. In an age when national map surveys are accepted as a normal and routine activity, it is difficult to imagine the momentous scale of Cassini II's undertaking. Without the use of modern surveying instruments or transportation, and without the understanding of the local community, even the basic tasks involved would prove extremely arduous. The teams began with a reconnaissance of the area to be surveyed, establishing physical and man-made features, and deciding where to measure base lines and angular distances. This immediately presented problems. Unlike the earlier surveys conducted in built-up areas of reasonably hospitable terrain, the geometers now confronted a landscape that proved unyielding to the progress of scientific accuracy. They needed to survey regions that were often bereft of notable landmarks from which to triangulate distances, or mountainous regions where siting equipment would prove extremely perilous. Working in the Vosges Mountains in the summer of 1743, they were suspected of being Anabaptists, and accused of inciting a revolt with their secretive encampments and mystifying behaviour; in the early 1740s one was hacked to death by the villagers of Les Estables in the Mezenc region, who suspected his instruments of bewitching the local crops.[26]

The teams also encountered tiny villages, populated by people with little connection to the wider world and with no idea as to why a group of strangers were marching round pointing strange instruments at the landscape, asking awkward questions. Even as the surveyors began to reconnoitre, their equipment was stolen, they were denied horses and guides and many were pelted with stones. Local knowledge proved difficult to obtain, as even those who understood what was happening remained opposed to the work, convinced (quite rightly) that the results would only lead to the imposition of higher tithes, rents and taxes.

When (or if) the basic reconnaissance of an area was completed, preparations were made to measure a base line. Micrometers and quadrants were used to calculate exact latitude. The base line could now be constructed, using wooden rods each measuring 2 *toises* laid end to end over the course of at least 100 *toises*. Only when the base line was correctly laid and measured could the process of triangulation begin. Having verified the distance between two points on the base line, the surveyors could now choose a third point to create a triangle. But even this presented problems. The surveyors had no way of measuring the altitude of the terrain; all they could try to do was triangulate points from a specific man-made vantage point – usually a church bell-tower. Having established this position, the angle to the third point was measured using a quadrant or graphometer. Turning to their trigonometrical tables, the surveyors measured all three angles and calculated the two new sides of the triangle. Having ascertained the three angular distances, the team could then construct a second triangle, and so on until the entire region was surveyed according to a network of adjacent triangles. As each triangle was completed, a plane table was used to sketch the beginnings of what would eventually become the precise map of the region.

The sheer physical task of moving such cumbersome equipment from place to place, before then taking and rechecking measurements and calculations to ensure their accuracy, was exceptionally gruelling, and the margins for error were legion. Not surprisingly, work was painfully slow. The surviving manuscript maps made in the field give an idea of the countless observations, readings and calculations that were made. Hardly any physical details are shown, apart from towns, villages and rivers. Instead, the maps were criss-crossed with innumerable angular lines representing triangulated measurements, which dominated entire sheets. As the survey gradually built up its vast databank of measurements, the realization began to dawn on those assessing the fieldwork back in Paris that Picard's original calculations were not as infallible as had been assumed. The survey had started by drawing triangles related to the original Paris meridian drawn by Picard. By 1740, having measured 400 triangles and 18 base lines, Cassini II and his young son, César-François, realized that Picard's original measurement of the position of the meridian was out by 5 *toises*, or 10 metres. The error was small, but if multiplied over the entire country, it would compromise all the original calculations. A complete recalculation of the established measurements was required. When they were completed in 1738, the results were once again bad news for Cassini II: the recalculated latitudes confirmed Maupertuis's Lapland measurements. Even the measurement of French soil now proved, once and for all, that Newton's theories were correct.

The influence of the Cassini lineage might have ended there but for the increasing involvement in the survey of Jacques's son, César-François Cassini de Thury, Cassini III – a better geographer than an astronomer, and a shrewd diplomat. He quietly accepted the triumph of Newtonianism, understood Orry's requirements for the new survey, and throughout the 1730s and 1740s deftly steered the arduous survey not only to its completion but to printed publication. As his disillusioned and increasingly aloof father appeared out of step with the shift towards a more professional approach to geography, Cassini III began planning the public dissemination of the survey's work.

In 1744, the survey was finally completed. Its geometers had completed an extraordinary 800 principal triangles and nineteen base lines. Cassini III had always envisaged printing regional maps as they were produced, and by 1744 the map was published in eighteen sheets. Its new map of France, on an appropriately small scale of 1:1,800,000, shows the country represented as a network of triangles, with virtually no expression of the land's physical contours, and with large areas such as the Pyrenees, the Jura and the Alps left blank. It was a geometrical skeleton, a series of points, lines and triangles following coasts, valleys and plains in connecting the key locations from which observations were carried out. Over it all lay the triangle, the new, immutable symbol of rational, verifiable scientific method.[27] On Cassini III's map the triangle almost takes on its own physical reality, a sign of the triumph of the immutable laws of geometry and mathematics over the vast, messy chaos of the terrestrial world. The Babylonians and the Greeks had revered the circle; the Chinese celebrated the square; the French now showed that it was the application of the triangle that would ultimately conquer the earth.

The publication of the 1744 survey represented the fulfilment of

Colbert and Orry's original plans. It was not in modern terms a national survey based on comprehensive topographical detail, but a geodetic survey which produced a positional illustration of places significant to the requirements of state planning. Cassini III admitted as much when he explained that his surveying teams 'didn't go off into each village, into each hamlet in order to survey the plan. We haven't visited each farm or followed and measured the course of every river ... such detail is required only of plans for some seigniorial land; the reasonable size at which one ought to fix the map of a country doesn't allow one to be able to mark so many things without great confusion.'[28] The logistics of completing not one but two surveys were simply too great; to complete a further, third survey of the entire topography of the country would have required a level of money, manpower and technical precision that Cassini III clearly regarded as unrealistic. As far as he was concerned, his work, and that of his family, was finished. Public and private individuals and organizations could now fill in the topographical blanks that were clear for all to see on his 1744 map. The commercial map trade in Paris had already produced a distinguished collection of mapmakers, including Alexis-Hubert Jaillot (1632–1712) and Guillaume Delisle (1675–1726), but a new generation, including Jean Baptiste Bourgignon d'Anville and Didier Robert Vaugondy (1723–86), now emerged to produce maps and atlases capitalizing on the cartographic opportunity presented by Cassini III.[29]

Though nobody, least of all Cassini III, regarded the survey of 1733–44 as a preliminary study for an even greater description of the country,[30] this is nevertheless precisely what happened and Cassini was asked to embark on yet another survey in the service of the Bourbon monarchy's dynastic and military ambitions. Just as his father had led France into one costly war over the dynastic succession to the throne of Spain, so Louis XV intervened in a similar dispute in 1740. This time it revolved around the Habsburg territories claimed by Austria on France's northern and eastern borders. The Austrian War of Succession (1740–48) led Louis into a series of bloody and expensive campaigns, which by the spring of 1746 saw his armies fighting in the Austrian Netherlands. Cassini III was invited to advise the French engineers on measuring base lines along the River Scheldt, and in October 1746 assisted in drawing up the topographic plans for the Battle of Rocoux outside Liège.

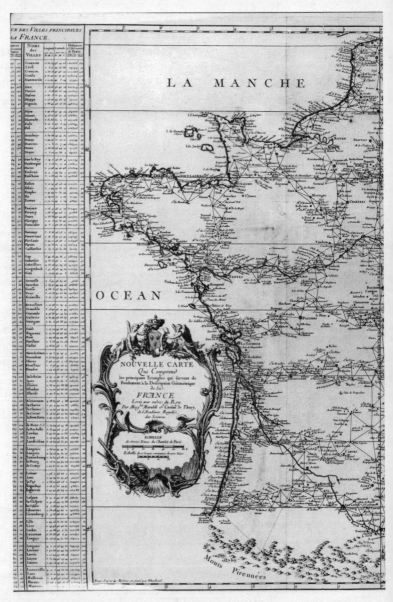

Fig. 28 César-François Cassini de Thury, 'New Map of France', 1744.

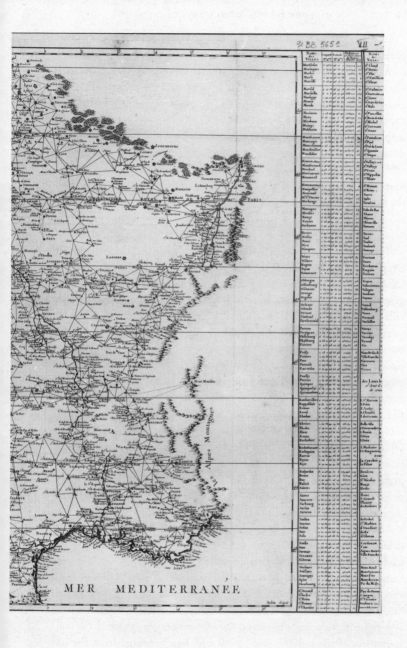

Following the French victory, Louis XV visited the region and compared its terrain with Cassini's maps. His comments would provide a turning point in the future of the national survey. 'The King,' recalled Cassini III, 'map in hand, found the country and the disposition of his troops so well represented that he had no question for his generals nor for the guides; and he did me the honour of saying to me: "I want the map of my kingdom to be done in the same way, and I charge you with doing it, inform [Jean-Baptiste de] Machault [the controller-general] of this".' After the labours of the previous eighty years, both king and cartographer knew that this would be no easy undertaking, but 'he did me the honour', Cassini continued, 'of asking me several times if the work would be easy of execution, and how much time it would require to bring it to perfection'.[31]

The pragmatic Cassini III soon had an answer to Louis's question. Despite his concerns about the feasibility of such a gigantic project, the chance to conduct yet another survey, this time covering every topographical feature in the country, from its rivers to its hamlets and villages, was too good to miss, and scientific immortality beckoned. He calculated that the survey would take eighteen years to complete. One hundred and eighty regional maps on a uniform scale of 1:86,400 would be required to cover the entire country, with ten maps produced every year, each costing 4,000 livres, which included the costs of equipment, surveying and printing. The annual budget of 40,000 livres would pay for ten teams of two engineers to measure and record the relevant information on the ground, which would then be sent back to the Paris Observatory where it would be checked prior to engraving and publication. Each printed map sheet would be sold for 4 livres on an estimated print run of 2,500 each. If all 180 sheets were sold at this figure, the project would raise revenue of 1,800,000 livres – an impressive return on a projected state investment of just 720,000 livres. Considering that a skilled worker could earn up to 1,000 livres a year, and that the king's cabinet-maker could charge 938,000 livres for ten years' work, Controller-General Machault appreciated that in purely financial terms Cassini's survey looked like an excellent investment.[32] He had been shocked at the depletion of the kingdom's coffers following the Austrian Wars of Succession, and was keen to reform the state's outdated tithe system by introducing a universal flat rate of taxation, much to the consternation of the nobility and clergy who had benefited from the old feudal arrangements. The new survey promised to assist him in his new scheme, and benefit many of those people who had so aggressively opposed its predecessors.

Cassini III grasped the opportunity to implement a new survey that would transform geographical understanding of France, but his methods would also change the entire practice of geography. He proposed to standardize the maps created from the survey by adopting what is now called a transverse equirectangular projection, which treats the globe as a cylinder projected onto a rectangle. The globe is rotated so that any given meridian acts just like the equator, ensuring that scale along this line, and anywhere at right angles to it, remains correct. The inevitable distortions at the northern and southern limits of such a projection were negligible for Cassini's purposes, as the regional areas to be surveyed were too small to be seriously affected. Otherwise he pointed out that, unlike in the first two surveys, no new scientific innovations would be required. Having established a geodetic framework, Cassini III now introduced a method that would allow its topographic detail to be filled in. In line with Orry's plan to educate a new generation of geographers in standard methods of surveying and mapmaking, Cassini III proposed to train his teams of engineers from scratch in the techniques of measurement and observation required to complete the survey. Each engineer would keep two logbooks. One would record topographical information, the location of villages, rivers, churches and other physical features, which would be verified by local priests and gentry. The other would record geodetic data on the measurement of triangulation in relation to established base lines and principal triangles, which would be sent to Paris and checked by members of the Observatory. Accuracy, uniformity and verifiability were established as central to the survey's political and financial success. Under Cassini III's guidelines, geography would now become a routine and continuous activity sanctioned by the state, its practitioners operating within strict guidelines determined by the authorities. The age of the learned savants uniting the arcane wisdom of astronomy, astrology and cosmography in the creation of their maps was coming to an end. Geographers were slowly but surely turning into civil servants.

Shortly after the conclusion of the Austrian War of Succession with the signing of the Peace of Aix-la-Chapelle in October 1748, the first instalment of money was paid to Cassini III to begin his new survey. Yet again, teams of engineers fanned out across the country, preparing to survey what Cassini III referred to as 'this innumerable quantity of cities, towns, villages, hamlets and other objects scattered across the whole extent of the kingdom'.[33] As usual, work started from the environs of Paris, following the tributaries of the Seine. Topography now preceded geometry, as Cassini's engineers endeavoured to put geographical flesh onto the triangulated skeleton of the first two surveys. The work was less specialized, but it would bring about an unprecedented portrayal of the impact of human settlement on the earth.

Cassini III had already established his reputation for tact and diplomacy. To this he now added an obsessive attention to detail and accuracy, characterized by his tireless micromanagement of every aspect of the survey, from personal involvement in the fieldwork, to overseeing the engraving of the plates for publication. Nothing was left to chance, as his account of a surveyor's average day in the field revealed:

> Located on the highest part of the bell-tower and accompanied by either the parish priest or syndic or other person able to provide knowledge of the country and to indicate to them the names of the objects they see, they had to spend part of the day becoming sufficiently familiar with the area to be able to represent it on the map, checking the condition of their instruments and the parallelism of the telescopes and taking and several times retaking the angles between the principal points, checking whether the angles taken in circling the horizon did not exceed 360 degrees, as good a proof of the precision of the angles composing the *tour d'horizon* as the observation of the third angle of a triangle. The work of the day would be followed by the work of the study: having acquired an idea of the layout of the area, it would be necessary to roughly draw the heights, the valleys, the direction of the roads, the course of the rivers, the nature of the terrain; to draw up, in fact, the map of the area while they were there and able to check that it was accurate, and to correct it if it were erroneous.[34]

Of equal importance to the work in the field was the paper trail it created; from the land to the study, Cassini's engineers were instructed to write up their observations and translate them into handdrawn sketch maps, correct them where necessary and then dispatch everything to Paris for another round of verification. Cassini III insisted that when the map was drafted, it was returned to the local dignitaries involved in initially checking the relevant topographical data. 'The geometrical part belongs to us,' proclaimed Cassini; 'the expression of the terrain and the spelling of the names are the work of the lords and priests; the engineers present the maps to them, profit from the information they provide, working under their orders, making in their presence the corrections to the map, which we publish only when it is accompanied by certificates' confirming the veracity of the information recorded.[35] It was an essential element in ensuring accuracy, but it had another consequence too: however reluctant the provincial nobility might have been to verify the observations made by the unwelcome engineers, they were now becoming part of the fabric of a national survey. Up to this point, local knowledge had been ignored in favour of the pure geometry of the triangulated framework of the survey; Cassini III now ensured that the visualization of the imagined community of France included the knowledge of those who lived and worked within it.

And it *was* slow. After eight of the eighteen years Cassini had initially estimated would be needed to complete the survey, only two maps were published, those of Paris and Beauvais. In the summer of 1756 Cassini III was granted an audience with Louis XV to present him with the map of Beauvais, fresh from the engraver. Initially, the meeting went well. Cassini recalled that the king 'seemed astonished by the precision of the detail' of the map. But then Louis dropped a bombshell. 'My poor Cassini,' he said, 'I am terribly sorry, I have bad news for you: my controller-general doesn't want me to go on with the map. There's no more money for it.'[36] The project was way behind, and costs had escalated, Cassini now estimating that each map would cost nearly 5,000 livres. Based on current progress, the entire project would not be completed until well into the next century. Machault's reforms had predictably foundered in the face of aristocratic opposition, and, given the parlous condition of the state's finances, his replacement, Jean Moreau de Séchelles, was clearly unprepared to sanction further expenditure. Whatever Cassini III's immediate reaction, he later recalled that his response to Louis's news was characteristically determined: 'The map will be made.'[37]

Cassini III was a geographer who insisted on absolute perfection, and the dilatory progress of the survey which resulted had put its existence in jeopardy. But, as a businessman, he now moved quickly to ensure its survival. Cassini had always hoped that the private sector would invest in his previous survey of 1733–44, and he now hit on a plan to test his belief and save the new survey. With Louis's backing, he formed the Société de la Carte de France, an association of fifty share-holders who were asked to provide 1,600 livres per year to support his estimate of the 80,000 livres per year which would be needed to complete the survey in just ten more years. In exchange, they received shares in the projected profits, as well as two copies of each completed map. Politically and financially, it was a brilliant move. Leading members of the nobility and the government publicly allied themselves with the project by subscribing, even including the king's mistress, Madame de Pompadour, and Cassini raised even more money than he actually needed. Despite this effective privatization of the enterprise, Cassini also stipulated that the Académie des Sciences would retain overall control of both the survey's management and the publication of its maps. Within a few weeks, Cassini III had rescued the survey from potential oblivion, ensured its future funding, and freed it from the interference of either the state or its shareholders.

Cassini's actions galvanized production of maps based on the survey. Within days he announced that the first maps of Paris and Beauvais were for sale, each costing 4 livres – considerably more expensive than other regional maps, which sold for as little as 1 livre. Cassini promised to release one map each month: Meaux, Soissons, Sens, Rouen, Chartres, Abbeville, Laon, Le Havre, Coutances and Châlons-sur-Marne quickly followed and within three years, 39 of the anticipated 180 maps had been published (although they were all concentrated in the northern and central areas around Paris). The print runs were substantial (500 for each sheet), and sales were impressive. By 1760, more than 8,000 copies of the first forty-five sheets printed had been sold.[38] By the end of the decade tens of thousands of individual sheets were in the hands of people living the length and breadth of the country. Although the number of printed maps published by Cassini was smaller than in Blaeu's *Atlas maior*, the cumulative circulation of all the published map sheets certainly eclipsed sales of the earlier, more expensive Dutch atlas. Maps were circulating on an unprecedented scale.

The publication of the first maps in 1756 made it clear for all to see that this was a staggering achievement. Having obsessively supervised every stage of their creation from survey to publication, Cassini III had created a series of maps that were unparalleled in their precision, detail, accuracy and standardization. Each map was produced using only the finest materials available. They included German black ink from

Frankfurt and *aqua fortis* which gave the maps their characteristically sharp but durable lines, as well as a soft, silvery aura. Cassini had demanded that the maps be 'laid out with a certain taste and clarity'. He understood that the 'public hardly judges but by this inconsiderable point'. The crisp exactitude of the finished product resulted in the kind of aesthetic beauty not normally associated with maps. Mapmaking might have become a science, but Cassini was also anxious for the public to regard it as an art.

Capitalizing on the public interest, Cassini took another innovative step: in February 1758, he offered a public subscription to the entire map of France. For 562 livres, subscribers would receive all 180 maps as they were published, representing a saving of 158 livres. One hundred and five subscribers took up the offer, rising to 203 by 1780.[39] Unlike the company's shareholders, very few of these subscribers were part of the Parisian elite. They included provincial farmers and businessmen, many drawn from the middling sector of French society that had previously been so opposed to the survey. Although their numbers were relatively few, these bourgeois subscribers represented an inadvertent 'nationalization' of the survey as a consequence of its putative 'privatization'. Just as Cassini III's obsessive demand for accuracy allowed local people to contribute to the national survey and regard it as representing their country, so his attempt to ensure its continued financial support enabled others to invest in a small piece of France.

Having delegated the funding of the project to the private sector, the state still retained an active interest in the progress of the reinvigorated survey. In 1764 a royal proclamation was issued requiring all unsurveyed regions to contribute to the outlay involved. The subsequent revenue provided Cassini III with nearly 30 per cent of the estimated costs for completing the entire survey. It was a timely boost for him. His original projections for finishing the survey had been wildly optimistic, as he must have realized when he had struggled to raise the revenue required to keep the project afloat. The new capital enabled him to employ a further nine engineers, but it was still not enough. Surveying and mapping the highly populated and unforbidding terrain in the country's central and northern regions was straightforward enough, but working in the vast, mountainous areas in the south and south-west was proving as difficult as ever. The expected completion of the survey in the late 1760s came and went; between 1763 and 1778, fifty-one further maps were published, mostly in the central and western regions, but that still left well over a third of the country to be mapped, including Brittany, where the conservative aristocratic authorities impeded what they regarded as the survey's centralizing demands.

Transferring the results of the survey's fieldwork onto engraved maps hundreds of kilometres away in Paris had the potential for endless minor errors. Cassini's response was the introduction of obsessive checks, tests and inspections. Even the paper on which the maps were printed was exactly measured to ensure that, on a standard sheet measuring 65×95 centimetres applying a uniform scale of 1:86,400, each discrete section would represent exactly 78×49 square kilometres. Having established a method of verifying the survey's results with the local population, Cassini then turned his attention to the problem of engraving. The florid italic of Mercator and Blaeu's maps and atlases was no longer sufficient to cope with the mass of data produced by the prescribed scale of 1:86,400. 'As regards the engraving of the sheets,' Cassini complained, 'one would not believe how this art, taken so far in France, has been so neglected in its geographical aspects.' In response, he was 'obliged to train engravers, to make a selection of models for them to follow in representing woods, rivers, and the conformation of the country'.[40] Still not satisfied, Cassini found it necessary to train two sets of engravers: the first to engrave the topographical plan, the second to complete the lettering. Pierre Patte, one of Cassini's main engravers, described how the black-and-white engraving aimed to reproduce the sensual detail of the natural world. 'As for the manner of expressing the different parts that make up a map,' he wrote in 1758,

> the whole art lies in grasping the general expression of nature and giving its spirit to what one wishes to be represented. From high on a mountain, consider the tone of the different objects that are on the surface of the surrounding terrain: all the woods seem to stand out in brown, bush-like, against a background that also seems a little brown . . . As regards mountains, unless they are peaks, they never seem to be clearly delineated, but their summits always seem on the contrary to be rounded, more or less elongated, and giving on the shadow side a velvety tone without crudity.[41]

Cassini and his engravers were building up a new grammar of mapping, developing signs, symbols and lettering that would translate the land's topography into a new language of cartography. The results can be seen in the most popular and iconic of all the survey's maps, the first sheet representing Paris. Its lack of decoration is immediately noticeable. There is no cartouche, no table of contents or explanation of symbols, and no extraneous artistic flourishes: just a topographical map of Paris and its environs. The triangular, geometric skeleton of previous Cassini maps has disappeared, subsumed by the wealth of local detail. The subliminal geometry of the map is hardly visible at all, just discernible along the Paris meridian which runs right down its middle, with the perpendicular forming a right angle at its dead centre: the Paris Observatory. But on this map, there is no great celebration of egocentric geography; instead, the eye is drawn to its precise toponymy and lovingly rendered topography.

Everything on the map is standardized. Cassini improved the established signs and symbols (such as the traditional hierarchy of city, town, parish, château and hamlet symbolized by different oblique perspectives), and added his own – an abbey was represented by a bell-tower with a crozier, a country house with a small banner, a mine by a small circle. Administrative divisions from the national to the regional level were distinguished by a variety of dotted and stippled lines, while hatching symbolized relief. In sheet after sheet, the same standard conventions and symbols were used. The message was unmistakable: whatever the terrain, every corner of the kingdom could now be mapped and represented according to the same principles. In a direct challenge to the country's defiant regionalism, the map established that nowhere was exceptional. It was a powerful message of unity echoed in the growing opposition to monarchical rule expressed by lawyers like Guillaume-Joseph Saige, who wrote in 1775 that 'there is nothing essential in the political body but the social contract and the exercise of the general will; apart from that, everything is absolutely contingent and depends, for its form as for its existence, on the supreme will of the nation'.[42]

Paradoxically, this 'will' was conveyed loudest through the maps' most obvious feature. In eighteenth-century France, the king's subjects spoke a diversity of languages, from Occitan, Basque, Breton, Catalan, Italian, German, Flemish and even Yiddish, to a variety of French dialects.[43] On Cassini's maps, the purely descriptive language – *ville*, *bourg*, *hameau*, *gentilhommière*, *bastide*, and so on – is written in Parisian French. With geographical standardization comes linguistic conformity. If everyone looking at the map is being asked to imagine the place where they themselves are as part of a unified France, they must do so in the language of its rulers in Paris.

The 1780s brought many momentous changes, both to the survey and to France. As Cassini approached his seventies, he was joined on the completion of the survey by his son, Jean-Dominique, comte de Cassini (Cassini IV). Efforts were redoubled to complete the work, but in September 1784 Cassini III contracted smallpox and died, aged 70. His achievements had been immense, from restoring the family's authority after the debacle over the shape of the earth in the 1740s, to the completion of yet another geometrical survey, first initiating then rescuing the most ambitious survey ever envisaged, and steering it towards a conclusion. Now the onerous task of completing the national survey was passed to Jean-Dominique, who also assumed the role of director of the Observatory. Named after his great-grandfather, born and brought up in the Paris Observatory that the family now saw as their home, and firmly established as part of the Parisian nobility since the 1740s, like Cassini I and II, Jean-Dominique regarded himself as an astronomer rather than a geographer (a title which, in the circles within which he moved, still brought far greater prestige). He chose to see himself as a patrician scientist and academician, loftily surveying the mechanical work done in the field by his engineers from the confines of what he believed to be his observatory. The completion of the survey, as his father acknowledged, would never produce the kind of scientific breakthroughs achieved by Cassini I and II. Assessing the impact of science on the study of geography, Cassini IV later wrote that:

> Thanks to the multiple voyages undertaken by educated men all over the world; thanks to astronomy's, geometry's, and clock making's easy and rigorous methods for determining the position of all places, geographers will soon find that they have neither uncertainty, nor choice, nor need of a critical faculty in order to fix the principal positions of the four parts of the globe. The canvas will fill itself bit by bit as time passes, imitating the procedure that we followed for the production of the general map of France.[44]

Geography was dismissed as a method rather than a science, bereft of 'a critical faculty', its practitioners reduced to painting by numbers, rather like Cassini's own engineers in the field. Cassini tacitly accepted that the survey would be a huge achievement, but for him, completing it simply consisted of a mechanical filling in of gaps, and needed to be finished as quickly as possible to allow him to pursue more ambitious astronomical research.

Ever mindful of his family's reputation, Cassini IV dutifully pursued the surveying and printing of the final maps, publishing another forty-nine throughout the 1780s. But as work continued and the decade drew to an end, larger political events began to overtake it. The bitterly cold winter of 1788–9 and subsequent drought sent food prices rocketing, leading to riots across the country. The monarchy was no longer able to trim its parlous financial situation, and turned the issue of political and fiscal reform over to the Estates-General, a representative assembly of three tiers – the Church, the nobility and commoners – which convened for the first time since 1614, at Versailles. When reforms foundered in the face of aristocratic opposition, opponents of the *ancien régime* finally took matters into their own hands. After being barred from a meeting of the Estates-General on 20 June 1789, the members of the Third Estate met to sign the 'Tennis Court Oath', which demanded a new, written constitution. It triggered the beginnings of a revolution, which would quickly see the creation of a new legislative assembly and a failed constitutional monarchy, culminating in the proclamation of a French Republic in 1792 and the execution of Louis XVI in 1793.

In the late 1780s the opposition to the king's rule demanded sweeping political reforms in language which repeatedly invoked the 'patrie' ('the fatherland') and the nation. Throughout the latter half of the eighteenth century, as the Cassini surveyors toiled across the country, both the royalists and their increasingly vocal political adversaries fought over the term 'patrie'. Initially, the king's supporters claimed that to be patriotic was to be royalist, but the opposition responded by referring to itself as the *parti patriote* from the early 1770s, and argued that, until the monarchy was swept away, France possessed no 'patrie', and could not truly be called a nation. The debate can be discerned in the titles of books: between 1770 and 1789, 277 works were printed with variations on the word 'patrie' in their title, and in the same period 895 titles used 'nation' or 'national'.[45] These ranged from pamphlets with titles like *Les Vœux d'un patriote* (1788) to Pierre-Jean Agier's anti-monarchical treatise *Le Jurisconsulte national* (1788), and the Abbé Fauchet's more conciliatory *De la religion nationale* (1789). As the supporters of the Third Estate seized the political initiative in 1789, their language repeatedly invoked a new idea of the nation. 'If the privileged order were removed,' wrote one of its supporters, 'the nation would not be something less, but something more.'[46] In a pamphlet entitled 'What is the Nation and What is France?' (1789), the writer Toussaint Guiraudet described the political situation as if he were looking down on Cassini's maps: 'France is not a compound of Provinces, but a space of twenty-five thousand square leagues.'[47] Another prominent supporter of the Third Estate, the abbé Emmanuel Sieyès, wrote of the need to make 'all the parts of France a single body, and all the peoples who divide it into a single body'. 'The nation', he argued, 'is prior to everything. It is the source of everything.' His book *What is the Third Estate?* presented its deputies as the nation's true representatives, and in June 1789 the Estate drew on Sieyès's rhetoric in declaring that 'the source of all sovereignty resides essentially in the nation'.[48]

As the political situation deteriorated daily, Cassini IV raced to complete the map of a country descending into revolution, adding the *Carte des Assemblages des Triangles*, increasing the project from 180 to 182 sheets. In August 1790, as the National Assembly began to reorganize the diocesan boundaries and departments surveyed by Cassini's engineers, Cassini presented a report to a meeting of the shareholders of the Société de la Carte de France. Fifteen maps remained to be published. The survey was finished, and the entire map agonizingly close to final publication. As the new regime prepared itself for war with its hostile neighbours, the army turned its attention to the map. The head of the military engineering corps, Jean-Claude Le Michaud d'Arçon, summarized the dilemma facing Cassini in assessing the dangers of publishing the last few remaining maps which contained potentially sensitive information about the vulnerable mountainous regions. 'It is essential that neither their strengths nor their weaknesses be indicated to the enemy, and it is of the very greatest importance that any knowledge of them profits us alone,' insisted d'Arçon. 'The privilege granted to M. de Cassini's engineers should exclude those parts of the frontiers knowledge of which should be reserved to us.' His conclusion captured Cassini's situation with brutal honesty. 'His map may be good or bad. If it is good, it will have to be banned, and if bad, it would hardly deserve favour.'[49]

But, as we saw at the beginning of this chapter, the map was not banned, but nationalized (or confiscated, depending on your political sympathies) by the National Convention in September 1793. Nationalization meant that the map was completely withdrawn from public circulation, its plates and published sheets confiscated by the Dépôt de la Guerre in the interests of the new nation. In December 1793, as the 'Reign of Terror' swept Paris, the company's shareholders were called to what would be its last general meeting. Cassini IV and his loyal assistant Louis Capitaine waited in vain. Finally, a solitary shareholder arrived. 'Gentlemen, believe me,' he announced, 'you may do as you wish, we all have plenty of other things to think of than maps. For myself, I shall bid you good day and find hiding where I can.'[50] The net was closing around Cassini: already stripped of his membership of the Académie (which was subsequently disbanded) as well as the directorship of the Observatory, he was thrown into prison in February 1794. Condemned by his students, he narrowly avoided the guillotine, a fate which befell his unfortunate cousin and fellow prisoner, Mlle de Forceville, as Cassini looked on helplessly.

As the Terror subsided in the summer of 1794, Cassini was released from prison, but he was a broken man. He turned his back on science, railing against the revolutionary reforms as 'overturning everything, of changing everything without need, and for the sole pleasure of destruction'.[51] He flirted with requests to join several academic societies, and supported Capitaine's attempts to compensate the shareholders of the Société de la Carte de France for their losses. When Philippe Jacotin, the head of the topographical department of the Dépôt de la Guerre, was delegated to assess what the state owed the shareholders (including Cassini), he simply calculated the changing metal value of the engraved copper map plates over twenty years, deducting the costs of maintaining them over the same period. He arrived at a figure of 3,000 new French francs (roughly equivalent to the old livre) for each share. Cassini was predictably outraged. 'It seems to me', he fulminated, 'that it isn't some colonel, the head of the topographic office, that one should turn to for such an expert opinion, but rather to a boiler-maker, who understands better than anyone the value of old copper.'[52] Fifty years of scientific labour by father and son were now valued according to the price of the copper used to make their map engravings. It was a miserable end to a glorious project. Disillusioned and scorned, Jean-Dominique retired to the family home in Thury, where he died in 1845, aged 97.

Technically speaking, the general map of France was never finished. Following its nationalization, everything related to the survey and its maps was transferred to the Dépôt de la Guerre. This included 165 finished sheets, eleven still to be engraved, and four sheets of Brittany, already surveyed but yet to be drawn. The Dépôt now had everything it needed to complete the map of France as it was originally envisaged way back in 1748: 180 map sheets of the entire country on a uniform scale of 1:86,400, with the addition of Cassini IV's *Carte des Assemblages des Triangles*. But yet again circumstances intervened. Even the most recent maps now needed correcting and updating to include new roads, as well as the administrative reforms of the Republic's departments. Reduced versions of the projected map of France were produced, but none matched the original plan. In 1790, prior to nationalization, Louis Capitaine drew up a reduced atlas based on the survey's work, designed to represent the National Assembly's reorganization of the regional departments. He also published the 'Carte de la France suivant sa nouvelle division en départements et districts'. The map was dedicated to the National Assembly and the shareholders of the Société de la Carte de France, in a valiant attempt to accommodate their divergent political and commercial interests. It was also the first map to represent the reformed departments. But it was still not the comprehensive survey covering every corner of the country envisaged by Cassini III and IV.

It is strangely appropriate that the individual who encouraged its completion and signalled its eclipse was both a revolutionary and an emperor: Napoleon Bonaparte. Having overthrown the republican authorities in 1799, he crowned himself Emperor Napoleon I in December 1804. Just weeks before his coronation, he wrote to his military chief of staff, Louis-Alexandre Berthier, concerning French troop movements across the Rhine. 'The *ingénieurs-géographes* are being asked to make cadastres [property maps] instead of military maps, which means that, twenty years from now, we shall have nothing,' he complained. He went on: 'If we had stuck to making maps on Cassini's scale, we should already have the whole Rhine frontier.' 'All I asked was that the Cassini map be completed.'[53] As far as Napoleon was concerned, the scale and detail of Cassini's maps were perfect instruments for military activity.

Ten years later, as his enemies closed in on him, a small incident shows how far the Cassini maps had by then permeated and shaped the national consciousness. In February 1814 Napoleon spent the night in the remote village of Her, in the Champagne-Ardennes region, preparing for the Battle of Arcis-sur-Aube, which would prove to be the penultimate battle before his abdication and exile to Elba. Lodging with the local priest, Napoleon and his officers settled down to dinner, at which point, Napoleon's faithful secretary Baron Fain recalled, 'it was with difficulty that our host comprehended how his military guests could be so well acquainted with its localities, and insisted upon our all being natives of Champagne. In order to explain the cause of his aston-ishment, we showed him some sheets of Cassini, which were in every one's pocket. He was all the more astonished when he found in them the names of all the neighbouring villages, so far was he from thinking that geography entered into such details.'[54] That virtually all of Napoleon's retinue owned copies of Cassini's maps is testament to their military application. But their almost magical revelation in front of the incredu-lous priest also shows how far they appeared able to bridge regional differences (regardless of reality); above all else, priest and soldier were 'French', regardless of their religious or ideological differences.

The Dépôt de la Guerre took over direct control of the publication and distribution of the remaining maps, appointing twelve engravers to update the confiscated plates and print new editions where necessary. The political and military importance of these maps ensured that state funding was always available, as Berthier had pointed out in a letter written to the Dépôt's director in 1806: 'with money, we shall lack for neither draughtsmen nor engravers.'[55] There was clearly a brisk market for these new maps, as single sheets sold for 4 francs. Finally, in 1815, the last sheets of Brittany were finished, completing the full set of 182 sheets of maps. But when, after sixty-seven years, the Cassini map of France reached its end it was already a thing of the past. Seven years earlier, in 1808, Napoleon had ordered a new map of France. A report highlighted the errors and mistakes that were now glaringly obvious on the *Carte de Cassini*:

> The Dépôt de la Guerre, in possession of the plates of the Cassini map, has had every opportunity to verify its exactitude. It has unfortunately identified major errors; localities placed a league distant from their true position; the impossibility of precisely determining longitudes from the Cassinis' data and calculations, etc. In addition, the Cassini plates, badly engraved to begin with, were almost worn out; a great many already retouched, many that will have to be engraved anew, an operation it makes no sense to undertake without making a great number of corrections, or, to be quite frank, a new survey.[56]

The Cassini survey and its maps were ultimately rendered redundant not by the edicts of a king, or the ideological demands of a republic, but by what any modern nation state undertakes with continuous regularity – just another survey. By 1818 the first trials for a new survey were undertaken, although it was not completed until 1866, with the final maps (273 in all) published in 1880. New methods of measuring elevation and relief, including the clinometer, which calculated angles of elevation relative to gravity, ensured that the new survey offered a level of accuracy that finally surpassed the technical achievements of the *Carte de Cassini*.[57]

One of the most enduring consequences of the Cassinis' endeavours was that they inspired the most famous of all national surveys, the British Ordnance Survey. In October 1783, less than a year before his death, Cassini III wrote to the Royal Society in London with a proposal to measure the difference in latitude and longitude between the observatories in Greenwich and Paris, using the methods of triangulation perfected by his engineers working across France, in the first truly cooperative international mapping project of its kind. Cassini's telescopic instruments were able to locate positions in England from France, and he now proposed a trigonometric survey across the sea, uniting the two old adversaries in a chain of precisely measured triangles.[58]

The proposal inevitably evoked old animosities between the two great European powers, and the Astronomer Royal, the Reverend Neil Maskelyne, grumbled at Cassini's cheek in suggesting that English estimates of Greenwich's geodetic position were inaccurate. But on this occasion, science overcame nationalism. The president of the Royal Society, Sir Joseph Banks, asked Major General William Roy to undertake the survey on the English side of the Channel, and in June 1784 Roy began work by painstakingly measuring the first base line of his survey on Hounslow Heath, west of London. Roy's base line provided the foundation for the Ordnance Survey's subsequent mapping of the whole of Britain, and it followed the same principles as Jean Picard's base line drawn 115 years earlier west of Paris. The instruments were new and improved (including the introduction of a monstrous 200 pound theodolite, capable of measuring vertical or horizontal angles), but the methods pursued by Roy and the Ordnance Survey throughout the rest of the nineteenth century were exclusively based on those developed by Picard and the Cassinis in France. For Cassini III, exporting his surveying techniques across the Channel was the culmination of a geodetic project that had stretched back over 120 years; for the English, it was the beginning of a national survey that would ultimately gain even greater fame than the *Carte de Cassini*.[59]

The *Carte de Cassini* was an unprecedented step in the history of map-making. It was the first general map of an entire nation based on geodetic and topographical measurement; 'it taught the rest of the world what to do and what not to do'.[60] Its pursuit of an '*esprit géométrique*', or 'quantifying spirit',[61] begun in the mid-seventeenth century, gradually transformed the practice of mapmaking over the next 150 years into a verifiable science, pursuing a standardized, empirical and objective method that could (and would) be extended right around the earth. The cartographer was now regarded as a disinterested engineer, capable of matching the map to the territory. Reduced to a series of geometrical triangles, the world became knowable and manageable.

But the claim of the Cassinis to pursue a disinterested, objective method of scientific enquiry was more of an aspiration than a reality. Looking back at his tenure as head of the Paris Observatory during his long retirement, Cassini IV wrote wistfully that '[e]nclosed in the Observatory, I thought that there I was in a port sheltered from all storms, beyond the sphere of jealousies and intrigues that we call the world. I saw in the movement of the stars only the noble and sweet contemplation of the marvels of the universe.'[62] This was in part a disillusioned response to his treatment at the hands of what he regarded as the ruthlessly instrumentalist attitude of the new republican regime, but it avoided the fact that for four generations his family had acted in response to the demands of an absolutist monarch. From the foundations of the Académie des Sciences in the 1660s, the surveying and mapping of France by the Cassinis provided a direct response to the political and financial requirements of the rule of first Louis XIV and then Louis XV. Successive controllers-general saw the surveys and maps as a tool for the effective management of the state. From Colbert onwards, ministers demanded a new kind of geography that could help map transport networks, regulate provincial taxation, facilitate civil engineering works and support military logistics. The Cassinis responded – often brilliantly – to such needs, rather than developing their surveying methods from neutral, disinterested scientific speculation.

The results of their methods were not as accurate and comprehensive as has sometimes been claimed. The sheer physical difficulties confronted by the engineers attempting to take precise measurements in adverse conditions using cumbersome and often limited instruments meant that, even after the conclusion of three surveys and the virtual completion of the *Carte de Cassini*, the Napoleonic authorities still found discrepancies in the position of locations, the omission of recently created roads and the measurement of longitude and altitude. The surveying was also highly selective in what it recorded. Those who bought individual sheets to see their local areas complained that certain features such as farms, streams, woodland and even châteaux were missing, even though the state wanted a 'locational diagram of significant places'[63] for specific administrative purposes, including taxation. Even Cassini III admitted that '[t]he topography of France was subject to too many variations for it to be possible to capture it in fixed and invariable measurements'.[64] Paradoxically, the limitations of both the surveys and the incomplete *Carte de Cassini* proved to be among their most significant legacies, because they showed that any national survey was potentially endless. The accumulation of topographical data resulted in a vast complexity, which overwhelmed the initial geometrical skeleton of the first surveys. When we see the Cassinis' engraved maps failing to record new roads, canals, forests, bridges and innumerable other man-made changes to the landscape, we realize that the land never remains static for very long, whatever the scientific claims to measure and map it accurately.

Ultimately, the *Carte de Cassini* was more than just a national survey. It enabled individuals to understand themselves as part of a nation. Today, in a world almost exclusively defined by the nation state, to say that people saw a place called 'France' when they looked at Cassini's map of the country, and identified themselves as 'French' citizens living within its space seems patently obvious, but this was not the case at the end of the eighteenth century. Contrary to the rhetoric of nationalism, nations are not born naturally. They are invented at certain moments in history by the exigencies of political ideology. It is no coincidence that the dawn of the age of nationalism in the eighteenth century coincides almost exactly with the Cassini surveys and that 'nationalism' as a term was coined in the 1790s, just as the Cassini maps were nationalized in the name of the French Republic.[65]

29. Bernard van Orley's remarkable tapestry 'Earth under the Protection of Jupiter and Juno' (1525), showing the Portuguese King John and his Habsburg wife Catherine, and the extent of the king's seaborne empire.

30. Diogo Ribeiro's world map (1525), the first in a series supporting the Castilian claim to the Moluccas (visible in both the far left- and right-hand corners), and offering a new outline of the North American coastline.

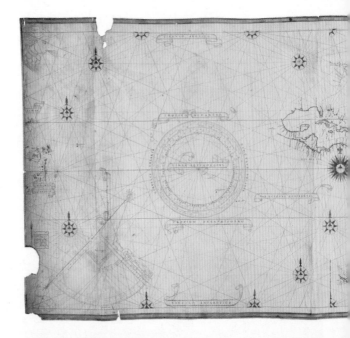

31. Ribeiro's third, and greatest, world map (1529), placing the Moluccas (again visible in both the far left- and right-hand corners) within the Castilian half of the globe in a brilliant act of cartographic manipulation.

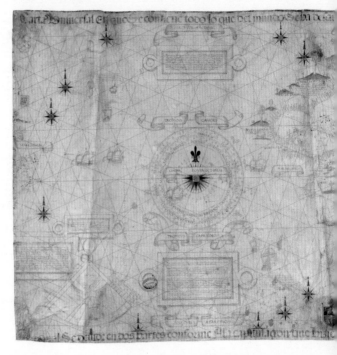

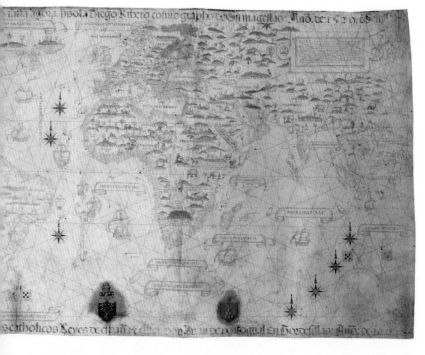

32. Gerard Mercator's map of the Holy Land (1538), showing surprising similarities to maps made by Luther's supporters.

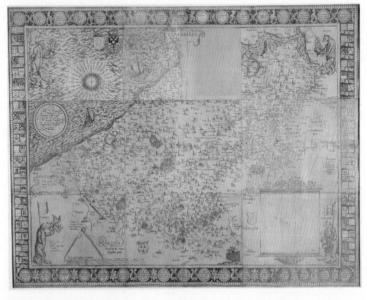

33. Gerard Mercator's incomplete wall map of Flanders (1539-40), a hasty attempt to head off a Habsburg occupation of Ghent in 1540, with unfinished sections.

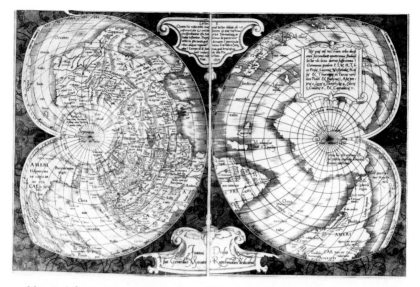

34. Mercator's first attempt at mapping the world (1538). The double cordiform ('heart-shaped') projection was only one of many alternatives available to him.

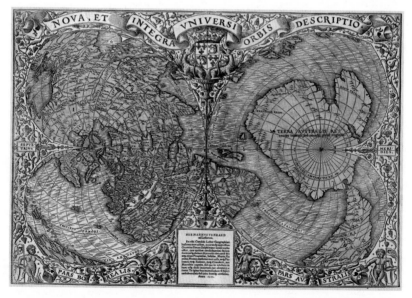

35. Oronce Finé's double cordiform world map (1531), copied by Mercator. Finé and many other mapmakers with occult and reformed religious beliefs chose this projection.

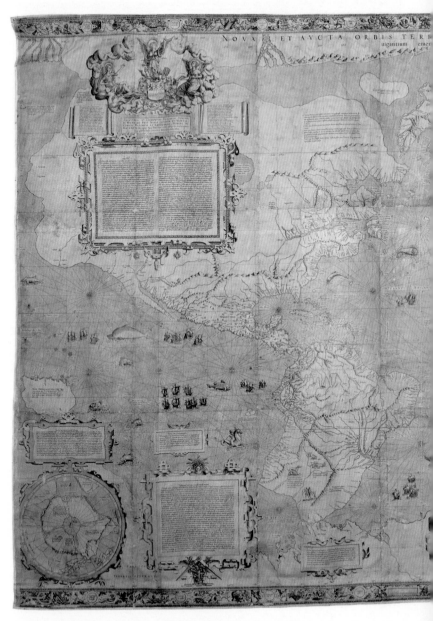

36. Gerard Mercator's map of the world on his famous 1569 projection.

NOVA TOTIVS TERR

40. Joan Blaeu's world map (1648), which celebrates the independence of the Dutch Republic and the global ambition of its East India Company. It also provides the first world map based on a heliocentric solar system. Just below the map's title, above where the hemispheres meet, is a diagram of the solar system labelled 'Hypothesis Copernicana', showing the earth revolving round the sun.

...RVM ORBIS TABVLA.

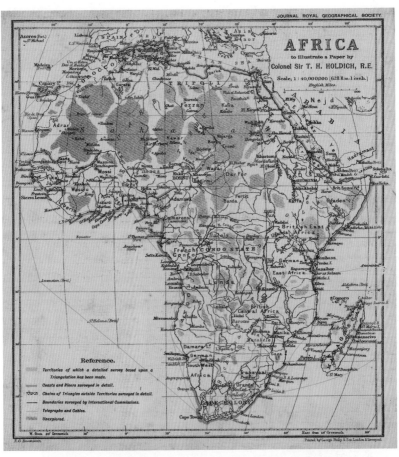

49. Colonel Sir Thomas Holdich's map of Africa (1901), showing the limitations of British imperial surveying in Africa. Red shows areas surveyed using triangulation, blue shows areas 'surveyed in detail'. Everywhere else, including large expanses of grey, is 'unexplored'.

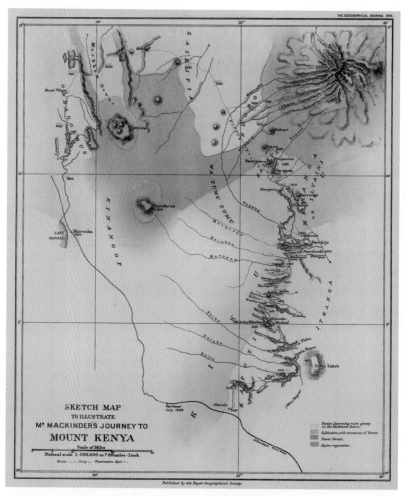

SKETCH MAP
TO ILLUSTRATE
Mr MACKINDER'S JOURNEY TO
MOUNT KENYA

Scale of Miles

Natural scale 1: 500,000 or 7·89 miles ·1inch.

*Published by the Royal Geographical Society.*

50. Halford Mackinder's map of his journey (in red) from Nairobi, at the bottom, to the summit of Mount Kenya, in the top right-hand corner (1900). To the west is Markham's Down, named after the Royal Geographical Society's president.

51. The first photograph of the whole earth, taken from space by the crew of Apollo 17 (1972), an iconic image of a fragile 'blue earth' that inspired the environmental movement.

52. A virtual world: the home page of Google Earth (2012).

53. An equal world? Arno Peters's world map on the Gall Orthographic Projection (1973).

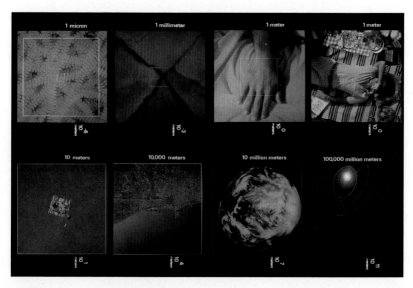

54. Early geospatial visualization: stills from *Powers of Ten*, a short film by Charles and Ray Eames (1968), a cult hit with computer engineers.

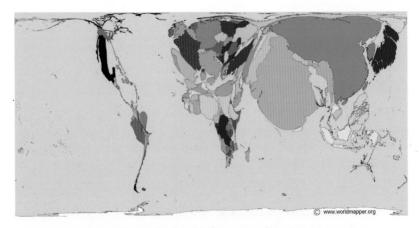

55. A cartogram showing the distribution of the human population in 1500 (2008). As the image of the world becomes increasingly familiar, demographic issues assume more importance than debates over geographical methods of projection.

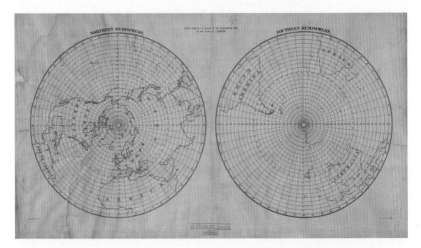

56. Index diagram of sheets for the proposed International Map of the World (1909) on the scale of 1:1,000,000.

In his classic study of the origins of nationalism, *Imagined Communities*, Benedict Anderson argues that the roots of national consciousness grew out of the long historical erosion of religious belief and imperial dynasties. As the certainty of religious salvation waned, the empires of the *ancien régime* in Europe slowly disintegrated. In the realm of personal belief, nationalism provided the compelling consolation of what Anderson calls 'a secular transformation of fatality into continuity, contingency into meaning'. At the level of political authority, the nation superseded the empire with a new conception of territory, where 'state sovereignty is fully, flatly and evenly operative over each square centimetre of a legally demarcated territory'. This is in direct contrast to empires, 'where states were defined by centres, borders were porous and indistinct, and sovereignties faded imperceptibly into one another'.[66]

The reasons for this shift lay in the transformation of vernacular languages and apprehensions of time. In the West, the rise of what Anderson calls 'print-capitalism' in the fifteenth century gradually signalled the ultimate decline of the 'sacred languages' of imperial and ecclesiastical authority, Greek and Latin, in favour of the vernacular languages spoken by a vast, new potential readership. The subsequent rise of the novel, the newspaper and the railway in Europe created a new perception of 'simultaneous' time, marked by 'temporal coincidence', and measured by the introduction of clocks and calendars. People began to imagine the activities of their nation taking place simultaneously across time and space, even though they were unlikely ever to visit or meet more than a tiny fraction of the places and people of which their nation is composed.

But in a typical example of what has been called the 'historian's strange aversion to maps',[67] Anderson initially failed to consider the most iconic of all manifestations of national identity. If changes in language and time 'made it possible to "think" the nation',[68] then a map's potential to alter perceptions of space and vision make it possible to visualize the nation. The *Carte de Cassini*, created during the same period that railways, newspapers and novels rose to cultural preeminence, was an image that allowed those who bought it to imagine the national space in one glance. Moving from their particular region to the nation as a whole, and reading the map in the standardized language of Parisian French (something which was standardized by the revolutionary authorities from the mid-1790s), the map's owners could identify with a topographical space and its inhabitants. As a result, the nation began the long and often painful process in the development of an administrative solidity and a geographical reality that helped to inspire an unprecedented emotional attachment and political loyalty from its subjects.

The Cassini surveys represented the beginnings of a new way of mapping a country, but its inhabitants needed an emotional attachment and political loyalty to something more than a geometrical triangle. Religion no longer provided the answer. Where Christ once presided over the map, looking downwards onto the world, the Cassini maps offered a horizontal perspective of the earth, from which every metre of territory (and by implication each of its inhabitants) had the same value. Political absolutism was also unsustainable. Despite its initial attempts to establish a way of mapping that could police and control the dynastic realm, the monarchy supported the map of a kingdom which unintentionally metamorphosed into the map of a nation.

Embedded in each and every one of its 182 sheets, the message of the *Carte de Cassini* would be easily appropriated by subsequent generations of national ideologues: one map, one language and one people, all sharing a common set of customs, beliefs and traditions. The *Carte de Cassini* presented its subjects with an image of a nation that was worth fighting for, and even dying for, in the endlessly repeated act of national self-sacrifice. It seemed a noble enough cause at the time, but the more intemperate consequences of such unwavering nationalism would be felt not only throughout France in the 1790s.

# 10

# Geopolitics

*Halford Mackinder, 'The Geographical Pivot
of History', 1904*

## London, May 1831

On the evening of 24 May 1831, a group of forty gentlemen met for
dinner at the Thatched House tavern in the St James's area of central
London. They were all experienced travellers and explorers, united by
their membership of one of London's growing number of private dining
societies: the Raleigh Travellers Club, named after the great Elizabethan
explorer Sir Walter Raleigh. The club had been founded in 1826 by the
traveller Sir Arthur de Capell Brooke and met every fortnight, when
each member took it in turns to provide a lavish banquet and impart
tales of travel and adventure. On this particular evening, the club's
dinner-card announced it had slightly different business. At a meeting
chaired by Sir John Barrow, Second Secretary to the Admiralty and himself a renowned traveller and statesman in China and South Africa, the
club's members 'submitted that, among the numerous literary and scientific societies established in the British metropolis, one was still wanting
to complete the circle of scientific institutions, whose sole object should
be the promotion and diffusion of that most important and entertaining
branch of knowledge, GEOGRAPHY'. It proposed that 'a new and useful
Society might therefore be formed, under the name of THE GEOGRAPH-
ICAL SOCIETY OF LONDON'.[1]

The club's members believed that the advantages of such a society
would be 'of the first importance to mankind in general, and paramount
to the welfare of a maritime nation like Great Britain with its numerous
and extensive foreign possessions'. It was therefore proposed that the
new society would 'collect, register and digest' all the 'new, interesting,
and useful facts and discoveries', 'accumulate gradually a library of the

of Janssonius and Blaeu had already questioned the validity of the cosmographer's task, but the impact of first Copernicanism and then Darwinism fatally compromised the traditional conception of cosmography as comprehending a universal image in a series of maps. As cosmography continued its decline at the beginning of the nineteenth century, so a new conception began to circulate that replaced it, and in the process offered a clearer description of the science of mapmaking: cartography. Karl Ritter, founder of the Berlin Geographical Society, first used the term 'Kartograph' in a paper written in 1828. Just a year later the French Société de Géographie began using the word 'cartographique'. In 1839 the Portuguese historian and politician Manuel Francisco de Barros e Sousa, viscount of Santarém, claimed to have coined the term in using the word 'cartographia'. Sir Richard Burton was the first Englishman to adopt it in 1859, on an RGS-sponsored expedition to explore the lakes of central Africa; 'cartographer' followed in 1863, and by the 1880s both words were firmly established in the lexicon.[7]

The rise of cartography gave the subjective act of mapmaking a degree of scientific expertise which enabled both its practitioners and political beneficiaries to represent it as a coherent discipline from which all geographical knowledge developed. It was increasingly regarded as an objective, empirical and scientifically verifiable field of study, separate from the extraneous disciplines of cosmography, navigation, surveying and astronomy with which it had been associated (and often subsumed by) for so many centuries.[8]

The idea was compelling, and it generated even greater advances within mapmaking. Developments in pure and applied mathematics inspired interest in map projections that went even further than the innovations of the sixteenth century. Between 1800 and 1899 an estimated fifty-three new map projections were proposed, more than three times the number developed throughout the eighteenth century. Mercator's projection, and the associated presumptions about projecting the globe onto a plane surface, was repeatedly challenged by a bewildering array of new mathematical projections which also responded to the need for medium- and small-scale maps to represent the increase in knowledge of the physical world. The combined study of calculus and geometry allowed mathematicians to propose increasingly complex projections that moved beyond the classical models of using cylinders

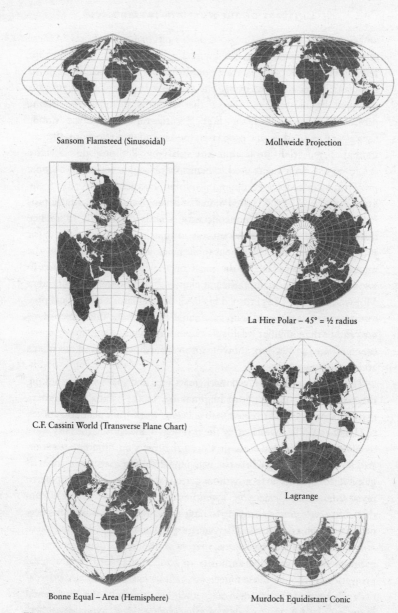

Sansom Flamsteed (Sinusoidal)

Mollweide Projection

C.F. Cassini World (Transverse Plane Chart)

La Hire Polar – 45° = ½ radius

Lagrange

Bonne Equal – Area (Hemisphere)

Murdoch Equidistant Conic

Fig. 29 Diagrams of eighteenth- and nineteenth-century map projections.

and rectangles to project the globe onto a piece of paper. Many of these new projections were proposed by amateurs motivated by a desire for self-promotion, but others were supported by geographical organizations and states eager to utilize the political and commercial insights provided by mapmaking. Those that endured included the Bonne projection, named after the French cartographer Rigobert Bonne (1727–95), a pseudo-conic projection used in topographic maps, the azimuthal perspective projection invented by Philippe de la Hire (1640–1718), which was used in hemispheric maps, and the polyconic projection created by Ferdinand Rudolph Hassler (1770–1843), the Swiss-born head of the United States's Survey of the Coast, whose projection used a series of nonconcentric standard parallels to reduce distortion, and which was so successful that it replaced the Mercator projection on official US topographic maps and coastal charts in the nineteenth century. One of the most prominent innovations came in 1805, when the German mathematician and astronomer Karl Brandan Mollweide (1774–1825) turned his back on Mercator's cylindrical projection to create a world map calculated to represent the depiction of area rather than angular fidelity. It became known as a pseudo-cylindrical equal-area projection, showing an oval earth with curved meridians and straight parallels.

These projections involved mathematicians and surveyors rethinking the possibilities and limits of mapmaking. In the 1820s the German mathematician Carl Friedrich Gauss began work on a geodesic survey of Hanover. While investigating the problem of measuring the curvature of the earth's surface, Gauss produced theorems on differential geometry, in which he argued that it was impossible to map the terrestrial globe onto a plane surface without serious distortion. He attempted to revise Mercator's projection, inventing the term 'conformality' (from the Latin *conformalis*, meaning to have the same shape), basing his new projections on correct shape around a particular point. Despite these and numerous other projections, there was no international geographical organization with the authority to adopt a standard geographical projection. Although most nineteenth-century atlases still used Mercator's projection on their world maps, their hemispherical and continental maps still drew from a variety of more than a dozen available projections.[9]

The consequence of all these changes was the emergence of a new genre, thematic mapping. A thematic map portrays the geographical nature of a variety of physical, and social, phenomena, and depicts the spatial distribution and variation of a chosen subject or theme which is usually invisible, such as crime, disease or poverty.[10] Although used as early as the 1680s in meteorological charts drawn by Edmund Halley, thematic maps developed rapidly from the early 1800s with the growth in quantitative statistical methods and public censuses. The development of probability theory and the ability to regulate error in statistical analysis allowed the social sciences to compile vast amounts of data, including national censuses. In 1801 France and England conducted censuses to measure and classify their populations. By the 1830s the Flemish astronomer Adolphe Quételet developed the statistical concept of the 'average man', inspiring 'moral' thematic maps measuring educational, medical, criminal and racial distributions.[11]

As well as contributing to the development of the social sciences, thematic maps also allowed the natural sciences to classify and represent data in a completely new way. Biology, economics and geology all exploited the new method to map the earth's atmosphere, its oceans and plant and animal life, as well as the land's surface. In 1815 William Smith combined geological analysis with statistical methodology to produce the first national thematic geological map of England, 'The Strata of England', and other scientists used these methods to create a new visual language of cartographic representation.[12] The rise of lithography led to lower costs and wider circulation. By the mid-1840s, printers in France were able to produce colour-printed lithographic maps of the geology of France at a cost of 3.5 francs each, in contrast to the usual engraved handcoloured copies, which cost 21 francs a copy.[13] Such maps were sufficiently cheap that they could be printed in their thousands rather than hundreds, creating a public market that dwarfed the circulation of Blaeu's *Atlas maior* or the *Carte de Cassini*.

Among these changes in mapmaking, most of which were brought about by individuals who did not even call themselves geographers, geography as a discipline found itself in ferment. Mapmaking's place within it seemed hopelessly confused, especially in Britain, where the inability of mapmaking to develop in an organized fashion had become a standard refrain in learned circles. As late as 1791, the president of the Royal Society, Sir Joseph Banks, complained that Bengal was mapped more accurately than England. 'I should rejoice could I say that Britons, fond as they are of being considered by surrounding nations as taking the lead in scientific improvements, could boast a general map of their island as well executed as Major Rennel's delineation of Bengal', a reference to James Rennel's East India Company-sponsored *Bengal Atlas* (1779).[14] Although 65 per cent of England had been surveyed at this time, the results were patchy. They lacked uniformity or standardization despite the formal creation of the state-sponsored Ordnance Survey in 1791, following Roy's initial survey work in 1784. Private maps of estates had been in use for centuries, but were usually made by local surveyors to serve the interests of landowners. As a result they used a variety of scales incompatible with the standardized aims of the Ordnance Survey, but which were often cheaper and more detailed. The prohibitive cost of a national survey meant that the Ordnance Survey left large tracts of land to be mapped by private surveyors. The result was a cartographic patchwork of uneven coverage.

In contrast to the Ordnance Survey's difficulty in providing standardized maps of England's complex and entrenched system of land ownership and management, the English East India Company assumed it would be much easier to survey overseas possessions like India by using new scientific techniques and simply ignoring local methods of mapping and owning land, notwithstanding the country's size. In the 1760s the company began providing financial support to individuals like Rennel for surveys that culminated in the Great Trigonometrical Survey of India. The survey was judged complete by 1843, but work carried on for decades, and, like the Cassini surveys, it has no decisive terminal date. In the words of Matthew Edney, the survey's most distinguished historian, the surveyors 'did not map the "real" India. They mapped the India they perceived and that they governed', and as a consequence created 'a *British* India'.[15] A similar process took place in Africa. When Joseph Conrad's protagonist Marlow peers at an imperial map in *Heart of Darkness* (1899), 'marked with all the colours of the rainbow', he is pleased to see 'a vast amount of red – good to see at any time, because one knows that some real work is done in there'.[16] In contrast to the French (blue), Portuguese (orange), Italian (green), German (purple) and Belgian (yellow) imperial possessions, the red patches of British dominion represent the pinnacle of Britain's civilizing imperial mission – at least for enthusiastic supporters like Conrad.[17] But as in the case of India, many of these maps showed imperial spheres of interest rather than direct colonial rule, little more than examples of the aspirational 'unofficial mind' of imperialism, which was driven by private initiatives like the RGS.

These organizations promoted a cartography that was more of an ideological projection based on apparently objective scientific principles than an administrative reality. Perhaps the most infamous example of Europe's use of mapmaking to lay claim to imperial territories is the Berlin Conference on Africa of 1884–5. It is still regarded as initiating the imperial 'scramble for Africa', with the assumption that its fourteen attending European powers proceeded to carve up the continent along the lines expressed in Conrad's *Heart of Darkness*. In fact, the minutes of the conference show it was convened to regulate European commercial access primarily in West Africa rather than partition the entire continent.[18] One British official expressed 'grave objections to [the conference's] definitions of the Congo which do not accord with geographical facts', while another protested that their geography was so confused that it was like drawing a map of the Rhine in the basin of the Rhône.[19] The conference did not produce any maps dividing Africa according to European interests, nor did it produce any binding statements on sovereignty, other than vague agreements to sanction subsequent claims to possession based on the principles of free trade, rather than political geography.

The RGS was particularly concerned with the haphazard nature of international mapmaking, especially in Africa. As late as 1901 Colonel Sir Thomas Holdich, a former surveyor of the Indian frontier and future president of the society, could publish an article in the society's journal bluntly entitled 'How Are We to Get Maps of Africa?'[20] Holdich complained that 'various surveys have been commenced in different parts of Africa under local administrations, which are unconnected with each other and have apparently no common basis of technical system or scale, from which it will be difficult eventually to compile a satisfactory and homogeneous first map of our African possessions'. He encouraged the adoption of more systematic mapping techniques such as common scales and base measurements, as well as the use of information gathered from local communities, or what he called 'native agency'. The map of Africa appended to Holdich's paper illustrated the problem: 6.75 million square kilometres of territory officially under British imperial control were still unmapped, and this figure excluded areas controlled by other European powers that still awaited survey. The map shows coastal areas of northern, eastern, western and southern Africa 'surveyed in detail', but the grey 'unexplored' regions overwhelmingly define the map, in stark contrast to the tiny red areas where 'a detailed survey based upon triangulation has been made'. Although political maps of the world might show nearly a quarter of its surface marked in British imperial red, physical maps of these regions told a far less convincing story of colonial mastery and domination.

Into this confused situation stepped an English academic named Halford Mackinder (1861–1947), who would almost singlehandedly both transform the study of geography in England, and create a whole new way of understanding and using the subject: geopolitics. Throughout the late nineteenth and early twentieth centuries Mackinder was one of the most influential figures in British academic and political life: one of the founders of the London School of Economics (1895), he was a Scottish Unionist Party Member of Parliament (1910–22), the British High Commissioner in Southern Russia (1919–20), and a keen amateur explorer, being the first European to climb Mount Kenya (1899). In 1920 he was knighted for his services as an MP, and in 1923 he became full professor in geography at the LSE.

Mackinder was born and educated in Gainsborough, Lincolnshire, where his interest in geography and politics began at an early age. Looking back over his life in 1943 at the age of 82, Mackinder recalled that 'my earliest memory of public affairs goes back to the day in September 1870 when, as a small boy who had just begun attendance at the local grammar school, I took home the news, which I had learned from a telegram affixed to the post office door, that Napoleon III and his whole army had surrendered to the Prussians at Sedan'.[21] At 9, Mackinder was already 'writing a history of the war in a notebook', as well as reading an account of Captain Cook's voyages, and delivering speeches to his family on geography, including one on Australia, which his father praised as 'delivery good, reception excellent'.[22] Such interests did not always endear him to his teachers. He later recalled that he 'had been caned at school for drawing maps instead of writing Latin prose'.[23] His boyhood games included being king of an island on which he 'civilised its usually backward inhabitants', and his adolescence coincided with the rise of British imperialism: in 1868 the Royal Colonial Society was founded, and in 1877 Queen Victoria was proclaimed Empress of India.

By the time he reached Oxford University in 1880, the belief in imperialism as a providential vocation was beginning to offer a viable alternative to the pursuit of organized religion, still reeling from the challenge of a variety of writers, most significantly the publication of Charles Darwin's evolutionary theories in *On the Origin of Species* (1859) and *The Descent of Man* (1871). As an Oxford undergraduate, Mackinder joined the Oxford *Kriegspiel* (or 'War Games') Society, which provided its members with training in military drill, manoeuvres and marksmanship. He also joined the Oxford Union, becoming its president in 1883, where he befriended some of the students who would formulate Britain's subsequent imperial policy. They included George Curzon (1859–1925), the future viceroy of India and Foreign Secretary, and Alfred Milner (1854–1925), who subsequently became High Commissioner for South Africa during the Boer War. Mackinder studied history and physical sciences, in the last of which he was influenced by Henry Moseley (1844–91), the Linacre Chair of Comparative Anatomy. Moseley had participated in the Challenger expedition (1872–6), an RGS-sponsored study of marine science that coined the term 'oceanography', and which discovered 4,717 new species on its 127,600 kilometre voyage around the world. Having been advised by Darwin, Moseley was a firm believer in evolutionary theory, but he also taught Mackinder the importance of geographical distribution: how geography affects biology in shaping the evolution of species.[24] This was a new kind of environmental determinism, which Darwin called 'that grand subject, that almost keystone of the laws of creation, Geographical Distribution'.[25]

Mackinder initially prepared to study international law at London's Inner Temple, and at the same time he also began to teach for the Oxford University Extension movement for adult education, aimed at widening educational access to those without the means to study at any of the established universities. Throughout the academic year 1886–7 Mackinder travelled hundreds of kilometres up and down the country, delivering lectures under the provocative series title 'The New Geography' in town halls and working institutes. He later recalled that he saw his task as that of 'gradually familiarising intelligent people throughout the country with the idea that geography consisted neither of lists of names nor of travellers' tales'.[26]

Having tirelessly championed the teaching of geography first in Oxford and then throughout the country, he went on to collaborate in the foundation of the Geographical Association in 1893, which was intended to address the absence of the study of human geography in schools. Just two years later his interest in reforming the study of geography alongside politics and economics led to his involvement in the birth of the London School of Economics, acting first as part-time lecturer in economic geography with the task of lecturing on 'Applications of Geography to Definite Economic and Political Problems', and then as the school's director from 1903 to 1908. Mackinder claimed that he was drawn to the school because it advocated 'the tearing to pieces of the old fashioned classical *a priori* political economy and the foundation of a group of specialists aimed at ascertaining the facts in the first place and then a generalisation from them in a really scientific spirit'.[27] He was appointed professor there in 1923 before his retirement in 1925. During this time he was also involved in setting up Reading University, acting as its principal from 1892 until it received university college status in 1903.

Mackinder also retained his affiliation with Oxford University, where his increasingly visible work represented a challenge to its dons. Mackinder knew that they were sceptical of the discipline of geography because of its novelty and its apparent lack of scientific rigour. These objections coalesced around the fact that rival universities in Paris and Berlin offered courses in geography, where its most famous proponents were Karl Ritter, the first professor of geography in Berlin, and Alexander von Humboldt (1769–1859), the great explorer and author of the enormously influential five-volume study *Cosmos: A Sketch of the Physical Description of the Universe* (published part-posthumously between 1845 and 1862). Humboldt's book singlehandedly redefined the possibilities of geography as a method of scientific enquiry, and its volumes offered no less than a complete account of the natural world and the physical universe.[28] As a result, Mackinder's lectures stressed the physical elements of geography, explaining how landscape, climate and environment acted upon and shaped human life. Today this approach to geography sounds obvious, even banal, but in the 1880s it was pioneering, and represented a bold attempt to convince the university authorities of the respectability of the subject as a science.

His lectures proved so successful that in 1887 the Royal Geographical Society invited Mackinder to present his ideas on geography to its Fellows. On 31 January, aged 25, Mackinder presented his first paper to the society. It was entitled 'On the Scope and Methods of Geography', a manifesto of Mackinder's new geography, although it took so long to deliver that the ensuing discussion was adjourned to the next meeting a fortnight later. The response to the talk was mixed, to say the least. Mackinder recalled that 'a worthy Admiral, a member of Council, who sat in the front row, kept on muttering "damn cheek" throughout the lecture'.[29]

Mackinder's opening question reflected the directness for which he was by then well known. 'What is geography?' he asked. He argued that there were two reasons for posing such a question. The first concerned 'the educational battle' being fought to enshrine the discipline within 'the curriculum of our schools and Universities', a battle which was, of course, being led by Mackinder. His second reason for posing the question was a direct challenge to the society. Geography was changing. 'For half a century,' he contended, 'several societies, and most of all our own, have been active in promoting the exploration of the world.' He continued, 'the natural result is that we are now near the end of the roll of great discoveries. The Polar regions are the only large blanks remaining on our maps. A Stanley can never again reveal a Congo to the delighted world.' Mackinder warned that 'as tales of adventure grow fewer and fewer, as their place is more and more taken by the details of Ordnance Surveys, even Fellows of Geographical Societies will despondently ask, "What is geography?"' In a cheeky swipe that probably incurred the wrath of the admiral in the front row, Mackinder conjured up the spectre of the society's closure unless it reformed, comparing it to 'a corporate Alexander weeping because it has no more worlds to conquer'.[30]

In the rest of his talk, Mackinder made a passionate call for geography, which he defined as 'the science whose main function is to trace the interaction of man in society and so much of his environment as varies locally', to be placed at the heart of English public and educational life. In attempting to unite physical geography with human (or what he called political) geography, Mackinder acknowledged the rival claims of history and the now wildly popular study of geology. 'Physical geography', he argued, 'has usually been undertaken by those already burdened with geology, political geography by those laden with history. We have yet to see the man who taking up the central, the geographical position, shall look equally on such parts of science and such parts of history as are pertinent to his inquiry.'[31] In pushing geography's case even further, Mackinder argued brusquely that 'the geologist looks at the present that he may interpret the past; the geographer looks at the past that he may interpret the present'.

What followed was an almost cosmographical survey of the earth's surface, starting with 'a geography of South-eastern England', and its chalk landscape, before swooping outwards even further to offer a god-like perspective across the entire surface of the earth. 'Imagine our globe in a landless condition,' Mackinder asked his audience, 'composed that is of three great concentric spheroids – atmosphere, hydrosphere and lithosphere [the outer shell of the earth].' At each turn he argued for the social and political development of a people, a nation, even a city, based upon its geographical environment. Building up his layers of geographically informed analysis, Mackinder insisted that 'everywhere political questions will depend on the results of the physical enquiry'. In concluding, Mackinder was clear about his ambitions for geography. 'I believe', he said, 'that on lines such as I have sketched a geography may be worked out which shall satisfy at once the practical requirements of the statesman and the merchant, the theoretical requirements of the historian and the scientist, and the intellectual requirements of the teacher.' It was a unification of what Mackinder called the scientific and the practical, and which, in one final claim that probably upset the admiral, he even suggested that geography might represent a 'substitute' for the study of the classics, and become 'the common element in the culture of all men, a ground on which the specialists could meet'.[32]

One of the society's councillors, the distinguished explorer and pioneering eugenicist Sir Francis Galton, responded with concerns about Mackinder's attempt to claim geography as a science. Nevertheless, he was sympathetic to the moves to adopt geography as an academic discipline, and remarked that, whatever the limitations of his paper, he was sure Mackinder 'was destined to leave his mark on geographical education'.[33] Galton knew more than he admitted: he was already in talks with the authorities at both Oxford and Cambridge universities to appoint an RGS-funded reader in the subject, a society aspiration that stretched back to the early 1870s, and had stage-managed Mackinder's invitation so that he would emerge as the most obvious candidate for any new post. On 24 May 1887, less than four months after Mackinder's talk, Oxford University agreed to establish a five-year Readership in Geography, supported by RGS funds. The following month Mackinder was formally appointed, on a yearly salary of £300.[34]

The creation of the new post was a huge coup for the RGS, which found a new mission to pursue, and a personal triumph for Mackinder. But the sceptics at Oxford were not so easily defeated. The subject was still not granted full degree status, and students attending Mackinder's lectures could only study for a one-year diploma. The results were predictable enough: after speaking around the country to halls of hundreds of people, Mackinder found that his first lecture at Oxford was not quite as popular. 'There was an attendance of three,' he recalled, 'one a Don, who told me that he knew the geography of Switzerland because he had just read Baedeker through from cover to cover, and the other two being ladies who brought their knitting, which was not usual at lectures at that time.'[35] Nevertheless, he struggled on, reporting back to the RGS at the end of his first year that he had delivered forty-two lectures on two courses. The scientific course, with its lectures on 'Principles of Geography', was less popular than the historical course, with its focus on 'The Influence of Physical Features on Man's Movements and Settlements'.[36] When in 1892 Mackinder came towards the end of his tenure and the Oxford authorities showed little interest in creating a full degree course in geography, he took up a post at the London School of Economics, where his interests turned increasingly towards politics and imperial adventure.

In September 1895, Mackinder gave a presidential address to the Geographical Association. His talk was entitled 'Modern Geography, German and English', and it provides a fascinating insight into his understanding of the evolution of geography and mapmaking throughout the course of the nineteenth century. Mackinder set out his case with typical fortrightness. 'As a nation,' he asserted, 'we may justly claim that for several generations we have been foremost in the work of the pioneer; nor need we view with dissatisfaction our contributions to precise survey, to hydrography, to climatology, and to biogeography.' Nevertheless, he continued, it 'is rather on the synthetic and philosophical, and therefore on the educational, side of our subject that we fall so markedly below the foreign and especially the German standard'.

Mackinder's concern was that, unlike German geographers, their English counterparts were unable to synthesize the practicalities of geographical research within an overarching theory of the discipline. 'What made the eighteenth century a transition age of such importance to geography', he believed, 'was the realization of new problems, which both Antiquity and the Renascence had either neglected or utterly failed to solve.' The great German geographers like Humboldt and Ritter had managed to overcome the age-old problem of seeing geography 'either as a discipline, or as a field of research'. The German philosophical tradition gave a very different perspective on the possibilities offered by the study of geography. Immanuel Kant's philosophical pursuit of a universal science, combined with Johann Wolfgang von Goethe (1749–1832) and Friedrich Schelling's (1775–1854) idealistic beliefs in a transcendental coordinating principle to explain nature, enabled Humboldt to celebrate geography as the greatest of all sciences, capable of synthesizing everything. The result was a school of geography that combined the scientific study of nature with an emotional response to its grandeur and beauty. In this tradition, August Heinrich Petermann (1822–78) established himself as one of Europe's most innovative cartographers, publishing a journal on new geographical studies, *Petermanns geographische Mitteilungen* ('Petermann's Geographical Communications', or 'PGM'); and Oscar Peschel (1826–75) and Ferdinand von Richthofen (1833–1905) pioneered geomorphology, the study of the form and evolution of the earth's surface. These German initiatives represented for Mackinder an 'exhaustive attempt to relate causally relief, climate, vegetation, fauna, and the various human activities' under the singular title of 'geography'.[37]

In lamenting the inadequacies of the English tradition, Mackinder produced a telling characterization of the role of maps in his new conception of geography:

There are three correlated arts (all concerned chiefly with maps) which may be said to characterize geography – observation, cartography, and teaching. The observer obtains the material for the maps, which are constructed by the cartographer and interpreted by the teacher. It is almost needless to say that the map is here thought of as a subtle instrument of expression applicable to many orders of facts, and not the mere depository of names which still does duty in some of the most costly English atlases. Speaking generally, and apart from exceptions, we have had in England good observers, poor cartographers, and teachers perhaps a shade worse than cartographers. As a result, no small part of the raw material of geography is English, while the expression and interpretation are German.

A map needed to offer more than just the empirical facts of observable place names: it required the expression and interpretation practised by German geographers of geomorphology, as well as 'biogeography' – the geography of organic communities and their environments – and 'anthropogeography' – the geography of men. For Mackinder, a map was not the territory it claimed to depict, but an interpretation of the geological, biological and anthropological elements which made up that territory.

In describing his 'ideal geographer', Mackinder observed:

In his cartographic art he possesses an instrument of thought of no mean power. It may or may not be that we can think without words, but certain it is that maps can save the mind an infinitude of words. A map may convey at one glance a whole series of generalizations, and the comparison of two or more maps of the same region, showing severally rainfall, soil, relief, density of population, and other such data, will not only bring out causal relations, but also reveal errors of record; for maps may be both suggestive and critical.

Unsurprisingly, the description of the ideal geographer was male, and bore a striking resemblance to the author. 'As a cartographer he would produce scholarly and graphic maps; as a teacher he would make maps speak; as an historian or biologist he would insist on the independent study of environment ... and as a merchant, soldier, or politician he would exhibit trained grasp and initiative when dealing with practical space-problems on the earth's surface.'[38]

It was another tour de force by Mackinder, insisting on the need for a 'modern geography'. Intellectual 'centralization' of geographical study was required in England to catch up with the German tradition, as a way of reiterating his belief that 'the geographical is a distinct standpoint from which to view, to analyze, and to group the facts of existence, and as such entitled to rank with the theological or philosophical, the linguistic, the mathematical, the physical, and the historical standpoints'.[39] It also anticipated an even more ambitious attempt to put Britain's geographers, and its explorers, at the forefront of international affairs.

In 1898 Mackinder hatched a plan to be the first European explorer to climb Mount Kenya in East Africa. Looking back on his decision in the 1940s, Mackinder admitted just how self-conscious his decision was to suddenly embark on a career as an explorer at the age of 37. 'To be generally regarded as the complete geographer,' he wrote, 'it was still necessary at that time for me to prove that I could explore as well as teach.'[40] His choice of Mount Kenya was based on a mixture of physical and political geographical considerations. Mackinder later wrote that it had become clear to him 'that when the Uganda railway had reduced the distance from the coast to Kenya by two-thirds, it should be possible, with no great expenditure of time, to convey a well-equipped expedition in a state of European health to the foot of the mountain, and that such an expedition would have a reasonable chance of completing the revelation of its alpine secrets'.[41] He wanted to get up the mountain before the railway brought other explorers, namely Germans, the great imperial rivals of the British in East Africa, and in particularly the German climber Hans Meyer, who had already scaled Mount Kilimanjaro, and announced in 1898 his intention of climbing Mount Kenya. The race was on between the two great imperial rivals in East Africa.

On 8 June 1899 Mackinder left England for Marseilles by train, where he met his team of six European guides and porters. On 10 June he set sail for Egypt, passing down the Suez Canal to first Zanzibar and then Mombasa, before arriving in Nairobi on the recently laid railway line in mid-June 1899. Here the expedition really began: 'we were six white men and our goods were carried on the heads of 170 natives, half of them stark naked, for at that time there were in East Africa no horses or draught oxen or mules, and of course no motor cars'. The 170 kilometre trek to the mountain was tough and delayed by various problems. 'The temper of the natives', according to Mackinder, 'was suspicious and dangerous', an accusation partly borne out by the murder of two of his Swahili guides, and the theft of most of their food just as they prepared their ascent of the mountain in late August.[42] Undeterred, Mackinder continued the climb, but had to abandon the ascent when the team ran out of rations. Having secured further supplies, he and two others set off again and spent a day scaling the mountain. It proved to be 'very steep and intensely hard', but finally, at noon on 13 September Mackinder reached the summit. He admitted that 'we dare, however, stay only forty minutes – time enough to make observations and to photograph', before the threat of storms forced them to descend. Mackinder only slightly overestimated the summit at 17,200 feet, or 5,240 metres (the actual height is 5,199 metres). It was an impressive physical achievement, matched by the precision of his scientific data. Mackinder returned home with 'a plane-table sketch of the upper part of Kenya, together with rock specimens, two route surveys along lines not previously traversed, a series of meteorological and hypsometrical observations, photographs by the ordinary and by the Ives colour processes, collections of mammals, birds, and plants, and a small collection of insects'.[43] Mackinder utilized Frederick Ives's new colour photographic techniques for the first time on a scientific expedition, and drew three beautiful maps of the mountain and his route, lithographically reproduced for his talk delivered at the RGS on 22 January 1900, two months after his return.

Fig. 30 Halford Mackinder on the summit of Mount Kenya, 1899.

The maps were classic examples of scientific imperial mapmaking. The first, illustrating Mackinder's journey, includes a scale of 1 : 500,000, a graticule, contour lines, and the route drawn in red. But it also shows the imprint of European exploration. His calculations were obtained by using a watch, a prismatic compass and a sextant. In the north-west Mackinder has named 'Markham Downs' in honour of the RGS's president, Sir Clements Markham, who co-sponsored the expedition. On the mountain itself 'Hausburg Valley' is named after the expedition's other sponsor, his wife's uncle, Campbell Hausburg. Mackinder also took the opportunity of leaving his own mark on the land: running north-east of the Hausburg Valley is 'Mackinder Valley'.

The news of the success of Mackinder's expedition was received with delight in London and dismay in Berlin. Returning home at the end of 1899, Mackinder immediately began writing up his exploits for presentation to the RGS Fellows in January 1900. What Mackinder later referred to as his 'holiday' in Kenya was greeted with unalloyed admiration by its vice-president Sir Thomas Holdich. On the evening of 22 January he introduced Mackinder as 'well known to all of us as a scientific geographer; tonight he comes before us as a most successful traveller, as the first man to ascend one of the principal peaks in East Africa, Mount Kenya'.[44] Having established himself as a distinguished geography teacher and an intrepid explorer, Mackinder was now sufficiently respected to define his subject as an intellectual discipline in late Victorian Britain, with the science of cartography at its centre. By announcing his new vision of geography as crucial to the protection of the British Empire (which went to war with the Boers in South Africa just as Mackinder returned from Kenya), his success was virtually assured, and his talk received none of the grumblings that met his previous efforts in the late 1880s.

The broader imperial aspirations of Mackinder's adventure also coincided with a decisive shift in his political views which propelled him into politics. Throughout the 1890s he still espoused a belief in international free trade, despite what he saw as the increasing threat that Germany posed to British manufacturing. But by the time he returned from Kenya, his beliefs were changing. In September 1900 Mackinder unsuccessfully contested a parliamentary seat in Warwickshire as a Liberal Imperialist. By 1903, increasingly swayed by the economic protectionist arguments of the Unionist Colonial Secretary Joseph

Chamberlain, he renounced free trade altogether and resigned from the Liberal Party to join the Conservatives, espousing a new theory of imperial protectionism based on a powerful British navy and tariffs to promote British overseas trade.[45]

But as a geographer, Mackinder's new political arguments presented him with a problem. How could his geopolitical thesis of imperial protectionism be represented on a map? He had already commented on the limitations of maps in showing basic topographical features such as relief. How were they to show his evolving world picture of economic protectionism and imperial authority? He addressed the problem at the height of his embrace of protectionism in *Britain and the British Seas*, published in 1902. In it he provided a familiar argument about how physical geography shaped the social world, but now with added political urgency. Geography had 'given to Britain a unique part in the world's drama', allowing it to become 'mistress of the seas' and develop a seaborne empire of unparalleled power and global authority.[46] But the changing balance of global power at the beginning of the twentieth century meant that such authority was now under threat.

In tracing the genealogy of Britain's seaborne power, Mackinder turned to maps. He began by examining the location of the British Isles on the Hereford *mappamundi*, to suggest that, before Columbus's voyages at the end of the fifteenth century, 'Britain was then at the end of the world – almost out of the world'. The subsequent discovery of America and opening of the Atlantic to its north, west and south, meant that 'Britain gradually became the central, rather than the terminal, land of the world'. But maps struggled to confirm his argument. 'No flat chart can give a correct impression of the North Atlantic,' he complained, as they only showed 'the mere lie of the coasts'. In a classic piece of egocentric geography, Mackinder blithely observed that the best way to understand Britain's new position on the globe after Columbus 'can best be realised by turning a terrestrial globe so that Britain may be at the point nearest the eye'. His illustrations of 'The Land Hemisphere' show the enduring problems of mapping the globe onto a plane surface – Australasia and half of South America have disappeared. In contrast a 'photograph' of the globe (an obvious misnomer in an age before space flight) clinches Mackinder's argument. On this image Britain is given a unique position, one from which 'the five historic parts of the world are accessible from its waters'.[47]

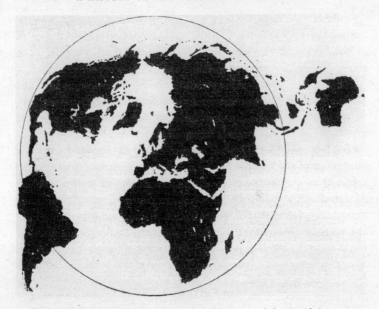

Fig. 31 'The Land Hemisphere', in Mackinder, *Britain and the British Seas*, 1907.

It was a brilliant act of global cartographic manipulation. By ignoring plane map projections of the earth and instead spinning his 'photographic' globe to place the British Isles at the centre, Mackinder was employing cartography in the service of a highly selective account of Britain's rise to seaborne and imperial dominance. Positioned on his map at the intersection of the great international maritime sea routes, but not connected to any continent, Mackinder claimed that Britain was 'possessed of two geographical qualities, complementary rather than antagonistic: insularity and universality'. It was '*of* Europe, yet not *in* Europe', allowing it to draw on the resources of the sea without the distraction of bordering neighbours.[48] But what made the empire great was also threatening to destroy it; without a renewed imperial drive to assimilate British colonies into a broader ideal of 'Britishness', its far-flung possessions risked absorption into the rising land-based empires of Russia, Germany and China. In the paternal language of so many imperialists, Mackinder ended his book by looking forward to a time when 'the daughter nations shall have grown to maturity, and the British Navy shall have expanded into the Navy of the Britains'.[49] It was an

Fig. 32 'Photograph of a Globe', in Mackinder, *Britain and the British Seas*, 1907.

almost mystical belief in the enduring power of the British Empire, and just two years later it would culminate in Mackinder's most famous and enduring theory.

On the evening of 25 January 1904, more than seventy years since its inception, the Royal Geographical Society opened the doors to its Fellows at its premises at 1 Savile Row in central London, to listen to yet another paper read by Mackinder. The society had led the way in funding and celebrating British imperial exploration since its foundation, supporting the colonial and missionary expeditions of public figures such as Sir Clements Markham, Dr David Livingstone, Sir Henry Morton Stanley and Robert Falcon Scott, as well as Mackinder's own adventure in Kenya. By the beginning of the twentieth century the society had turned its attention towards geography's more philosophical and educational dimensions, an interest that had already benefited individuals like Mackinder.[50] Its politically influential Fellows were also struggling with restoring the tarnished reputation of the British Empire following the disastrous Boer War (1899–1902), which had cost Britain more than £220 million, as well as the loss of 8,000 troops killed in action and a further 13,000 to disease. Of the estimated 32,000 Boers who died, the vast majority were women and children who died in British 'concentration camps', the first time such methods had been used in modern warfare. International condemnation had been virtually unanimous, and, in the face of Germany's aggressive policy of colonial expansion and armament, Mackinder's prediction of the British Empire's growing diplomatical isolation, military vulnerability and economic decline appeared increasingly prescient: though it had generated over 25 per cent of the world's trade in 1860, by the time Mackinder spoke the figure had dropped to just 14 per cent, with France, Germany and the United States quickly catching up.[51]

As a long-standing Fellow of the society, a successful explorer and now a passionate advocate of imperial protectionism, Mackinder was guaranteed a warm reception, but neither he nor his audience could have possibly anticipated the impact of his talk. The title of his paper was 'The Geographical Pivot of History'. It began by sketching a vast panorama of global history. He told his audience yet again that they were coming to an end of what he called 'the Columbian epoch', a 400-year period of intense seaborne exploration and discovery, in which 'the outline of the map of the world has been completed with approximate accuracy, and even in the polar regions the voyages of Nansen and Scott have very narrowly reduced the last possibility of dramatic discoveries'. This was a shrewd topical reference to Scott's first successful Antarctic expedition, funded by the RGS, remnants of which were struggling home even as Mackinder spoke. 'But the opening of the twentieth century', Mackinder continued, 'is appropriate as the end of a great historical epoch.' This was a moment, he believed, when 'the world, in its remoter borders, has hardly been revealed before we must chronicle its virtually complete political appropriation'. Foreshadowing twenty-first-century debates over the effects of economic and political globalization, Mackinder argued: 'Every explosion of social forces, instead of being dissipated in a surrounding circuit of unknown space and barbaric chaos, will be sharply re-echoed from the far side of the globe, and weak elements in the political and economic organism of the world will be shattered in consequence.'[52] For Mackinder, everything was connected, and the only way to trace these connections was through the society's and his own particular field of study: geography.

As far as Mackinder was concerned, to understand and even influence the changes which had recently come about in the world required a renewed geographical understanding of history and politics. 'It appears to me, therefore,' he continued,

> that in the present decade we are for the first time in a position to attempt, with some degree of completeness, a correlation between the larger geographical and the larger historical generalizations. For the first time we can perceive something of the real proportion of features and events on the stage of the whole world, and may seek a formula which shall express certain aspects, at any rate, of geographical causation in universal history.

He concluded: 'If we are fortunate, that formula should have a practical value as setting into perspective some of the competing forces in current international politics.'[53] This was a call not just for the importance of geography as an academic discipline that had characterized Mackinder's public pronouncements for years: it was now a demand for the discipline's insights to shape international diplomacy and imperial policy.

Having established the importance of geography, Mackinder then came to his central thesis. He claimed that, contrary to prevailing British imperial ideology, it was central Asia, or what he called 'Eurasia', which stood as 'the pivot of the world's politics'. Such a claim challenged the complacent assumptions of many of his audience, and Mackinder knew it. 'I ask you, therefore,' he requested, 'for a moment to look upon Europe and European history as subordinate to Asia and Asiatic history, for European civilization is, in a very real sense, the outcome of the secular struggle against Asiatic invasion.' It was a startling assertion, but one that Mackinder proceeded to defend through a vast, synoptic account of the physical geography of central Asia. It was a region, he argued, that throughout history had produced nomadic, warlike communities that repeatedly threatened the settled agricultural communities and maritime societies at the margins of the vast, landlocked plains of what he called 'Euro-Asia', which he described as

> a continuous land, ice-girt in the north, water-girt elsewhere, measuring 21 million square miles, or more than three times the area of North

America, whose centre and north, measuring some 9 million square miles, or more than twice the area of Europe, have no available water-ways to the ocean, but, on the other hand, except in the subarctic forest, are very generally favourable to the mobility of horsemen and camelmen.[54]

Moving towards the present, Mackinder asked: 'is not the pivot region of the world's politics that vast area of Euro-Asia which is inaccessible to ships, but in antiquity lay open to the horse-riding nomads, and is to-day about to be covered with a network of railways?' By this point, Mackinder was explicit about the kind of imperial map of the world he envisaged. 'Russia replaces the Mongol Empire,' he warned, and its 9,000 kilometres of railway from Wirballen in the west to Vladivostok in the east created the conditions for the mobilization and deployment of a vast military and economic machine that drew on such extensive landlocked natural resources that it would eclipse the seaborne power of maritime empires like the British. This, he predicted, 'would permit of the use of vast continental resources for fleet-building, and the empire of the world would then be in sight'. In a direct address to current British foreign policy, he warned that this might happen 'if Germany were to ally herself with Russia'.[55] These two great empires would effectively control the geographical pivot area of the entire world, stretching from western Europe to the Chinese Pacific coast and reaching as far south as central Persia and the borders of India. It was a timely observation. As Mackinder spoke, Japan was mobilizing its armies in response to Russia's imperial claims to Korea and Manchuria. Russian expansion in the Far East threatened British imperial interests in Hong Kong, Burma and even India. Two weeks after Mackinder delivered his paper, on 8 February, war broke out.[56]

Mackinder demonstrated his new world order by illustrating his talk with maps on lantern slides. After some regional maps of Eastern Europe and Asia, the later sections of the paper included a world map that provided a graphic explanation of Mackinder's argument, and which would come to be regarded as 'the most famous map in the geopolitical tradition'.[57] Entitled 'The Natural Seats of Power', the map shows three distinct zones. The first, the dotted pivot zone, covers most of Russia and central Asia and is exclusively landlocked (Mackinder makes the point by marking its northern extremes as abutting onto what he labels 'Icy Sea'). Beyond this zone are two concentric crescents. The first, labelled the inner or marginal crescent, is shown as partly continental,

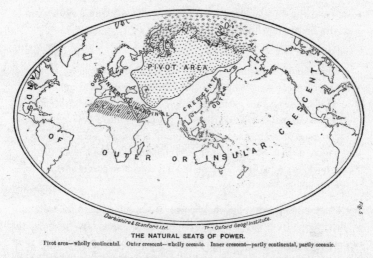

THE NATURAL SEATS OF POWER.

Pivot area—wholly continental.　Outer crescent—wholly oceanic.　Inner crescent—partly continental, partly oceanic.

Fig. 33 Halford Mackinder, 'The Natural Seats of Power', world map, in Mackinder, 'The Geographical Pivot of History', 1904.

partly oceanic, and is composed of Europe, North Africa, the Middle East, India and part of China. The outer or insular crescent, which was predominantly oceanic, included Japan, Australia, Canada, the Americas, South Africa and Britain.

Acknowledging the unfamiliar appearance of southern and northern America on the map's eastern and western borders, Mackinder argued that 'the United States has recently become an eastern power, affecting the European balance not directly, but through Russia, and she will construct the Panama Canal to make her Mississippi and Atlantic resources available in the Pacific [the United States had just been granted rights to begin work on the canal, which began four months later in May 1904]. From this point of view the real divide between east and west is to be found in the Atlantic Ocean.'[58] To an audience used to looking at a world map that placed the Americas in the western hemisphere, and which usually regarded the cultural and geographical division between east and west as falling somewhere in today's Middle East, Mackinder's map and the argument it supported was a shocking revelation, as was its implication for the military threat to the future of the British Empire.

Mackinder's remarks reflected his ambitions for the place of geography in political life. 'I have spoken as a geographer,' he said. But in what followed he suggested a new role for his academic profession in line with his changed political beliefs.

> The actual balance of political power at any given time is, of course, the product, on the one hand, of geographical conditions, both economic and strategic, and, on the other hand, of the relative number, virility, equipment, and organization of the competing peoples. In proportion as these quantities are accurately estimated are we likely to adjust differences without the crude resort to arms. And the geographical quantities in the calculation are more measurable and more nearly constant than the human. Hence we should expect to find our formula apply equally to past history and to present politics.[59]

For Mackinder, geography was the only discipline able to measure and predict the shifting balance of international politics. His 'formula' for understanding the geographical pivot of history was all that could limit what he regarded as the inevitable military confrontations or 'crude resort to arms' that would result from any significant shift in the balance of global power.

The audience's response to his paper was on this occasion decidedly mixed. The society's Fellows were not used to such sweeping conceptual arguments (which they tended to regard as the preserve of foreigners) and certainly not ones which, despite the prevailing political climate, suggested the British Empire was in imminent peril. His first respondent, Mr Spencer Wilkinson, lamented the absence of Cabinet ministers among the audience. They could learn, he suggested, from Mackinder's explanation that 'whereas only half a century ago statesmen played on a few squares of a chess-board of which the remainder was vacant, in the present day the world is an enclosed chess-board, and every movement of the statesman must take account of all the squares in it'. His scepticism towards 'some of Mr Mackinder's historical analogies or precedents' was shared by many other Fellows, who refuted his suggestion that the British Empire was under threat, insisting as Wilkinson did that 'an island state like our own can, if it maintains its naval power, hold the balance between the divided forces which work on the continental area'.[60] Surely, they thought, whatever the brilliance of Mackinder's argument, the empire's naval might was unassailable?

Wilkinson also worried about Mackinder's world map, 'because it was a map on Mercator's projection, which exaggerated the British

Empire, with the exception of India'.[61] It was indeed an odd map, but made absolute sense when compared with Mackinder's previous carto- graphic efforts. Alongside his Kenyan map presented to the RGS almost exactly four years earlier, the world map illustrated the transformations in geography and mapmaking that took place throughout the course of the nineteenth century. There are obvious differences between these two maps, but their overall approach to mapping and imperial policy is the same. The Kenyan map is a straightforward example of chorography – regional mapping – using standard cartographic conventions and symbols to chart the territory depicted and lay claim to it.

In contrast, the geographical pivot world map operates at a global level, and is extremely simplistic. After the great surveys of Cassini and the Ordnance Survey, Mackinder's map is noticeably bereft of the estab- lished attributes of regional or global cartography. Unlike the regional Kenyan map, there is no scale or graticule of latitude and longitude. It lacks even a basic toponymy: the oceans, countries, even continents are unlabelled, and for a map that sustains such an overtly political thesis, it is strange to see absolutely no divisions of territory along national, imperial, ethnic or religious lines. Even its peculiar oval shape was obso- lete, having been all but abandoned by cartographers since the sixteenth century. Although it drew on the shape of world maps like Mollweide's, which challenged Mercator, Mackinder still chose to use the 1569 pro- jection, even though the oval frame only amplifies the distortion which Mercator admitted affected his world map.

The world map also represented the culmination of the arguments rehearsed in *Britain and the British Seas*. The image Mackinder designed was in effect a thematic map, using highly charged political 'data' he had built up over the previous two decades as a teacher, explorer and politician. It drew on the physical and moral thematic maps that domi- nated so much nineteenth-century mapmaking to produce the foundational image of geopolitics, a compelling but ideologically loaded map of the world as a giant imperial chessboard. Some geographers would question its status as a map: it certainly stretched the definition of thematic mapping, using no verifiable data to argue its case. But its moralizing force was unquestionable, even though virtually every phrase on it was purely interpretative. Apart from its description of 'desert' and 'icy sea', Mackinder's image was composed of 'pivots' and 'crescents' that bore no relationship to any previous geographical language.

Like the images used in *Britain and the British Seas* that struggled to make his case for England's global position, Mackinder pushed the parameters of the 'map' as far as he could to give his thesis maximum graphic power and authority. Like Spencer Wilkinson, he knew the limitations of Mercator's projection. Nevertheless, he chose to use it because of its iconicity, and because it emphasized the eastern and western hemispheres in such a way that suited his imperial mentality: like Mercator, Mackinder had no interest in the North and South poles, which in Mercator are mapped to infinity, and in Mackinder's map are not even shown. By plotting the projection within an oval frame he could stretch his continents to show the range of his outer insular crescent, and present an image of a mutually interconnected world. This also allowed him to overcome the awkward plane maps and global 'photographs' created just two years earlier. The map which resulted looked both strikingly modern and strangely archaic. Though its argument was emblematic of a new geopolitical world order, the image itself is purely geometric, evoking the emblematic lines of imperial partition drawn across early sixteenth-century maps and globes as Spain and Portugal claimed to split the world in two, even though their influence stretched over only a fraction of the known earth. An even stronger visual and intellectual predecessor is a medieval map like the Hereford *mappamundi*, which Mackinder used as the first illustration in *Britain and the British Seas*. It was an image that Mackinder would return to in 1919, as he outlined his theory of the 'heartland' in *Democratic Ideals and Reality*. He described the Hereford *mappamundi* as 'a monkish map, contemporary with the Crusades', on which 'Jerusalem is marked as at the geometrical centre, the navel, of the world'. 'If our study of the geographical realities, as we now know them in their completeness, is leading us to right conclusions, the medieval ecclesiastics were not far wrong.' He concluded that '[i]f the World-Island be inevitably the principal seat of humanity on this globe . . . then the hill citadel of Jerusalem has a strategical position with reference to world realities not differing essentially from its ideal perspective of the Middle Ages, or its strategical position between ancient Babylon and Egypt'.[62]

To Mackinder, the Hereford *mappamundi* was defined not by theology, but by the geopolitics of the Crusades and the westward shift of empire from Babylon to Jerusalem. *Mappaemundi* were therefore simply an early confirmation of his central thesis: the enduring conflict between empires for control of a heartland. With the benefit of historical distance, we can see that Mackinder's 1904 map is in fact a manifestation of the same kind of ideological geometry that inspired the Hereford *mappamundi*: the providential mission of empire had replaced the pursuit of organized religion, but both of them sought to reduce the plurality and complexity of the world to a series of timeless truths. The belief now was that geography could disclose an ultimate reality upon which its creators could predict the political future. The two maps, made 700 years apart, look very different, but they were both inspired by the imperative to create a particular image of the world based on a prescriptive ideological geometry.

Throughout his life Mackinder returned to the geographical pivot thesis, revising it in response to the First and Second World Wars. In 1919 he published *Democratic Ideals and Reality: A Study in the Politics of Reconstruction*, written in the aftermath of the Armistice of the previous year and intended to influence the peace negotiations in Versailles. It modified the 'pivot' theory into an enlarged 'heartland', stretching from eastern Europe across central Asia. Mackinder warned against any diplomatic resolutions that allowed either Germany or Russia to take control of the 'heartland', and by implication the space he termed the 'World-Island', the conjoined space connecting Europe with Asia and North Africa. He summarized his argument in what would become one of the most infamous slogans of modern geopolitical thinking:

> Who rules East Europe commands the Heartland:
> Who rules the Heartland commands the World-Island:
> Who rules the World-Island commands the World.[63]

Following the outbreak of the Second World War, as the geopolitical world map changed once more, Mackinder had to amend his theory again. In July 1943, as the tide of the war turned in favour of the Allies, Mackinder published an article entitled 'The Round World and the Winning of the Peace'. His enduring fear of an alliance between Germany and Russia had finally happened in 1939 with the Nazi–Soviet Pact, although neither state ultimately dominated the 'pivot' or 'heartland'. Hitler's invasion of the Russian 'heartland' in 1941 was a further confirmation of Mackinder's thesis; its failure provided Mackinder with the basis of his plans for 'winning the peace' once the conflict ended.

The creation of a strong naval presence in the Atlantic and a dominant military power in central Asia would confront 'the German mind with an enduring certainty that any war fought by Germany must be a war on two *unshakable* fronts'.[64] It was a brilliant restatement of the strategic importance of the 'heartland' and a remarkably prescient account of the post-war geopolitical world of NATO and the Soviet Bloc that proposed a model of geopolitical checks and 'balances' in containing the inevitable post-war influence of the Soviet Union.

Anticipating the creation of NATO, Mackinder argued for the importance of a new transatlantic military alliance across the North Atlantic, or what he called 'the Midland Ocean'. It would involve 'a bridgehead in France, a moated aerodrome in Britain, and a reserve of trained manpower, agriculture and industries in the eastern United States and Canada'. On Mackinder's post-war world map, geopolitics would be reduced to an abstract geometrical ideal, in which a 'balanced globe of human beings' would be 'happy, because balanced and thus free'.[65] It was perhaps idealistic, but it prefigured the Anglo-American Cold War rhetoric that would come to dominate international politics for most of the second half of the century, and would influence subsequent US foreign policy towards what it regarded as the containment of the Soviet Union and South-east Asia. According to the political theorist Colin Gray, 'the most influential geopolitical concept for Anglo-American statecraft has been the idea of a Eurasian "heartland", and then the complementary idea-as-policy of containing the heartland power of the day within, not to, Eurasia. From Harry S Truman to George Bush, the overarching vision of US national security was explicitly geopolitical and directly traceable to the heartland theory of Mackinder'. Gray believes that 'Mackinder's relevance to the containment of a heartland-occupying Soviet Union in the cold war was so apparent as to approach the status of a cliché'.[66]

Although it is always difficult to identify precisely how ideas translate into direct policy, the pronouncements of a range of statesmen throughout the 1990s show the pervasiveness of Mackinder's thinking. In 1994 Henry Kissinger, the former National Security Advisor and Secretary of State to Richard Nixon and Gerald Ford, wrote that 'Russia regardless of who governs it, sits astride what Halford Mackinder called the geopolitical heartland, and is the heir to one of the most potent imperial traditions'. As late as 1997 Zbigniew Brzezinski, another former National Security Advisor, argued that 'Eurasia is the world's axial supercontinent', situated at the heart of a 'geopolitical chessboard'. He concluded that a 'glance at the map also suggests that a country dominant in Eurasia would almost automatically control the Middle East and Africa'.[67] Ostensibly, Mackinder's political geography was based on a stated desire to keep the peace. In reality, it was predicated on perpetual military conflict and international warfare, as the various pieces on his global chessboard vied with each other for increasingly scant resources. It also contributed to a post-war American geopolitical strategy that pursued military intervention both covertly and openly on nearly every continent across the globe.

Looking back in 1942 on the reception of Mackinder's original 1904 talk, the German political scientist Hans Weigert wrote that it must have 'seemed shocking and fantastic' to many Englishmen. However, by the time of his death in 1947 Mackinder's argument was firmly established as one of the most influential political theories of its time. Some of the most famous (and reviled) politicians of the twentieth century drew on his ideas, from George Curzon and Winston Churchill to Benito Mussolini. The German academic Karl Haushofer (1869–1946) adopted Mackinder's ideas in the development of the Nazi theory of geopolitics, which he regarded as 'geography in the service of world-wide warfare'.[68] Haushofer was a close friend of Rudolf Hess, deputy leader of the Nazi Party; Hitler's speeches throughout the 1930s on the Russian threat to Germany repeatedly used the language of Mackinder.[69] The geographical pivot even resonated throughout George Orwell's 1948 novel, *Nineteen Eighty-Four*, with its world divided into three great military powers, Oceania, Eurasia and Eastasia, perpetually at war with one another in an attempt to resolve the enduring conflict described by Mackinder between oceanic and landlocked states. By 1954, seven years after Mackinder's death, the prominent American geographer Richard Hartshorne argued that Mackinder's original model was 'a thesis of world power analysis and prognosis which for better or worse has become the most famous contribution of modern geography to man's view of his political world'. Coming from the founder of the geography division of the Office of Strategic Services (OSS), the wartime predecessor of the Central Intelligence Agency (CIA), this was praise indeed.[70]

Mackinder's argument had not just aspired to transform the status of geography as an academic discipline: he had effectively defined a whole new field of study in the English-speaking world – geopolitics, although the 1904 lecture never actually used the term. Variously defined as 'an attempt to draw attention to the importance of certain geographical patterns in history', 'a theory of spatial relationships and historical causation' and 'the study of international relations from a spatial or geographical perspective',[71] geopolitics has now become a ubiquitous part of our political vocabulary. The first person to use the term was the Swedish politician and social scientist Rudolf Kjellén (1864–1922), who in 1899 defined it as 'the theory of the state as a geographical organism or phenomenon in space'.[72] In the United States, the naval strategist Alfred Mahan (1840–1914) was also developing a similar geopolitical vocabulary. In his book *The Influence of Sea Power upon History* (1890), Mahan advocated 'the use and control of the sea' in response to what he regarded as the threats faced by the United States' 'weakest frontier, the Pacific'.[73] In 1902 he also coined the term the 'Middle East' in a paper on the 'Persian Gulf and International Relations'.[74] In Germany, the geographer Friedrich Ratzel (1844–1904) was also developing a geopolitical theory based on the expansion of the German state. In his *Political Geography* (1897), Ratzel argued that the struggle for human existence was a perpetual fight for geographical space. The 'conflicts of nations' he wrote, 'are in great part only struggles for territory'.[75] In his 1895 lecture on 'Modern Geography', Mackinder greatly admired Ratzel's 'anthropogeography', but it was also based on the superiority of the German race. Ratzel would extend his arguments into the theory of national struggles for 'living space', or *Lebensraum*, which Hitler believed provided the justification for much of his foreign policy throughout the 1930s, and which ultimately led to the outbreak of war in 1939.[76]

Each of these writers, Mahan and Ratzel in particular, developed a theory of geopolitics justifying the apparent inevitability of global warfare. They all had an effect upon their native country's foreign policy, but it was Mackinder's formulation of geopolitics that would have the greatest impact. And at the heart of his theory sat a world map, endlessly reproduced by subsequent geographers and politicians, giving a graphic shape and form to the idea of geopolitics. The terms Mackinder's associates and their followers contrived – 'heartlands', 'Middle East', 'iron curtain', 'third world', and the more recent 'evil empire' and 'axis of evil' – are all examples of the ideologically loaded language of geopolitics. At the beginning of the twentieth century, such ideas were still only implicit in either geography or politics. It was Mackinder's great achievement to change all that, and in the process to play a part in establishing both modern geography and mapmaking's relationship to politics and empire. To judge from the volume of recent academic publications on both Mackinder and the geopolitical research he inspired, it is a legacy with which geography is still coming to terms today.[77]

In April 1944, as Allied forces made preparations for the invasion of Normandy, Mackinder, then aged 83, was awarded the Charles P. Daly Medal for services to geography at the American Embassy in London. Delivering an address to the ambassador, he reflected on the extraordinary influence of his talk on 'The Geographical Pivot of History':

> I am grateful to you, in the first place, for the testimony you have borne to my loyalty to democracy, since absurd as it may seem I have been criticised in certain quarters as having helped to lay the foundations of Nazi militarism. It has, I am told, been rumoured that I inspired [Karl] Haushofer, who inspired Hess, who in turn suggested to Hitler while he was dictating *Mein Kampf* certain geo-political ideas which are said to have originated with me. Those are three links in a chain, but of the second and third I know nothing. This however I do know from the evidence of his own pen that whatever Haushofer adapted from me he took from an address I gave before the Royal Geographical Society just forty years ago, long before there was any question of a Nazi Party.[78]

Mackinder was understandably horrified at the implication that his geopolitical ideas influenced the rise of Nazism and Europe's descent into world war. The connection was not inevitable – but it is understandable. Mackinder's ultimate legacy was to ensure that during his lifetime the study of geography was established as what has been called 'the science of imperialism *par excellence*',[79] and out of this marriage of geography and imperialism geopolitics was born. In contrast to Nazi or Soviet ideologues, Mackinder never incited conflict or open warfare in his writings, but they were based on the inevitability of imperial conflict over terrestrial space and the need to exert force in the maintenance of political authority or, to use his own language, 'the winning of the peace'.

Mackinder's 1904 map represented the ultimate version of a globe seemingly bereft of collective agency, where the messy reality of the world is reduced to enduring warfare between cultures for ever determined by their physical location and quest for increasingly scarce resources. It was an indispensable part of Mackinder's extraordinarily successful mission to elevate the study of geography to a hitherto unknown stature, and situate it within the cartographic imagination of international political relations. But it was a double-edged legacy. The impact of decolonization after the Second World War has slowly led geographers and mapmakers to question the ease with which their discipline surrendered to the established political powers. Although many reaped the benefits of Mackinder's legacy, others became deeply uncomfortable about the enhanced authority of geography.

The world view of Mackinder's map continues to influence foreign policy across the globe. In an article written for the US Army War College's journal *Parameters* in the summer of 2000 entitled 'Sir Halford Mackinder, Geopolitics and Policymaking in the 21st Century', Christopher Fettweis argued that 'Eurasia, the "World Island" to Mackinder, is still central to Amercian foreign policy and will likely continue to be so for some time'. Today, as Fettweis points out, the 'heart of the Heartland is floating on top of a sea of oil'.[80] The first Gulf War of 1990–91 is already regarded by many political observers as the first of a series of 'resource wars' launched to ensure US control over global oil supplies. Writing in the *Guardian* newspaper in June 2004, Paul Kennedy, a distinguished professor of history at Yale University and an expert on Mackinder, wrote that '[r]ight now, with hundreds of thousands of US troops in the Eurasian rimlands and with an administration constantly explaining why it has to stay the course, it looks as if Washington is taking seriously Mackinder's injunction to ensure control of "the geographical pivot of history"'.[81] It is a disturbing fulfilment of Mackinder's original predictions, and current US involvement in the Gulf shows it will not be the last international conflict over increasingly scarce physical resources. It is a sobering reminder that, although Mackinder's world map is virtually obsolete, the world view that it expressed continues to affect people's lives right across the globe.

# I I

# Equality

*The Peters Projection, 1973*

*India, 17 August 1947*

In June 1947 the British Government commissioned Sir Cyril Radcliffe, a lawyer and former Director-General of the Ministry of Information, to travel to India for the first time in his life to produce a report partitioning the subcontinent. His mission was to divide the country along religious lines, separating Hindus from Muslims in the creation of India and Pakistan. Over just three months the Radcliffe Boundary Commission was required to create a 6,000 kilometre geographical boundary dividing 90 million people living in a region covering over 400,000 square kilometres. Without any experience of India, and with no inclination to commission updated geographic surveys or revised boundary demarcations, Radcliffe set about using outdated census reports in partitioning the country to 'demarcate the boundaries of the two parts of the Punjab on the basis of ascertaining the contiguous majority areas of Muslims and non-Muslims'.[1] His so-called Award was published on 17 August 1947, just two days after the official declaration of the independent states of India and Pakistan. The Indian artist Staish Gurjal remembered the chaos involved in communicating news of the partition. 'Curiously,' he recalled, 'the news of such magnitude was conveyed to us not by newspapers (which had ceased publication) but by posters pasted on the walls.'[2] The consequences of Radcliffe's map of partition were swift and disastrous. It sparked the largest migration in history, with between 10 and 12 million people moving across the newly established borders of Punjab and Bengal. The new border areas descended into bloody violence, with as many as a million people murdered in communal massacres.[3]

Radcliffe's Award satisfied nobody. The mainly Islamic Kashmir joined India, while Muslim minorities remained, and by late 1947 India and Pakistan were at war over the contested borders. Further wars followed in 1965 and 1971, and the tensions between the two states continue to this day, although now with the added threat of nuclear confrontation. Never before had the drawing of a line on a map led to such terrible human consequences.

The catastrophic geographical partition of India was a logical, if not necessarily inevitable, consequence of the ambitious but incomplete mapping projects of the eighteenth and nineteenth century, and their preoccupations of nation-building and imperial expansion. In France, as we have seen, several generations of the Cassini family had created ambitious but imperfect mapping techniques that played a part in shaping a distinctly French national consciousness. Their cartographic methods were soon adopted throughout Europe as the political geography of the continent slowly evolved from a group of disparate empires and monarchies into a series of sovereign nation states. In England, the gulf between the claims to cartographic practice and their reality in the administration of Britain's imperial dominions in Africa, India, South Asia and the Middle East meant that any partition of a country like India would inevitably lead to conflict. The legacy of Mackinder's geopolitical version of a world order, underpinned by imperialism and vividly illustrated by his infamous 1904 world map, showed how mapmaking could be appropriated by a range of political ideologies with little interest in its claims to scientific objectivity and impartiality.

The ease with which political power used cartographic expertise is a recurrent theme of twentieth-century history. As the century progressed and Europe descended into global conflict, maps became more explicitly politicized than ever before, and in some cases transformed into servants of what is now very familiar political propaganda. Even before the outbreak of the Second World War, the Nazis had grasped the power of maps to convey their political message. An infamous map of 1934 purports to show the danger posed to German sovereignty by Czechoslovakia, a manufactured threat that would ultimately provide the pretext for the Nazi invasion in March 1939. Lacking a proper scale or toponymy, the image hardly qualifies as a map in a technical sense, but its use of light and shade creates a contrast between the passive, blank space of Germany and the more menacing outline of Czechoslovakia.

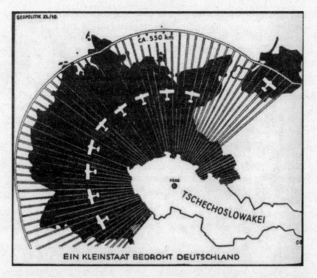

Fig. 34 'A small state threatens Germany!' Propaganda map, Germany, 1934.

The crude approximation of a fan-shaped graticule suggests the threat of airborne bombing (despite the minute size of the Czech air force). As one commentator wrote during the Second World War, in propaganda maps such as these, 'geography as a science and cartography as a technique become subservient to the demands of effective symbol manipulation'.[4] Although crude in its execution and message, this map exemplified the systematic political distortion of German maps and geographical textbooks throughout the 1930s, as the racial and ethnic message of Nazism appropriated the supposedly objective and scientific methodology of geography.[5]

The process of cartographic manipulation reached new and tragic heights during the Second World War, when the Nazis used maps in the pursuit of their 'Final Solution', the systematic mass murder of European Jews. In 1941 Nazi officials drew up an ethnic map of the puppet state of Slovakia based on official statistics of population distribution based on ethnicity. The map is a highly accurate representation of Slovakia, but its clusters of black circles betray its more sinister function: they depict the location of Jewish ('Juden') and Gypsy ('Zigeuner') communities. Labelled 'For official use only', this map was used with the

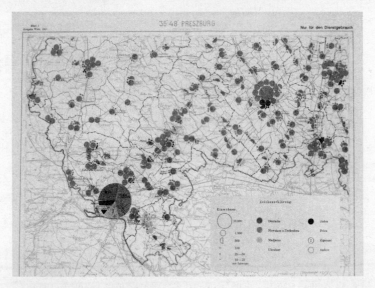

Fig. 35 Ethnic map of Slovakia, 1941.

support of the sympathetic Slovak authorities to round up Jews and Gypsies the following year, who were deported to extermination camps, where most met their death.

The appropriation of maps throughout the Second World War quickly translated into the political brinksmanship of the Cold War, exemplified by *Time* magazine's map of 'Red China' published in 1955. The illustration suggests the global stakes of the post-war military confrontation between the Soviet Union and the United States, with its depiction of China, Japan, Korea and Vietnam, with American possessions in the Pacific vulnerably positioned in the foreground. It mimics cartographic 'accuracy' in implying a geopolitical subtext which Mackinder would undoubtedly have understood: the fear of the spread of 'Red' Communism throughout South-east Asia and the threat to American interests in the Pacific.

As Cold War strategists on both sides of the ideological divide used 'persuasive cartography' to play on the anxieties of their fearful publics, geography also inevitably found itself tracing the collapse of European imperial dominion in Africa and South Asia. Having imposed arbitrary cartographic lines dividing ethnic, linguistic and tribal groups across entire continents like Africa in the nineteenth century, the former colonial powers were required to unravel these prescriptive cartographies in the post-war period of decolonization. The results, as in India's case, were rarely convincing, and often fatal for those who found themselves, literally, on the wrong side of a line.

The impact of political influence and manipulation on mapmaking also led to new developments to its medium, which could sometimes lead to different, more positive perspectives on the world. One of the most momentous shifts in twentieth-century perceptions of the earth began on 7 December 1972, when the three astronauts on board NASA's Apollo 17 spacecraft took a series of photographs of the earth with a handheld camera. One of the photos, taken at more than 33,500 kilometres above the earth's surface, was released by NASA following the safe return of the mission on Christmas Day. It became one of the most iconic images of not only a new age of space travel and exploration, but also of the earth itself. Since the time of Ptolemy, earthbound mapmakers had speculated and projected imaginative visions of the appearance of the world as seen from space. Historically, most map projections adopted such a perspective. But implicit in such projections was the assumption that no human would ever actually witness the earth from such a position. Now, for the first time, the whole earth, the subject of the study of geography since its inception, was finally captured for all to see, not on a map or through the skills of a mapmaker, but by a photograph taken by an astronaut.[6]

The Apollo 17 photograph, in its depiction of both the sublime grandeur and exquisite beauty of a singular blue world floating in the dark abyss of empty, inhospitable space, inspired wonder and also indignation at the state of 'our' world. The language of religious awe that accompanied the photograph's reception was quickly superseded by political and environmental reflections on the fragility of a world that united all its inhabitants, regardless of creed, colour or political orientation. The impact of the image found its way into 'The Brandt Report', a commission chaired by the former West German Chancellor Willy Brandt, which was published in 1980 to address the problems of economic development between the northern developed world and the southern developing nations. The report's authors wrote that 'from space, we see a small and fragile ball dominated not by human activity and edifice but by a pattern of clouds, oceans, greenery and soils.

Humanity's inability to fit its doings into that pattern is changing planetary systems fundamentally.'[7] Indeed, the whole-earth photograph had a significant influence on the growth of thinking about environmentalism and climate change. As this is the only world we possess, reasoned this new strand of ecological thinking, we had better look after it, and transcend our petty, earth-bound disputes in favour of a more holistic approach to the environment. It also had an impact on James Lovelock, who was developing his 'Gaia' hypothesis of the earth as a self-sustaining organism when he worked for NASA in the 1960s (but did not publish it until 1979), and gave new impetus to the Canadian thinker Marshall McLuhan's invention of the idea of a 'global village' in the early 1960s. Such sentiments echoed the transcendent global image that ran throughout the history of mapmaking, from Ptolemy through Macrobius to Mercator, although now with an added political urgency.

A further consequence of the Apollo earth photographs was their impact on global cartography. If it was now possible to photograph the whole earth rather than produce partial maps of its surface based on unsatisfactory projections, who needed mapmaking at all? One answer was of course that photographs from space were still limited to showing the earth as a disc, not a globe or a map on a plane surface (and the Apollo 17 photograph was centred on eastern Africa and the Persian Gulf, with no sign of the Americas or the Pacific Ocean). Another would be provided by the rapid improvement of Geographic Information Systems (GIS) which merged aerial and satellite photographic imagery with electronic database technologies to begin the rise of online mapping, examined in the final chapter.

Less than six months after the release of the Apollo 17 earth photographs, a world map was unveiled in Germany that claimed to turn its back on the selective political mapmaking of the twentieth century and to present an image of the world that promised equality to all nations. In May 1973, the German historian Arno Peters (1916–2002) called a press conference in Bonn, then capital of the Federal Republic of West Germany. In front of an assembled gathering of 350 international reporters, Peters announced a new map of the world based on what he called the Peters Projection. It was an immediate sensation, and it quickly made international front-page news. In the United Kingdom, the *Guardian* newspaper ran a story entitled 'Dr Peters' Brave New World', heralding the new map and its mathematical projection as 'the most honest projection of the world yet devised'.[8] *Harper's Magazine* even went as far as to run an article on Peters's projection entitled 'The Real World'.[9] For those who first saw the map in 1973, its novelty lay in its appearance. To those used to Mercator's projection, the northern continents appeared radically reduced in size, while Africa and South America took on the appearance of enormous teardrops sliding down towards Antarctica, or as one reviewer infamously put it, 'the land-masses are somewhat reminiscent of wet, ragged, long winter underwear hung out to dry on the Arctic Circle'.[10]

Peters claimed that his new world map offered the best alternative to the 400-year-old hegemony of Mercator's 1569 projection, and the sup-posedly 'Eurocentric' assumptions that lay behind it. In unveiling his map, Peters believed that the 'usual' map of the world by his German-speaking forebear, with which his audience were so familiar, 'presents a fully false picture particularly regarding the non-white-peopled lands', arguing that 'it over-values the white man and distorts the picture of the world to the advantage of the colonial masters of the time'. In explaining the technical innovations of his own map, Peters pointed out that Mer-cator put the equator nearly two-thirds of the way down his map, effectively placing Europe at its centre. On Mercator's projection the land masses were subject to distortion, leading to an inaccurate increase in the size of Europe and the 'developed' world and a subsequent decrease in the size of what Peters called 'the third world', in particular Africa and South America. Peters insisted that his own map provided what he called an 'equal area' projection that accurately retained the 'correct' dimen-sions of countries and continents according to their size and area. It therefore rectified what he regarded as the Eurocentric prejudice of Mer-cator and offered 'equality' to all nations across the globe.[11]

The impact of Peters's projection and his attack on Mercator was extraordinary. Over the next two decades it became one of the most popular and bestselling world maps of all time, rivalling the American cartographer Arthur Robinson's 1961 projection reproduced in the international bestselling Rand McNally and National Geographic Soci-ety's world atlases, and even Mercator's ubiquitous projection. In 1980 it adorned the cover of the Brandt Report, and in 1983 it appeared in English for the first time, in a special issue of the global devel-opment magazine *New Internationalist*. Praising what it called a 'remarkable new map', the magazine reproduced Peters's claims that

Mercator's map 'shows the ex-European colonies as relatively small and peripheral', while his own map 'shows countries according to their true scale', which, it believed, 'makes a dramatic difference to the portrayal of the Third World'.[12]

In the same year the British Council of Churches distributed thousands of copies of the map, which was also endorsed by OXFAM, Action Aid and more than twenty other agencies and organizations. Even the papacy praised its progressive agenda. But the United Nations was the most passionate advocate of Peters's map. UNESCO (the Educational, Scientific and Cultural wing of the organization) adopted it, and UNICEF (the United Nations Children's Fund) issued an estimated 60 million copies of the map under the slogan 'New Dimensions, Fair Conditions'. The map was so successful that Peters issued a manifesto in German and English outlining his approach. It was published in English in 1983 as *The New Cartography*, and was soon followed in 1989 by *The Peters Atlas of the World*. More than 80 million copies of the map have probably now been distributed across the world.[13]

But if the media and progressive political and religious organizations quickly accepted the map and Peters's cartographic methods, the scholarly community reacted with horror and disdain. Geographers and practising cartographers queued up to launch a bitter and sustained attack. The projection's claims to greater 'accuracy', they countered, were inaccurate: Peters, untrained in cartography, lacked an understanding of the basic principles of map projection; as a target, Mercator was a straw man, his influence unnecessarily overstated; Peters's skilful marketing of his map and the subsequent atlas looked like someone cynically exploiting an ignorant public to promote his own personal and political ends.

This response, even by academic standards, was vicious. In one of the first English language reviews of Peters's projection, published in 1974, the British geographer Derek Maling condemned it as 'a remarkable act of sophism and cartographic deception'.[14] Another British geographer, Norman Pye, dismissed the publication of Peters's *Atlas* as 'absurd', and complained that 'only the cartographically naïve will be deceived and fail to be exasperated by the pretentious and misleading claims made for the atlas by the author'.[15] Reviewing *The New Cartography*, the prominent British cartographer H. A. G. Lewis wrote that '[h]aving read this book many times in German and in English, I still marvel that the author, any author, could write such nonsense'.[16]

The most damning review of Peters's projection came from Arthur Robinson. In 1961 Robinson had created a new projection with the explicit aim of offering a compromise between conformal and equal-area projections. He used evenly spaced, curving meridians which did not converge onto a single point, limiting distortion at the poles, which allowed for a relatively realistic representation of the whole earth as a globe. The projection is also known as orthophanic (from the French for 'correct speaking'), although Robinson's colleague John Snyder captured its inherent compromises when he described it as providing 'the best combination of distortions'.[17] Nevertheless, with the backing of the Rand McNally publishing house and the National Geographic Society, millions of copies of the projection were circulated, and it finally eclipsed Mercator's as the most popular and widely distributed map of the world. Reviewing Peters's work in 1985, Robinson was unsparing in his attack on his German rival. *The New Cartography* was 'a cleverly contrived, cunningly deceptive attack' on the discipline of cartography, but its method was 'illogical and erroneous', 'absurd', 'the arguments spurious and in some instances just plain wrong'. Echoing Lewis's review, Robinson concluded that '[i]t is difficult to imagine how anyone who claims to be a student of cartography can write such things'.[18]

Even in Germany, the attacks continued. Following the release of Peters's projection in 1973, the German Cartographical Society felt compelled to issue a statement condemning it. 'In the interests of truthfulness and of pure scientific discussion', the society decided to intervene in what it called 'the continuing polemic propaganda by the historian Dr Arno Peters'. Invoking 'the mathematical proof that the projection of a spherical surface to a plane surface is not possible without distortions and imperfections', the society's statement went on: 'If Mr Peters, in the "catalogue of world map qualities" produced by him, maintains that his world map possesses only positive qualities and no shortcomings, then this contradicts the findings of mathematical cartography and arouses doubts regarding the author's objectivity and the usefulness of his catalogue.' Having systematically dismantled most of Peters's claims, the statement concluded, 'the Peters map conveys a distorted view of the world. It is by no means a modern map and completely fails to convey the manifold global, economic and political relationships of our times!'[19]

Despite such ferocious responses, Peters's supporters continued to champion the map through government and aid organizations. By 1977 the West German government's Press and Information Office were circulating press releases endorsing Peters's new map, much to the consternation of many cartographers. When one of the releases was published in the bulletin of the American Congress of Surveying and Mapping (ACSM), its members responded in November 1977 with an article entitled 'American Cartographers Vehemently Denounce German Historian's Projection'. The article was even more intemperate than the response of the German Society. Written by Arthur Robinson and John Snyder, two of the organization's most distinguished members, it savaged Peters as having little 'good sense', and his projection, which was 'ridiculous and insulting to dozens of other inventors' of more valid map projections.[20]

From the academic response to Peters's map, it would be easy to say that he created a flawed projection, and that his conclusions were wrong. But it is never that simple with maps. Both sides in the controversy claimed that objective truth was on their side, but invariably this objectivity quickly unravelled to disclose more subjective beliefs and vested personal and institutional interests. Gradually the debate turned into a deeper reflection on the nature of mapmaking. Were there established criteria for assessing world maps, and, if so, who should establish them? What happened when a map was accepted by the public at large but rejected by the cartographic profession, and what did this say about people's ability to read (or misread) maps? What was an 'accurate' map of the world, and what was the role of maps in society?

Initially, such questions were ignored in the professional condemnation of Peters's projection because most technically trained cartographers were so busy falling over themselves to dismiss the projection as 'bad' and Peters's claims as 'wrong'. There was indeed much to criticize. Of greatest concern was that Peters seems to have simply got his calculations wrong when drawing up his world map. Having measured the graticule on Peters's projection, one of his earliest critics noticed that his parallels were out by up to 4 millimetres, which on a global scale was a serious distortion, and meant that, technically speaking, 'Peters' projection is not equal area'.[21] Peters's claim that scale and distance were correctly represented on his projection was also mathematically impossible, as any plane map that attempts to replicate the distances between two points on the globe must adopt a scale relative to the curvature of the earth's surface. The argument that his projection dramatically reduced territorial distortion and correctly represented those countries colonized by the European powers was also not borne out by closer inspection. Reviewers claimed that on his map Nigeria and Chad both appeared twice as long as they should be, while Indonesia was represented at twice its north–south height and half its actual breadth east to west.[22] These were serious mistakes, but, when challenged, Peters stuck to his calculations, and refused to accept he had made any. Ironically, the distortions of shape which affected his projection were at their greatest in Africa and South America, two of the continents he argued suffered so greatly from European 'misrepresentation'. In contrast, regions predominantly covering the middle latitudes, including most of North America and Europe, suffered very little distortion. These errors and contradictions were only compounded with the subsequent publication of *The Peters Atlas of the World* in English in 1989. Here Peters altered his standard parallels, and also contradicted his claim to use one universal projection for every regional map: in his polar maps he adopted two of the more traditional projections (including Mercator's) which had been summarily dismissed in his *New Cartography*.

As well as exaggerating his map's accuracy, Peters also failed to practise what he preached. If he was so eager to reorient the cartographic tradition of putting Europe at the centre of the map and distorting colonized nations, then why, asked his critics, did he reproduce Greenwich as his central meridian when somewhere in Africa, China, or the Pacific could have been easily adopted? Another problem identified by the critics was the political dimensions of his projection. 'Since area alone is neither the cause nor the symptom of division between the North and South,' wrote David Cooper, 'does this map improve our understanding of the problems of the world?'[23] By producing a map which ostensibly offered equality of surface area in its projection, Peters implied that it was possible to address political inequality. Size, at least for Peters, did matter. But, as another critic asked, did a more accurate representation of the size of Indonesia really address that country's exceptionally high infant mortality rate, or only further obscure it? To some extent it was a valid question, but Peters's point was that perceiving Indonesia according to its actual relative size was an important step in establishing its place in the wider geopolitical world. Such criticisms suggested the need for a debate (not pursued for several years) as to how *any* world map could meaningfully address statistically derived social inequalities in graphic form.

Nearly all of Peters's critics questioned his attack on Mercator to the exclusion of almost all other projections. To ascribe 'Eurocentrism' and complicity in the subsequent colonization of large sections of the globe to Mercator appeared anachronistic, and conceded far greater power and authority to the map than it actually possessed. Many reviewers pointed out that the technical limitations of Mercator's projection had been acknowledged from the eighteenth century, and that its influence in maps and atlases had been on the wane since at least the late nineteenth century. Mercator was too easy a target to condemn as producing an 'inaccurate' world map to allow the promotion of Peters's 'accurate' map with its depiction of equal-area over all other elements. It was a grossly simplistic opposition that ignored countless other projections, but one which, in its visual clarity, would quickly capture the public imagination.

More than thirty years after its first publication, the Peters projection still causes consternation among the cartographic profession, and curiosity in the media. In 2001 the acclaimed US television political drama series *The West Wing* featured the fictional 'Organization of Cartographers for Social Equality', lobbying presidential staff to 'support legislation that would make it mandatory for every public school in America to teach geography using the Peters projection map instead of the traditional Mercator [map]'.[24] Following the episode's release the Peters projection experienced a fivefold increase in sales. The distinguished American geographer Mark Monmonier remained unimpressed. In 2004, two years after Peters's death, Monmonier revisited the controversy in his book *Rhumb Lines and Map Wars*, a social history of the Mercator projection. He castigated Peters for offering 'a ludicrously inapt solution' to the problem of how to revise Mercator's methods, and argued that 'the Peters map is not only an equal-area map but an exceptionally bad equal-area map that severely distorts the shapes of tropical nations its proponents profess to support'.[25]

By the time Monmonier made his reflective but still hostile criticism of Peters, the map and its projection were no longer used in atlases, and were already becoming objects of historical curiosity. In reassessing now both the technical and political controversy of the Peters projection, it is possible to see it as what has been called a 'defining moment' in the history of mapmaking. Peters's methods were suspect and his world map made unsustainable claims for greater accuracy, but his work revealed a more important truth about mapmaking: by arguing that all maps and their projections are either deliberately or inadvertently shaped by their social and political times, the 'map wars' ignited by Peters forced mapmakers to concede that their maps had never been, and never could be, ideologically neutral or scientifically objective 'correct' representations of the space they claimed to depict. Peters asked both cartographers and the general public to confront the fact that all maps are in some way partial and, as a consequence, political.

This turn to politics was a direct consequence of Peters's personal experience of a century that witnessed the political appropriation of maps for the purposes of military conquest, imperial administration and national self-definition. But in the phenomenal impact of the Apollo 17 earth photographs he also saw the power of the image of the whole earth to inspire awareness of the environment and of the baleful effects of inequality across the globe. If Peters made one mistake above and beyond his questionable cartography, it was in failing to acknowledge that his own map was just another partial representation of the world, and was subject to the same interplay of political forces that he identified throughout the course of Western cartography. Now, nearly forty years after it was first published, we can see more clearly the place of Peters and his world map in the history of mapmaking.

Despite his antagonism towards Mercator's projection and the historical gulf that separated them, Peters's own life reveals that he and Mercator had more in common than he probably liked to admit. Like Mercator, Peters was born in the German-speaking lands east of the Rhine during a time of political and military conflict. Growing up in the Weimar Republic of the 1920s and the Nazi Germany of the 1930s, and building his career in the post-Second World War context of a politically divided West and East Germany, Peters understood better than most how geography could be used to divide nations and people. He was born in Berlin in 1916 into a family of labour and union activists, and his father was imprisoned by the Nazis for his political beliefs. The teenage Peters was educated in first Berlin and then the United States, where he studied film production, writing his Ph.D. thesis on 'Film as a Means of Public Leadership' as Europe descended once again into total war (it was this interest in propaganda that many of his later critics would seize on when claiming his 'manipulation' of cartography). Recalling the origins of his politicization in the 1970s, Peters wrote that 'it was here in Berlin, three decades ago, that my basic criticisms of our historic-geographical view of the world crystallised'. Having witnessed the widescale manipulation of cartography throughout the Second World War, Peters concluded that his critique would be aimed subsequently at 'the narrowness of our European-oriented – nay German-oriented – view of the world and the realisation of its incongruity with the broad, all-embracing manner of regarding the world and life in our epoch'.[26]

In the late 1940s Peters worked as an independent scholar, receiving funding from the German regional government and the US military to write a textbook on global history that could be used in both East and West Germany. The result, the *Synchronoptische Weltgeschichte*, or 'Synchronoptic World History', was published in 1952. A synchronop-tic perspective involves displaying several timelines concurrently, and this is what Peters created in attempting to avoid the traditional linear, written accounts of history focused on Western achievements. Using noticeably geographical language, Peters complained that, in concen-trating on European history, 'the remaining nine tenths of the occupied earth' gets ignored. A good example of his revisionist approach can be seen in his account of the Middle Ages: 'six hundred years of Greco-Roman flowering are stretched in our world histories to make it seem as though human civilization began with them. After the decline, history books move rapidly again. As is well known, the so-called Middle-Ages are "Dark Ages" in Europe, and therefore in our history books. But for the rest of the world, these thousand years were an age of flowering.'[27] In an attempt to provide equal weight to each slice of history, Peters abandoned a written narrative and instead described the period from 1000 BC to AD 1952 through a series of tables 'of eight colours divided into six bands: economics, intellectual life, religion, politics, war and revolution'.[28] Central to its creation was, Peters argued, the 'idea of charting time in the same manner as space is charted on our maps'. Describing the genesis of his book, Peters recalled, 'I took a sheet of blank paper and first entered time as such to scale. Each year got a ver-tical strip, one centimetre in breadth.' As a result, 'the map of time was born'.[29]

The right-of centre German magazine *Der Spiegel* described the book as 'the biggest scandal of the last two weeks'.[30] Peters's later critics would leap upon the controversy to suggest that decades before publication of his geographical projection he had already been manipulating academic information for personal and political ends. In December 1952 the right-wing American magazine *The Freeman* published an article entitled 'Official Misinformation', in which it reported with indignation that US officials in Germany had 'with the laudable motive of "democratizing" that country', commissioned Peters and his wife to write their 'World History', but 'only after they had spent $47,600 on the project and distributed 1100 of 9200 copies received, did they learn that the authors of the book were Communists and the book itself pro-Communist, anti-democratic, anti-Catholic and anti-Semitic'. Such lurid accusations were hardly justified by the text of Peters's book, which went on to become a bestseller, but that did nothing to abate *The Freeman*'s wrath. 'So the American taxpayers are not merely the victims of a £47,600 swindle,' it thundered; 'they have been gravely injured by incompetent and disloyal officials who used their funds to finance enemy propaganda.'[31]

But *Der Spiegel* took a more emollient approach to the controversy. Its main objection to the book lay not with its content, but with the revelation that it was partly financed by a member of the SPD (German Socialist Party). *Der Spiegel* praised the book as a laudable but unsuccessful attempt to provide a comprehensive account of world history. Peters claimed that his book was trying 'to bring equality and balance to the treatment of history', but within the context of the polarized world of US–Soviet Cold War politics, such progressive initiatives by academics like Peters were inevitably prey to ideological attacks by not just right-wing publications like *The Freeman*, but also left-wing authorities like the SPD, who argued that simply allocating space to huge periods of prehistory during which, as far as they were concerned, nothing really happened, seemed absurd. As a result, the book was partially withdrawn from circulation.

It is ironic that Peters should develop his subsequent geographical projection as a result of working on a history of the world (as he later acknowledged), in much the same way that his *bête noire* Gerard Mercator had compiled an innovative chronology of world history as he completed his famous map projection. Their intellectual and ideological influences were of course very different, but both produced their histories in accordance with deeply held personal convictions. For Mercator, this was the righteousness of biblical scripture; for Peters, it was the equality of all nations and races. Both men produced books that required a different spatial approach to world history through the use of columns and tables, and both realized that their universal histories led them to a reconsideration of how to portray global geography. Mercator's preoccupations were shaped by the theological and commercial imperatives of his time, which led him to create a map that allowed people to navigate (practically and even spiritually) across the world. In contrast, Peters appreciated that accurate navigation was no longer the aim of a global projection. For him, living in an age he called 'the post-colonial period', defined by global warfare, nationalism and decolonization, questions of land distribution, population control and economic inequality were central to the study of geography and the practice of mapmaking.

Following the publication (and subsequent withdrawal) of his 'World History', Peters spent the late 1950s and 1960s editing the German Socialist magazine *Periodikum*, as his interests became more focused on space and cartography. 'During the preparation of an atlas volume to accompany my synchronoptic world history,' he wrote, 'it became clear to me that existing global maps were worthless for an objective representation of historical situations and events.' He went on, 'the quest for the causes of arrogance and xenophobia has led me repeatedly back to the global map as being primarily responsible for forming people's impression of the world seen from their standpoint.'[32] It was a compelling statement about the power of maps, and its ramifications would come to dominate the rest of Peters's career. When he disseminated his new map within the academic community, Peters's projection was just one among a bewildering variety of others; by turning to the world's media and announcing a 'new map of the world' at his press conference in Bonn, Peters changed dramatically the terms on which both the public and academia understood the role of world maps.

There is an immediate difficulty in providing an objective account of Peters's aims, because his own arguments were so steeped in the kind of myths, ideological presumptions, scientific errors and self-aggrandizement that he was quick to identify in earlier cartographers. It is also difficult to distinguish his claims of cartographic accuracy from his reactions to the prejudicial and often highly personal criticism that quickly followed, and which often led him to change the terms of the debate. We can, however, now piece together his published statements

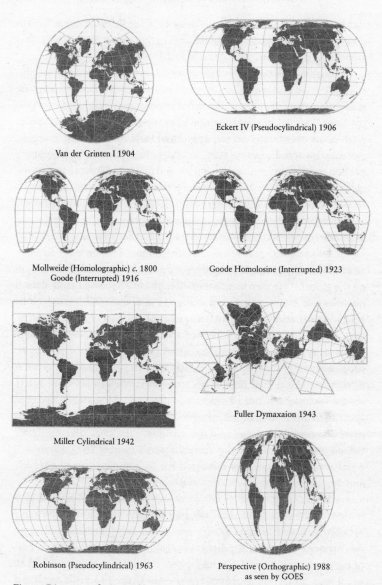

Van der Grinten I 1904

Eckert IV (Pseudocylindrical) 1906

Mollweide (Homolographic) *c.* 1800
Goode (Interrupted) 1916

Goode Homolosine (Interrupted) 1923

Miller Cylindrical 1942

Fuller Dymaxaion 1943

Robinson (Pseudocylindrical) 1963

Perspective (Orthographic) 1988
as seen by GOES

Fig. 36 Diagrams of twentieth-century map projections.

and lectures over two decades to describe what Peters thought he was doing, before assessing the avalanche of argument and debate that greeted his projection.

Throughout the twentieth century, developments in communications, transportation and global strategy and related innovations in surveying methods, statistical analysis and aerial photography produced new uses for maps. This led to a proliferation of new projections and revisions to established ones based on the appropriateness of particular mapping methods for specific practical applications. For example, as Mercator's projection became increasingly questioned as a way of representing the globe, it gained a new lease of life as a method of regional surveying.[33] In *The New Cartography*, Peters both described and responded to the increasing diversity of map projections by explaining what he regarded as a series of 'myths' that sustained traditional cartography, or what he called 'half truths, irrelevancies and distortions'. He summarized these as the myth 'that Europe dominates the world from a central position on the globe'.[34] He then went on to offer 'the five decisive mathematical qualities and the five most vital utilitarian aesthetic qualities' which he believed were necessary for an accurate modern map of the world. The five decisive qualities were fidelity of area, axis, position, scale and proportionality; the five 'vital' qualities were universality, totality, supplementability, clarity and adaptability.[35] In providing an overview of eight historical map projections from Mercator's to his own, Peters scored his map ten out of ten, while the 1569 Mercator projection, Ernst Hammer's 1892 equal-area projection, and J. Paul Goode's elaborate 1923 projection that split the world into six lobes, all lagged well behind with a poor four out of ten. Peters's nearest rival, Hammer's equal-area projection, was dismissed for its complex curved parallels and apparent lack of universality and adaptability.

For Peters, what he called 'fidelity of area' was central to his new projection: it should ensure that 'any two selected areas are in the same proportion to one another as they are on the globe', because 'only with this property can the real proportion of the sizes of various continents of the earth be achieved'. Cartographers call this particular method an equal-area projection as it retains the equivalence in size of territorial areas. Like Mercator's map, it is based on wrapping a flat map around a cylinder, but the crucial difference is that whereas Mercator's projection retains conformality, the correct shape around a particular point, an equal-area projection retains equivalence according to relative area. To achieve this, Peters had to find a different way of spacing his parallels and meridians.

Based on established measurements of the globe's circumference, Peters drew standard parallels at 45° N and 45° S, where minimal distortion occurred in transferring the globe onto a flat map. He plotted parallels of latitude which are all the same length as the equator. He then halved the scale from east to west running along the equatorial line, while doubling the scale running north to south at the equator to create a rectangular frame. It is no surprise that, whereas Mercator was influenced by the need to move across the sixteenth-century globe according to the exigencies of trade from east to west, Peters plotted a projection according to the north–south economic and political preoccupations of the second half of the twentieth century. The result of this north–south elongation and east–west compression is quite obvious on Peters's map: tropical areas in the southern hemisphere such as Africa and South America are long and thin, while the increasing compression towards the poles makes regions like Canada and Asia appear squat and fat. Even though the particular shapes of these areas were distorted through relative compression or elongation, such distortions allowed Peters to transfer relative surface area from the globe onto the map more accurately.[36]

The concern with area was central to Peters's political argument over the significance of map projections. For Peters, the relative failure to represent the world according to area, culminating in Mercator's conformal projection, was a basic act of political inequality. Looking only at the representation of territorial areas, Peters had a point: on the Mercator projection, Europe, at 9.7 million square kilometres, appears considerably larger than South America, which is nearly twice the size, 17.8 square kilometres; at 19 million square kilometres, North America is represented as considerably larger than Africa, 30 million square kilometres. Although China covers 9.5 million square kilometres, on the Mercator map it is dwarfed by Greenland, which is just 2.1 million square kilometres. A similar point can be made by looking at most atlases published prior to Peters's projection. The geographer Jeremy Crampton surveyed a range of twentieth-century atlases and found that, despite covering 20 per cent of the earth's land area, Africa was usually represented by just three maps on a scale of 1:8,250,000. In contrast the United Kingdom, covering just 0.16 per cent of the earth's land area, is shown on three maps using a more detailed scale of at least 1:1,250,000.[37] Such inequalities were summarized in the Brandt Report (1980), which divided the world between the developed northern hemisphere, covering just over 30 million square kilometres, and the developing southern hemisphere, covering over 62 million square kilometres.

Although the calculation of equal-area was central to the political and mathematical definition of Peters's projection, his *New Cartography* also laid out what he regarded as further requirements of any new map of the world. He dismissed any global projection that adopted curved meridians (of which there were many, both before and after Mercator) by invoking his second decisive quality: fidelity of axis. 'A map has this quality', Peters claimed, 'if all points, which on the globe lie north of any selected reference point, lie exactly vertically above it and all points to its south lie exactly vertically below it.' According to Peters, this quality aids 'orientation' and the accurate imposition of international time zones across its surface. In effect, it meant imposing a uniform rectangular grid of parallels and meridians across the earth's surface, like Mercator's – or his own.

Next came fidelity of position. This, according to Peters, is achieved when 'all points which exist at an equal distance from the equator are portrayed as lying on a line parallel to the equator', a quality that again can only be achieved through a graticule of right-angled parallels and meridians. Fidelity of scale 'reproduces the original (the surface of the globe) with quantifiable accuracy'. Because of his concern with 'absolute fidelity of area', Peters rejected the usual scales (for example, 1:75,000,000), and adopted a scale which, in the case of his projection, was 1 square centimetre to 123,000 square kilometres.[38] Finally came what Peters called 'proportionality'. Any map 'on which the longitudinal distortion along its upper edge is as great (or as small) as along its lower edge' is proportional. His projection certainly complied with this principle, but Peters also admitted that proportionality was required to minimize what was still an inevitable 'degree of distortion', in the transfer of the globe onto any flat map projection of the world. At least, he added with deft understatement, the apparent proportionality of his own map ensured 'an even distribution of errors'.

Each of Peters's other five 'vital' qualities ultimately denigrated rival

projections at the expense of his own. Universality, totality and adaptability emphasize the need for one uninterrupted projection of the world
that can be used for a variety of geographical purposes, while 'supplementability' and 'clarity' allowed for a comprehensive perspective of the
earth. Most of these categories were aimed at dismissing another group
of equal-area world maps, constructed along the lines of so-called 'interrupted' projections. As their name suggests, these maps attempted to
minimize distortion by 'interrupting' or dividing the globe up into discrete sections. Peters took as his example J. Paul Goode's equal-area
projection, invented in 1923. Goode fused various projections to come
up with a map that divided the earth into six peculiarly shaped lobes,
which looked like a peeled and flattened orange. It was a sign of the
impossibility of achieving both conformality and equivalence of the
globe on a flat map that Goode needed to resort to such a contorted and
discontinuous shape to come to a closer approximation of the spherical
earth in two dimensions.

Peters was quick to point out that such 'interrupted' maps lacked
universality, totality or clarity at either a technical or even aesthetic
level. Nor could they be easily adapted for more detailed mapping of
local regions. For Peters, these projections came closest to challenging
Mercator's cartographic dominance because 'they have fidelity of area,
but they bought this quality at the price of abandoning important qualities of Mercator's map' such as clarity and supplementability, 'and
could therefore not supplant it'. In one deft move, Peters dismissed all
earlier map projections, apart from Mercator's, which was ideologically
partial in its apparent 'Eurocentrism', and motivated by its inventor
'following the old, naïve practice of placing his homeland in the centre
of the map'. Ultimately, the only world map able to achieve what Peters
called 'the objectivity necessary in this scientific era'[39] was his own.

Despite these claims about the originality and accuracy of his map,
Peters's critics quickly spotted what they regarded as another example
of his opportunism and unreliability: his projection was not new at all.
It had been invented over a century earlier, by a Scottish evangelical
minister, the Reverend James Gall (1808–95), who presented his new
map at an address to the British Association for the Advancement of
Science (BAAS) in 1855, and labelled it 'Gall's Orthographic Projection'. To all intents and purposes it is identical to Peters's projection, so
much so that many cartographers now refer to it as the 'Gall–Peters

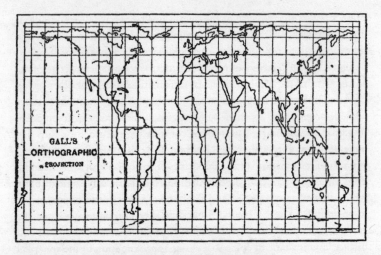

Fig. 37 James Gall, 'Gall's Orthographic Projection', 1885.

Projection'. In fact Gall's projection itself had been previously attributed to Marinus of Tyre (*c.* AD 100) by Ptolemy.

Peters always denied that he knew of Gall's projection, which is surprising considering his immersion in the history of map projections. Gall and Peters had much in common, although the contrasting response to their 'new' projections reveals a great deal about the state of geography in their respective eras. Like Peters, Gall was an amateur cartographer, and a prolific writer. He was a classic Victorian gentleman-scholar: deeply religious, highly learned, passionate about social welfare and slightly eccentric. His publications ranged from religion to education and social welfare; they included books on a triangular alphabet for the blind and *The Primeval Man Unveiled* (1871), in which he claimed that Satan and his demons were a pre-Adamic race of men who lived on the earth prior to Creation. His books on astronomy were particularly popular, and included the *People's Atlas of the Stars*, and *An Easy Guide to the Constellations* (1866).

It was this last book that led Gall to invent his new map projection. In trying to find a suitable method for depicting the stars, Gall realized that by 'representing only one constellation in each diagram', he was 'able to present it on a large scale, and without appreciable distortion, which could not be done if a large portion of the heavens were mapped on the same sheet'.[40] In a move that was strikingly reminiscent of the great Renaissance cosmographers, Gall later explained how he transferred his astronomical projection onto a comprehensive vision of the earth below. 'It then occurred to me', he wrote in 1885, 'that the same, or a similar projection, would give a complete map of the world, which had never been done before; and, on drawing a projection with the latitudes rectified at the 45th parallel, I found that the geographical features and comparative areas were conserved to a degree that was very satisfactory.'[41]

Gall's presentation to the BAAS meeting held in Glasgow in September 1855 was entitled 'On Improved Monographic Projections of the World'. Arguing that only cylindrical projections 'can represent the whole world in one diagram', Gall explained that such projections, including Mercator's, inevitably sacrificed some qualities (such as area and orientation) in favour of others. 'The best projection', he concluded, 'is that which will divide the errors, and combine the advantages' of a range of different qualities.[42] With this aim, he proceeded to offer not one but three different world projections – not just the orthographic, but also a stereographic and isographic projection (a variation on the equirectangular projection). Ironically, considering Peters's later adoption of the orthographic projection, Gall concluded that 'the Stereographic is best of all; for although it has none of the perfections of the others, it has fewer faults, and combines all the advantages of the others in harmonious proportions'. However, he still believed that there was a limited place for the orthographic projection. It was, he argued, 'a valuable map for showing the comparative area occupied by different subjects, such as land and water, as well as many other scientific and statistical facts'. He conceded that 'the geographical features are more distorted on this than on any of the others, but they are not distorted so as to be unrecognisable; and so long as that is the case, its advantages are not too dearly bought'.[43]

Even Gall's map was not the first of its kind. The very first equal-area map of the world on a rectangular projection and based on reproducible mathematical calculations was invented as early as 1772 by the Swiss mathematician Johann Heinrich Lambert. By using the equator as his standard parallel, Lambert produced a map which retained equal-area properties, but suffered serious distortion north and south. Like Gall, Lambert acknowledged the impossibility of producing a world map that was both conformal and equal-area, and went on to produce a conformal map on a conical projection, to demonstrate the options available between the two methods. Gall appears not to have known of Lambert's projection, but effectively reproduced it with the important modification of two standard parallels either side of the poles.[44]

Unlike Peters, Gall did not receive immediate condemnation for the unreliability or duplication of his new projection. There were several reasons for this. Gall disseminated his findings within a Victorian institution that was amenable to his goals and philosophy. The BAAS was founded in the same year as the Royal Geographical Society, but with a different purpose. It was a more peripatetic organization, which held meetings in provincial cities across the country which were designed to educate and enlighten middle-class laymen in the practical application of science for the betterment of Victorian society.[45] Queen Victoria's husband, Albert, the Prince Consort, was an honorary member, and its speakers included such luminaries as Charles Darwin, Charles Babbage and David Livingstone. Rather than challenging the moral and intellectual ethos of Victorian society, Gall was energetically involved in delineating it by his talks and publications on religion, education and science. He acknowledged that his projections were limited, and never claimed they could play a part in anyone's moral improvement. Reflecting on their impact in an article written thirty years later, Gall could have been addressing Peters when he observed that 'it is always difficult to introduce changes when long established custom has created a rut'. He also confessed sadly that over the subsequent twenty years 'I was the only person that used them'.[46] This, of course, was not a fate which befell Peters's projection.

For many of Peters's critics, the fact that his 'new' projection was almost identical to Gall's was at best poor scholarship; at worst it betrayed an opportunistic plagiarism. Placed alongside Gall's more modest claims for the significance of his orthographic projection, and his wider understanding of the partiality of all projections, Peters's claims for the radical and universal status of his projection look ridiculously inflated. But it also revealed the gulf that had opened up between professional and public perceptions of mapmaking between the 1850s and the 1970s. While Gall was broadly in step with the aims of the Victorian institutions that disseminated his ideas, Peters represented a direct challenge to the late twentieth-century cartographic profession and the ideological imperatives which he believed underpinned it.

By the end of the 1970s, the battle lines were clearly drawn. On one side, the cartographic profession and its institutions closed ranks to condemn Peters's projection at a technical level, according to its own rules and methods of mapmaking. On the other, political and aid organizations embraced the projection's explicitly social and ideological aims. While these organizations were understandably reluctant to engage in the debate over the projection's technical mistakes, the cartographic profession was equally unwilling to acknowledge Peters's insistence that all world maps (apart from his own) are partial and predisposed to subjective, ideological interests. The problem was compounded by the silence of many of Peters's critics about their own vested institutional interests. Although Arthur Robinson's technical criticisms of Peters's projection were widely accepted, he failed to acknowledge that Peters's world map posed the first serious challenge to his own projection, which in the 1970s was being distributed in atlases worldwide, thanks to US publishers. At the same time, as professional cartographers continued to attack Peters, they began to sound increasingly patrician, portraying the general public as a gullible mass, unable to read maps and see how Peters was deceiving them.

The gulf between Peters's supporters and his detractors was caused by more than just an argument over the mathematical accuracy of map projections. The changing political climate of the 1960s, exemplified by the political protests that took place in France in May 1968, represented among many other things a radical reassessment of the status of the humanities and social sciences within society. While subjects like history and philosophy were leading the way in criticizing established political orthodoxy, others that were deeply embedded in social policy and state organization, like geography, were understandably more reluctant to react to such changes. Standing on the margins of geography, politically active individuals like Peters were able to provide a version of cartography in step with the times, more radical than its leading practitioners, many of whom had vested political and institutional interests in upholding the political status quo.

Peters's rhetoric also chimed with the political debates of the early 1970s. There was a growing political awareness of the need to address inequality in response to the widening economic and political gap between the developed Western world and the developing southern world. In the early 1970s the World Bank estimated that 800 million people in the developing world were living in absolute poverty, with only 40 per cent in the same region able to secure the most basic necessities of life. The Brandt Report highlighted the gulf between the developed north and the developing south: it demanded that an 'action programme must be launched comprising of emergency and longer-term measures, to assist the poverty belts of Africa and Asia and particularly the least developed countries'. The report's authors had a vested interest in addressing the problem, arguing that 'whatever their differences and however profound, there is a mutuality of interests between North and South. The fate of both is intimately connected.' It called for a wholescale transfer of funds from the former to the latter, representing 0.7 per cent of the GDP of the countries involved, rising to 1 per cent by 2000 (neither figure was met).[47]

The developed north was not without its own problems: the 1970s witnessed a fall in economic growth of nearly 50 per cent from the 1960s, and by the end of the decade the thirty-four developed countries that made up the Organization for Economic Co-operation and Development (OECD) were experiencing inflation, recession and cumulative unemployment of 18 million people. The United States was also experiencing what the economist Paul Krugman has called 'the great divergence' in economic and political inequality. Although average American workers began to double their output, they suffered a simultaneous decline in wages, while the top 0.1 per cent of American society became seven times richer during the course of the second half of the twentieth century. This led to income inequality higher than at any time since the 1920s, which, according to Krugman, has been responsible for the subsequent polarization of American political culture.[48]

Few geographers were equipped to acknowledge these complex but profound levels of global inequality, but Peters was different. Having lived under the iniquities of both Nazism and the Stalinist regime of the German Democratic Republic, he was well placed to voice the rhetoric of inequality, and to propose equality as a virtue. Geography could play a role in tackling inequality, and even expressing its opposite on a map.

The late 1970s also saw a shift in the study of geography and the history of cartography. Philosophers like Gaston Bachelard and Henri Lefebvre had already started to ask basic questions about how we understand and live within space. Bachelard's *Poetics of Space* (first published in French in 1958) alerted readers to how the most intimate phenomena of spaces – attics, cellars – shaped our lives (as well as our dreams); Lefebvre's *The Production of Space* (1974) took a more Marxist approach to explain how the creation of our public environments helped to enable (or constrain) personal identity. Others soon followed in arguing that space had a history. Within geography and the history of cartography, one of the most important advocates of this new approach was the English scholar J. B. Harley. Having trained in the traditional, positivist approach to mapmaking, and published extensively on the history of the English Ordnance Survey throughout the 1970s, Harley performed a remarkable volte-face in the early 1980s. Having digested the work of Bachelard, Lefebvre and other influential French thinkers including Michel Foucault and Jacques Derrida, Harley published a series of ground-breaking articles which called for a complete reconsideration of the historical role of maps. In one of his most influential articles, published in 1989, entitled 'Deconstructing the Map', Harley voiced his 'frustration with many of the academic cartographers of today, who operate in a tunnel created by their own technologies without reference to the social world'. Claiming that 'maps are too important to be left to cartographers alone', he argued that 'we should encourage an epistemological shift in the way we interpret the nature of cartography'.[49]

Harley claimed that 'from at least the seventeenth century onward, European map-makers and map users have increasingly promoted a standard scientific model of knowledge and cognition'. He went on:

> The object of mapping is to produce a 'correct' relational model of the terrain. Its assumptions are that the objects in the world to be mapped are real and objective, and that they enjoy an existence independent of the cartographer; that their reality can be expressed in mathematical terms; that systematic observation and measurement offer the only route to cartographic truth; and that this truth can be independently verified.

This was indeed the prevailing view of mapmaking, an Enlightenment belief in the transparent, objective reality of the map. As a description of cartographic practice, Harley's account would undoubtedly have been accepted by both Arno Peters and his most vociferous critics.

But Harley went further. He invited his readers to 'pick a printed or manuscript map from the drawer almost at random'. What stands out 'is the unfailing way its text is as much a commentary on the social structure of a particular nation or place as it is on its topography. The map-maker is often as busy recording the contours of feudalism, the shape of a religious hierarchy, or the steps in the tiers of social class, as the topography of the physical and human landscape.' Harley's contention was not, as many of his critics claimed, that all maps lie, but that they contained historical conventions and social pressures that produced what he called a 'subliminal geometry'.

Coming from such a respected member of the cartographic profession, Harley's arguments for what he would later call 'the new nature of maps' marked a sea-change in the understanding of cartography. The impact on geography's understanding of itself as an academic discipline was soon affected by Harley's work, as it began to reflect on its own historical involvement in the endorsement of the ideologies of nationalism and imperialism.[50] However, practising cartographers still remained sceptical about Harley's adoption of Alfred Korzybski's dictum that 'the map is not the territory'.[51]

Matters came to a head in 1991. Harley had just completed another important article developing his earlier work by asking the question 'Can there be a cartographic ethics?' If maps can never be neutral, and are always subject to power, political authority and ideology, then is it possible for academic and professional cartographers to develop and sustain an ethical position in relation to their work? It was almost inevitable that Harley would invoke the controversy over the Peters projection to prove his point, although the consequences of doing so in this particular article only emphasized the problem he was trying to address. 'The *cause célèbre* of the Peters projection', wrote Harley, 'led to an outburst of polemical righteousness in defence of "professional standards".' Nevertheless, as he went on:

Ethics demand honesty. The real issue in the Peters case is power: there is no doubt that Peters' agenda was the empowerment of those nations of the world he felt had suffered an historic cartographic discrimination. But equally, for the cartographers, it was their power and 'truth claims' that were at stake. We can see them in a phenomenon well-known to sociologists of science, scrambling to close ranks to defend their established way of representing the world.

What followed was a startling accusation: 'They are still closing ranks. I was invited to publish a version of this paper in the ACSM [American

Congress on Surveying and Mapping] *Bulletin*. After submission I was informed by the editor that my remarks about the Peters projection were at variance with an official ACSM pronouncement on the subject and that it had been decided not to publish my essay!'[52] Nearly two decades after the map's release, the ACSM was still fighting a rearguard action by forbidding discussion of Peters's world map that was anything other than negative.

But Harley was as much concerned with the question of institutional power as with competing claims over cartographic 'accuracy'. The Peters projection *was* inaccurate by any standards: even its own author's account of the history of cartography was highly selective and its claims to objectivity seriously exaggerated. As a cartographic historian Harley understood this, and knew that its longevity was limited. The broader problem that the controversy inspired was how to produce an ethical cartography once the profession accepted that *all* maps were partial and ideological representations of the space they purported to depict.

It is a telling sign of the nature of these debates over Peters that virtually none of those involved seriously discussed the ways in which his world map was understood or used by the many organizations that enthusiastically adopted it throughout the 1970s and 1980s. In a survey in 1987 of forty-two of the United Kingdom's leading national non-governmental organizations (NGOs) working primarily in Third World development issues, the geographer Peter Vujakovic found that twenty-five had adopted the Peters world map. Of this group, fourteen organizations admitted previously using world maps based on Mercator's projection. When asked a series of questions about the use of world maps, nearly 90 per cent of the NGOs who responded agreed that map-making played a vital role in informing the public about Third World issues.[53] Peters's marketing campaign and the political arguments put forward for adopting his world map had apparently achieved remarkable success.

When the fourteen NGOs who adopted the Peters map were questioned more closely on the reasons for their choice, the results were more mixed. Asked to explain what they saw as the map's advantages over others, 48 per cent cited its equal area projection; 36 per cent cited its distinctive appearance, believing that it 'provokes reaction and thought'; 32 per cent cited its rejection of a 'Eurocentric' world view; 24 per cent claimed it provided 'a better representation of the relative importance of the Third World countries'; and just 4 per cent thought the map was 'a political statement in itself'. When questioned about the map's disadvantages, the responses overwhelmingly fell into two categories: the public's unfamiliarity with the map (32 per cent), and its distortion (32 per cent). It is noticeable that, apart from the map's claims to equal-area representation and a non-Eurocentric perspective, none of Peters's claims for it being a superior map projection are even mentioned. Nobody cited accuracy or objectivity as their reasons for adopting the map.

How then was the map used? Of the NGOs consulted, most admitted to using it as a logo for design purposes in published reports, documents and pamphlets, intending that the image's unfamiliarity would provoke surprise and debate. Others used it for educating people about development issues. This involved using the map to identify the location of overseas projects by selecting limited areas from the larger global projection to create regional maps. In most of these cases, the graticule was removed, making any discussion of scale or proportion (central to Peters's argument) irrelevant. Although such surveys are inevitably selective, they do reflect at least some understanding of Peters's world map by the agencies involved in disseminating it in such vast numbers, but their responses suggest a limited level of cartographic literacy. The Peters world map's ideological claims to enhance the geographical representation of developing countries simply offered a more attractive symbol of the political issues at stake for development agencies than any other cartographic projection currently available. The survey raises a question about the use of world maps, not just today, but throughout history: if mathematical accuracy and cartographic issues such as conformality or equal-area representation are of little interest to groups who use a world map, does anyone within the general public consider such questions in the world maps they use in their everyday lives?

The applications of the Peters projection throw into stark relief the fact that, ever since Ptolemy, individuals and organizations have used world maps for their own symbolic and political ends, regardless of the cartographer's claims to comprehensiveness and objectivity. Such claims have also been subject to appropriation, and used to further the ideological agenda of the map's users, rather than representing an end in their own right. Although modern cartographers may have a better understanding of the mathematical impossibility of comprehensively projecting the globe onto a flat map, such knowledge continues to make little difference to how people understand and use world maps today.

The release of the Peters projection in 1973 sparked a controversy within the world of cartography that reached far beyond the map's purported accuracy. The projection *was* questionably executed, its claims to accuracy and objectivity were wildly exaggerated, the championing of 'Third World' countries laudable in many ways but ultimately limited, and its attack on Mercator's projection clumsy and misguided. But Peters captured a perceptible shift in Western intellectual culture's understanding of mapmaking, the realization that all maps of the world are inevitably selective, partial representations of the territory they claim to represent, and that such representations are always subject to personal prejudice and political manipulation. The eighteenth-century belief in the ability of mapmaking to offer transparent, rational and scientifically objective images of the world, exemplified by the Cassini surveys, had slowly unravelled from the late nineteenth century onwards, as the political dictates of nationalism, imperialism and a range of ideologies appropriated cartography to produce persuasive but selective maps designed to legitimate their particular political versions of the world. Once dissident thinkers and political activists began to question such maps, it was almost inevitable that a figure like Peters would challenge the established cartographic hegemony. The resulting controversy unwittingly revealed the terminal limits of traditional world mapping and took cartography right to the brink of its next great evolution: the virtual world of online mapping.

Today, the Peters projection is no longer used, but the thematic sections in the *Peters Atlas of the World* (1989) which addressed social and economic issues such as population, economic growth and social issues have been absorbed into most early twenty-first century atlases. In the provocatively entitled *Atlas of the Real World: Mapping the Way We Live* (2008), Daniel Dorling, Mark Newman and Anna Barford dispensed altogether with maps according to physical size, and produced 366 world maps drawn according to demographic issues ranging from population growth to military spending, immigration, infant mortality, endangered species, and deaths from war. Their *Atlas* uses computer software to represent statistical data according to its geographical distribution on a world map. The cartogram of the world's population in 1500, for example, shows the relative insignificance of the Americas. These cartograms represent many of the issues of today's global world – population, environmentalism, poverty, inequality and conflict – but not one of them attempts to show the world according to either equal area or conformal principles.

The problem with Peters's world map lay not in his technical limitations in drawing a map, but in persisting with the belief that it was still possible to create a more 'accurate' and scientifically objective map of the world. Having convincingly argued that the history of cartography has always explicitly or implicitly reproduced the prevailing cultural values of its time, Peters still clung to the Enlightenment belief that his own world map could transcend such conditions, and be truly objective. In being so wrong, both technically and intellectually, Peters and the controversy that surrounded his projection inadvertently illustrated a deeper truth about mapping the world, that any map of the world is always partial and inherently selective, and that as a result it is inevitably prey to political appropriation. Cartography is still digesting these lessons, not in spite of Arno Peters, but to some extent because of him.

# 12

## Information

*Google Earth, 2012*

From 11,000 kilometres above its surface, the planet earth spins into view out of the black void of deep space. The sun's rays illuminate its surface, which appears free of clouds and water, although its ocean floors still sparkle ultramarine blue, the continents a beguiling patchwork of greens, browns and pinks. North Africa, Europe, the Middle East and central Asia curve round in a crescent through the right-hand half of the globe. The Atlantic Ocean dominates the bottom left, giving way to the tip of North America, with the brilliant white sheet of Greenland nearly crowning the planet's apex, looming over the North Pole. This is a vision of the world as Plato imagined it nearly two and a half thousand years ago in the *Phaedo*, a gleaming, perfect sphere, 'marvellous for its beauty'. It is the *oikoumenē* that Ptolemy projected on his geometrical grid in the second century AD, the globe that Mercator plotted onto a rectangle nearly 500 years ago, and the earth that NASA captured in the first extraterrestrial photograph of the whole planet taken in the last decades of the twentieth century. This is the geographer's ultimate object of study, an image of the whole earth.

But this is not an omniscient fantasy of the earth imagined from some god-like perspective. It is an image of the earth as seen from the home page of Google Earth.[1] Launched in 2005, the application, alongside Google's Maps, is now the world's most popular geospatial application (a combination of geographical data and computer software). In April 2009 Google edged past its mail rival MapQuest.com with just under 40 per cent of the market share of online visits to mapping websites.[2]

Since then its share of the market has continued to grow, and, despite efforts by other rivals such as Yahoo! Maps and Microsoft's Bing Maps, it is now virtually synonymous with online mapping. By November 2011 Google's market share in the United States was over 65 per cent, with Yahoo! its nearest rival trailing way behind with just 15 per cent.[3] Globally, Google's dominance is even more pronounced, with its share of the global online search market at around 70 per cent.[4] Of an estimated 2 billion people currently online globally, more than half a billion have downloaded Google Earth, and the figure continues to rise.

The application's benefits and subsequent popularity are obvious to anyone who has used it. As well as drawing on the iconic image of the blue planet suspended in space popularized by NASA in the 1970s, Google Earth offers its users a level of interaction with the earth unimaginable on printed paper maps or atlases. The application's display allows the world to be tilted, panned and rotated; geographical places and physical objects can be clicked to provide more information, and even to introduce time in the form of video streaming; other data can be integrated and 'layered' onto its surface, from political boundaries to historical maps depicting the same region; users can zoom down through its layers of data, or enter any location on the planet, and within seconds go from thousands of kilometres above the earth to within a few metres of its surface, presented with photo-real, three-dimensional images of recognizable neighbourhoods, streets, buildings and houses. Because Google Earth's application programming interfaces (or APIs) are free to anyone with internet access, individuals and companies can now create their own virtual maps within a computer-simulated environment that currently allows them to appropriate Google's geographical data and repackage it for their own use.[5] Not only has Google released a mountain of geographical data online for free; it has endorsed the ways in which its applications have been used by a variety of non-governmental organizations to support a range of environmental campaigns and humanitarian responses to natural catastrophes and civil wars across the globe.

The sheer scale of information that stands behind that first image of the earth is unprecedented, and staggering when compared to that of a traditional paper map. The viewer is looking at an extraordinary ten petabytes of potential geographical information distributed across the globe's surface. A byte is a unit of data representing a single 8-bit value of data in a computer's memory; in Western languages, a bit can be used to hold a single alphabetical or numerical character, such as the letter A, or the number 0. A standard 80 gigabyte hard drive contains approximately 80,000,000,000 bytes; one petabyte represents one million gigabytes, with a capacity to store 500 billion pages of printed text. At this size, Google Earth is able to call on a volume of digital data equivalent to six month's worth of the BBC's total programming output, any byte of which can be retrieved in seconds as the online viewer enters their coordinates and hurtles down towards the earth. As the image refreshes at up to fifty frames per second (FPS), Google Earth's technology is able to produce the highest definition of all its online competitors, giving a crisp, flicker-free image that simulates flight, and which has ensured its dominance in the world of online mapping.

In less than a decade, Google Earth has not just set the standard for these applications, but has led to a complete re-evaluation of the status of maps and the future of mapmaking, allowing maps to appear more democratic and participatory than ever before. It seems that anywhere on the earth can potentially now be seen and mapped by anyone online, without the inevitable subjective bias and prejudice of the cartographer. And as the cartographic limits of what it is possible to create online are expanded, so are the definitions of a map and its maker. If we use the established definition of a map as a graphic representation that provides a spatial understanding of the world, many geographers would not categorize Google Earth as a map at all (even its creators are cautious about using the term, preferring 'geospatial application' instead). Based on the manipulation of satellite and aerial imagery, the application produces a photographic realism free of the usual graphic signs and symbols that now define modern maps. Its makers are no longer even formally trained in geography or even cartography. The technological breakthroughs that inspired these geospatial applications were made by computer scientists, and those who work in virtual mapping today are usually called 'geospatial technologists', rather than 'cartographers'.

By its supporters, Google and its applications are spoken about with reverential awe. The computer scientist John Hennessy hails Google as 'the largest computer system of the world', while David Vise, the author of *The Google Story*, claims that '[n]ot since Gutenberg has any new invention empowered individuals, and transformed access to information, as profoundly as Google'.[6] Others are not quite so enthusiastic.

Some complain that Google's caching of content as it crawls across the web infringes all manner of copyrights; others (recently vindicated) argue that the company's ability to save individuals' search history represents a violation of privacy – a criticism intensified by Google's Street View initiative, which captures photographic imagery of everyday life. The company has also been attacked by civil liberties groups for censoring content, particularly in cooperation with the Chinese government, although in January 2010 the company took the decision to stop taking down material deemed sensitive by the Chinese. They continue to receive criticism from states such as Iran, North Korea and even India for displaying militarily sensitive locations within its geospatial applications. In December 2005, V. S. Ramamurthy, secretary to India's federal Department of Science and Technology, worried that Google's data 'could severely compromise a country's security'.[7] In most cases surrounding copyright and privacy, Google has successfully fought its case in the US courts. The newly evolving field of 'spatial law' is trying to keep up with Google as it releases ever more sophisticated technological applications that test the boundaries of what is legally permissible.[8]

As a result, many working in academic geography and professional cartography regard Google Earth with suspicion, even alarm. For some it signals the end of the traditional print-based map industry and the death of paper maps. For others it is a retrograde step in the quality of mapmaking: personalized maps made by 'amateurs' appear basic and lack the usual protocols of professional verification and review. Google Earth also faces accusations of homogenizing maps by imposing a singular geospatial version of the world in an act of cyber-imperialism.[9] Speaking in 2008, the British Cartography Society's president Mary Spence summarized many of her colleagues' concerns when she argued that online mapping (and Google Maps in particular) remains some way from matching the detail and comprehensiveness of state-sponsored reference maps like the Ordnance Survey, because it is not designed to represent medium-scale data of the kind seen on traditional reference maps.[10] Others question the application's innovativeness, claiming that Google acts simply as a data aggregator, using relatively basic programming to piece together licensed material from a variety of satellite imagery providers. Google does not reveal fully which companies provide specific data, and this makes it almost impossible to evaluate the quality of the data, or how it has been rendered.[11]

There is also a paradox that the free circulation of virtual maps and the ability to appropriate them for other uses online is led by some of the wealthiest multinational net-based corporations currently trading on the NASDAQ, many like Google generating vast revenues from advertising and sponsored links on their websites (in the third quarter of 2011 the company's net income rose 26 per cent to $2.73 billion). It is impossible to predict the future of such applications. Anything approaching a history of them is necessarily still to be written, as the technology continues to evolve on an almost daily basis, but this chapter is the first attempt to provide a printed account of Google Earth and incorporate it within the wider history of mapmaking.[12]

Each map in this book has constructed a particular cultural world view as much as it represents one, and nowhere is this process more evident than in the rapidly evolving development of geospatial applications like Google Earth. The application's ability to draw on ten petabytes of potential geographical data within seconds is just one of the most dramatic manifestations of the current and ongoing transformation in information technology, a change so profound that the Spanish sociologist Manuel Castells has called it 'the beginning of a new age, the Information Age'.[13]

Writing in 1998, Castells argued that we were experiencing 'a technological revolution, centred around information',[14] which he calls a networked society, where social behaviour is organized around electronically processed information networks.[15] Such a society generates a 'spirit of informationalism', where information and its processing become paramount in economic organization. Castells believes that the circuit of instantaneous electronic exchanges – telecommunications, computer processing, microelectronic devices – is creating a new spatial environment, what some commentators are calling 'DigiPlace',[16] in which networked individuals navigate their way through an apparently endless flow of virtual information. It offers its users the promise of understanding their place in the world by encouraging them to move between spaces that are partly physical, but increasingly virtual, from finding their way around a city to shopping and gaming.[17] All of this sounds like the stuff of dystopian science fiction, where the 'real' world is replaced by a virtual, digital world. But Castells points out that 'all reality is virtually perceived', because we apprehend the world through a variety of signs and symbols. The network society represents a new communication system which generates what Castells calls 'real virtuality'. This is a system in which 'reality itself (that is, people's material/symbolic existence) is entirely captured, fully immersed in a virtual image setting', and where 'appearances are not just on the screen through which experience is communicated, but they become the experience'. At the heart of the network society is information. According to James Gleick, 'information is what our world runs on: the blood and the fuel, the vital principle'. For modern physicists, the whole universe is now approached as 'a cosmic information-processing machine'.[18] No company better exemplifies the rise of the network society and this 'spirit of informationalism' than Google, with its defining mission statement 'to organize the world's information and make it universally accessible and useful'.[19] To grasp how applications like Google Earth have changed irrevocably the terms of mapping involves understanding the monumental changes to the theory and practice of communicating information that took place in the second half of the twentieth century.

During the late 1940s a group of American mathematicians and engineers began to develop ways of predicting what they called stochastic processes – events which appear random and indeterminate. Individuals like Norbert Wiener (1896–1964) and Claude Shannon (1916–2001) were employed during the Second World War to work on stochastic problems like firing mechanisms and cryptography. They started to propose complex 'control systems' that could decode and predict apparently arbitrary communicative acts between humans – and between machines. In 1948 Wiener wrote that 'we have decided to call the entire field of control and communication theory, whether in the machine or in the animal, by the name *Cybernetics*', a term taken from the Greek *kybernetes*, meaning a steersman of a ship, which is also used to define control or governance.[20]

Wiener was convinced that 'the brain and the computing machine have much in common',[21] and in a paper also published in 1948 called 'A Mathematical Theory of Communication' Shannon took this idea a step further. He argued that there were two connected problems in any act of communication: the act of defining the message, and what he called the 'noise' or interference that affected its transmission from one source to another. For Shannon, a message's content was irrelevant: to maximize the effectiveness of its transmission he envisaged communication as a conduit. The message originates from a source, enters a transmission device, and is then transmitted across a specific medium, where it encounters a variety of irrelevant 'noise', before reaching its intended destination where it is interpreted by a receiver. This metaphor drew on a functional account of human language, but it could also be applied to mechanical messages like the telegraph, television, telephone or radio. Shannon showed that all these messages (including speech) could be digitally transmitted and measured, through sound waves composed of ones and zeros.[22] 'If the base 2 is used,' argued Shannon, 'the resulting units may be called binary digits, or more briefly *bits*' – thus introducing the term as a unit of countable information.[23] Shannon went on to develop a theory of probability using complex algorithms that showed how to maximize the performance of the signal (or units of information) and minimize the transmission of unwarranted errors or 'noise'.[24]

Today, Shannon's paper is widely regarded by many computer engineers as the Magna Carta of the information age. He provided a theory of how to store and communicate digital information quickly and reliably, and how to convert data into different formats, allowing it to be quantified and counted. Information was now *fungible*, a commodity capable of quantification and mutual substitution with other commodities. The impact of such a theory on the field of computing hardware would be enormous, and it would also affect other disciplines – including cartography. Over the next two decades cartographers began to adopt Shannon's theory to develop a new way of understanding maps, based on the so-called 'map communication model' (MCM). In 1977 Arno Peters's great adversary Arthur Robinson proposed a radical reassessment of the function of maps to reflect what he described as 'an increased concern for the map as a medium of communication'.[25] Traditionally, any theory of maps had ended with their completion: the interest was purely in the cartographer's struggle to impose some kind of order on a disparate, contradictory (or 'noisy') body of information that was incorporated into the map according to the cartographer's subjective decisions. Drawing on Shannon's theory of communication, Robinson now proposed that the map was simply the conduit across which a message travels from mapmaker to its user, or what he called the percipient.

The effect upon the study of mapmaking was decisive. Instead of analysing the subjective and aesthetic elements of map design, Robinson's map communication model demanded a new account of the functional and cognitive aspects of maps. The result was an examination of mapping as a *process*, explaining how mapmakers collected, stored and communicated geographical information, and which then studied the percipient's understanding and consumption of maps. Used alongside Shannon's theories for maximizing communicative performance and minimizing noise, Robinson's map communication model addressed a conundrum at least as old as Herodotus and Ptolemy: how to accommodate a mass of noise and disparate geographical *akoē* (or hearsay) into an effective and meaningful map. Adapting Shannon's theories of 'noisy' interference in transmitting information, Robinson aimed to minimize obstacles in what he designated as a map's effective transmission. This meant avoiding inconsistent map design (for instance in the use of colour or lettering), poor viewing conditions (focusing again on the percipient) and ideological 'interference' (an enduring problem that took on greater resonance as Robinson continued his attack on Peters throughout the 1970s). Having directly incorporated both Shannon's theory of communication and Robinson's map communication model into subsequent computer technology, digital geospatial applications like Google Earth appear to fulfil the dream of producing maps where form and function are perfectly united, and geographical information *about* the world is communicated instantaneously to the percipient at any time or place *in* the world.

Claude Shannon's theories changed the perception of the nature of information and its electronic communication, and would provide the foundation for the development of subsequent computerized technology. The spectacular growth of information technology (IT) and graphic computer applications like Google Earth are indebted to Shannon's mathematical and philosophical propositions. To put Shannon's theory of communication into practice in the 1940s required a degree of computing power that only began to emerge in subsequent years with vital breakthroughs in electronic technology. The invention of transistors (semiconductors, or what we now call 'chips') at the Bell Laboratories in New Jersey in 1947 predated Shannon's ideas, and in theory enabled the processing of electrical impulses between machines at a hitherto unimaginable speed. But it needed to be made from a suitable material to optimize its usage. In the 1950s a new process manufacturing transistors was developed using silicon, which was perfected in 1959 by a company based in what became known as Silicon Valley, northern California. In 1957 integrated circuits (ICs, known more commonly as 'microchips') were invented by Jack Kilby and Bob Noyce, enabling lighter, cheaper integration of transistors. By 1971 these developments culminated with the invention of the microprocessor – a computer on a chip – by the Intel engineer Ted Hoff (also working in Silicon Valley).[26] The electronic vehicles required to test Shannon's theories were now a reality.

Due to their exorbitant cost at the time the initial impact of these technological developments was limited beyond their use in governmental military and defence, but some geographers were already beginning to use Shannon's ideas in developing new ways of representing data. The most important practical innovation for subsequent geospatial applications was the emergence of geographical information systems (GIS) in the early 1960s. GIS are systems that use computer hardware and software to manage, analyse and display geographical data in solving problems in the planning and management of resources. To ensure standardization, the results are referenced to a map on an established earth-coordinate system which treats the earth as an oblate spheroid.

In 1960 the English geographer Roger Tomlinson was working with an aerial survey company in Ottawa, Canada, on a government-sponsored inventory to assess the current use and future capability of land for agriculture, forestry and wildlife. In a country the size of Canada, to cover agricultural and forest areas alone would require over 3,000 maps on a scale of 1:50,000, even before the information could be collated and its results analysed. The government estimated that it would take 500 trained staff three years to produce the mapped data. But Tomlinson had an idea: he knew that the introduction of transistors into computers allowed for greater speed and larger memory. 'Computers', Tomlinson recalled, 'could become information storage devices as well as calculating machines. The technical challenge was to put maps into these computers, to convert shape and images into numbers.' The problem was that the largest machine then available was an IBM computer with just 16,000 bytes of memory, costing $600,000 (more than £4 million today), and weighing more than 3,600 kilograms.[27]

In 1962 Tomlinson put his plan forward to the Canada Land Inventory. Showing the demonstrable influence of Shannon and Robinson's theories of communication, he called it a geographic information system in which 'maps could be put into numerical form and linked together to form a complete picture of the natural resources of a region, a nation or a continent. The computer could then be used to analyse the characteristics of those resources ... It could thus help to devise strategies for rational natural resource management.'[28] His proposal was accepted, and the Canada Geographic Information System (CGIS) became the first of its kind in the world. The ability of the resulting maps to represent colour, shape, contour and relief was still limited by printing technology (usually dot matrix printers), but at this stage it was their capacity to collate huge amounts of data that really mattered.

The CGIS was still active in the early 1980s, using enhanced technology to generate more than 7,000 maps with a partially interactive capability. It inspired the creation of hundreds of other GIS systems throughout North America, as well as substantial US government investment in the foundation of the National Center for Geographic Information and Analysis (NCGIA) in 1988. These developments in GIS marked a noticeable change in the nature and use of maps: not only were they entering a whole new world of computerized reproduction, but they promised to fulfil Shannon's model of noise-free communication, facilitating new and exciting ways of organizing and presenting geographical information.[29]

In the early days of implementing the CGIS, Tomlinson allowed himself a brief flight of fantasy: would it not be wonderful if there was a GIS database available to everyone that covered the whole world in minute detail? Even in the 1970s, the idea was still the preserve of science fiction, as computing power was simply unable to match Tomlinson's aspiration. It was at this point that computer science began to take over from the geographers. Shannon had provided a theory of communicating countable information; the development of integrated circuits and microprocessors had led to a profound change in the capacity of computerized data; one of the challenges now was to develop hardware and software with the capability of drawing high resolution graphics composed of millions of Shannon's binary 'bits' of information, and which could then be distributed across a global electronic network to a host of international users – in other words, an Internet.

The Internet as we know it today was developed in the late 1960s by the US Defense Department's Advanced Research Projects Agency in response to the threat of a nuclear attack from the Soviet Union. The department needed a self-sustaining communication network invulnerable to a nuclear strike, even if parts of the system were destroyed. The network would operate independently of a controlling centre, allowing data to be instantly rerouted across multiple channels from source to destination. The first computerized network went online on 1 September 1969, linking four computers in California and Utah, and was named ARPANET.[30] In its first years, its interactivity was limited: public connection to ARPANET was expensive (between $50,000 and $100,000), and using its code was difficult. But gradually, technological developments throughout the 1970s began to open up the network's possibilities. In 1971 the American computer programmer Ray Tomlinson sent the first email via ARPANET, using the @ sign for the first time to distinguish between an individual and their computer. The invention of the modem in 1978 allowed personal computers to transfer files without using ARPANET. In the 1980s a common communication protocol was developed that could be used by most computerized networks, paving the way for the development of the World Wide Web at CERN (the European Council for Nuclear Research) in Geneva in 1990. A team of researchers led by Tim Berners-Lee and Robert Cailliau designed an application that was capable of organizing Internet sites by information rather than location, using a hypertext transfer protocol (HTTP, a method of accessing or sending information found on web pages), and a uniform resource locator (URL, a method of establishing a unique address for a document or resource on the Internet).[31]

These developments in information technology went hand in hand with the profound restructuring of Western capitalist economies that took place between 1970 and 1990. The worldwide economic crisis of the 1970s described in the last chapter led governments in the 1980s to reform economic relations through deregulation, privatization and the erosion of both the welfare state and the social contract between capital and labour organizations. The aim was to enhance productivity and globalize economic production, based on technological innovation. As Castells argues, the relations between a reinvigorated capitalism and electronic technology were mutually self-reinforcing, characterized by 'the old society's attempt to retool itself by using the power of technology to serve the technology of power'.[32] In contrast to Arno Peters's 1973 projection, which was a direct response to the economic crisis and political inequalities of the 1970s, the next generation of geospatial applications emerging in the early 1980s were born out of the economic policies of Reaganism and Thatcherism.

The results of this economic change can be seen in the rise of computer graphics companies in California's Silicon Valley throughout the 1980s, which began developing user-friendly graphics that would characterize the future of online user experience. In the late 1980s, Michael T. Jones, Chris Tanner, Brian McClendon, Rémi Arnaud and Richard Webb founded Intrinsic Graphics to design applications that could render graphics at a previously unimaginable speed and resolution. Intrinsic was subsequently acquired by Silicon Graphics (SGI), which had been founded in 1981, and specialized in 3D graphics display systems. SGI understood that the most compelling way to demonstrate their new technology was by visualizing it geographically.

One of SGI's inspirations was a nine-minute documentary film, *Powers of Ten*, made in 1977 by Charles and Ray Eames. The film opens with a couple picnicking in a park in Chicago filmed from just 1 metre away. It then zooms out by a factor of ten, to $10^{+25}$ or a billion light years away to imagine the perspective from the very edge of the known universe. The film then tracks back to the couple in the park, into the man's hand, right down through his body and molecular structure, and finally ending with a view of subatomic particles of a carbon atom at $10^{-17}$.[33] For its producers, the film's message was one of universal connectedness, derived from the graphic visualization of mathematical scale. It quickly attained cult status within and beyond the scientific community. SGI's challenge was to take the principle explored in *Powers of Ten* and unify satellite imagery and computerized graphics to zoom seamlessly between the earth and space very quickly – without being locked into the power of ten (or any other particular multiplier). They needed to mask the obvious intervention of technology in an attempt to simulate perfectly the experience of flight above the earth and deep into the cosmos.

By the mid-1990s, SGI were starting to demonstrate its new capabilities. They began working on hardware called 'InfiniteReality', which used an innovative component called a 'clip-map' texture unit.[34] A clip-map is a clever way of pre-processing an image that can quickly be rendered on the screen at different resolutions. It is a technological refinement of a MIP map (from the Latin 'multum in parvo', 'many things in a small space'). It works on the basis of creating a large digital image – like a map of the United States – to a resolution of 10 metres. The dimensions of the image would be in the region of 420,000 × 300,000 pixels. If the user pans out to view the image on a 1,024 × 768 monitor, each pixel of data would correspond to thousands of pixels on the map. Clip-maps create a slightly larger source image by including extra pre-processed data for lower levels of resolution for the image. When the computer renders lower resolution versions of the image it avoids the need to interpolate each pixel from the full-size image but instead uses the pre-processed lower level resolution pixel data, arranged rather like an inverted pyramid. Using an innovative algorithm, all the clip-map needs to know is where you are in the world; it will then extract the specific data required from the larger virtual 'texture' – all the information which represents the world – 'clipping' off the bits you don't need. So as you zoom down towards the earth from space, the system is supplying the screen with the information central to the user's view, and everything else is discarded. This makes the application extremely economical in terms of memory, allowing the application to run quickly and efficiently on home computers. As Avi Bar-Zeev, one of the early employees at Intrinsic Graphics puts it, the application is 'like feeding an entire planet piecewise through a straw'.[35] In Claude Shannon's terms, clip-mapping allows for the uploading of as little data as possible onto a graphics processing unit, to maximize speed and enable animation in real time of complex realities – like physical geography.

Mark Aubin, one of SGI's engineers, recalls that 'our goal was to produce a killer demo to show off the new texturing capabilities' that drew on commercially available satellite and aerial data of the earth. The result was 'Space-to-your-face', a demonstration model that Aubin reveals was inspired more by computer gaming than by geography. After looking at a flipbook of *Powers of Ten*, Aubin remembers, 'we decided that we would start in outer space with a view of the whole Earth, and then zoom in closer and closer'. From there, the demo focused in on Europe,

> and then, when Lake Geneva came into view, we'd zero in on the Matterhorn in the Swiss Alps. Dipping down lower and lower, we'd eventually arrive at a 3-D model of a Nintendo 64 [video game console], since SGI designed the graphics chip it uses. Zooming through the Nintendo case,

we'd come to rest at the chip with our logo on it. Then we'd zoom a little further and warp back into space until we were looking at the Earth again.[36]

SGI's 'killer demo' was impressive and enthusiastically received by those who saw it, but more work was needed on both the software and the data. They needed to move quickly, because the bigger corporations were already beginning to see the potential of developing such applications. In June 1998 Microsoft launched TerraServer (the forerunner of Microsoft Research Maps, or MSR). In collaboration with the United States Geological Survey (USGS) and the Russian Federal Space Agency *Sovinsformsputnik*, TerraServer used their aerial photographic imagery to produce virtual maps of the United States. But even Microsoft did not fully grasp the significance of the application. Initially it was developed to test how much data its SQL Server could store without crashing. The content was secondary to the sheer size of its data, which within less than two years was over 2 terabytes.[37]

As TerraServer grew, SGI made a vital breakthrough. When one of their engineers, Chris Tanner, invented a way to do clip-mapping in software for PCs, some of the group founded a new software development company in 2001 called Keyhole, Inc. Keyhole's intention was to take the new technology and try to find applications for it, and to answer the question that many of the team, including Mark Aubin, kept asking, and which could have been posed of Claude Shannon's theory of communication: 'What was it really good for?'[38] In Shannon's theory, the content of his units of information was irrelevant; all that mattered was how to store and communicate them. At this stage the fact that SGI's developments took geographical data as their focus seemed almost incidental. Aubin understood that the capability to rapidly render graphic information on a globe was something that people found mesmerizing, and which went beyond the technical wizardry. The company attracted interest in the new application, which was obviously an innovative tool even if it still lacked what one of its creators would later call an 'actionable application platform'.[39] The data could be quantified and counted, but according to what value of use? Rather like late fifteenth-century printers, computer scientists at companies like SGI and Microsoft responded to the technical challenge of rendering geographical information in a new medium, but with little foresight as to how the new form would also change the content of maps.

These computer engineers were beginning to realize that they were tapping into one of the most enduring and iconic graphic images in the human imagination: the earth as seen from above, and the ability to swoop down on it from a seemingly omniscient, divine location beyond terrestrial time and space. The technological ability to offer yet another perspective on this transcendent view of the globe was given an enormous boost by two specific political interventions made by the Clinton administration in the final years of the twentieth century. In January 1998 Vice President Al Gore delivered a talk at the California Science Center in Los Angeles entitled, 'The Digital Earth: Understanding our Planet in the 21st Century'. Gore began by arguing that a 'new wave of technological innovation is allowing us to capture, store, process and display an unprecedented amount of information about our planet and a wide variety of environmental and cultural phenomena. Much of this information will be "georeferenced" – that is, it will refer to some specific place on the Earth's surface.' Gore's aim was to harness this information within an application he called 'Digital Earth': a 'multi-resolution, three-dimensional representation of the planet, into which we can embed vast quantities of geo-referenced data'.

Gore asked his audience to imagine a young child entering a museum and using his Digital Earth program.

> After donning a head-mounted display, she sees Earth as it appears from space. Using a data glove, she zooms in, using higher and higher levels of resolution, to see continents, then regions, countries, cities, and finally individual houses, trees, and other natural and man-made objects. Having found an area of the planet she is interested in exploring, she takes the equivalent of a 'magic carpet ride' through a 3-D visualization of the terrain. Of course, terrain is only one of the many kinds of data with which she can interact.

Gore admitted that 'this scenario may seem like science fiction', and that 'no one organization in government, industry or academia could undertake such a project'. Such an initiative, if it could be realized, would have progressive global ramifications. It could facilitate virtual diplomacy, fight crime, preserve biodiversity, predict climate change and increase agricultural productivity. In pointing the way forward, Gore acknowledged the challenges of integrating and freely disseminating such a vast body of knowledge, 'especially in areas such as automatic interpretation of imagery, the fusion of data from multiple sources, and intelligent agents that could find and link information on the Web about a particular spot on the planet'. Nevertheless, he believed that 'enough of the pieces are in place right now to warrant proceeding with this exciting initiative'. He then proposed: 'we should endeavour to develop a digital map of the world at one meter resolution.'[40]

The Clinton administration's grasp of the need to open up online information did not end there. Since its development in the 1960s, Global Positioning Systems (or GPS) had been controlled by the US Air Force, through dozens of satellites orbiting the earth. The GPS signal allowed US military receivers to pinpoint any location in the world to an accuracy of less than 10 metres. Any member of the public prepared to spend thousands of dollars on a GPS receiver could pick up this signal. But for what were deemed reasons of national security the government filtered the signal for public consumption, using a programme called Selective Availability (SA). This degraded signal could only locate a position to within a few hundred metres, making it virtually useless for practical purposes. The Clinton administration faced increasingly vociferous lobbying by various business interests, including the automobile industry which wanted the deregulation of SA to allow the improved signal to support a range of commercial spin-offs such as in-car navigation systems.

As a result, and primarily because of Al Gore's advocacy, the Clinton administration turned off Selective Availability at midnight on 1 May 2000. The result was a much stronger and consistent GPS signal. Commercial businesses immediately grasped the potential of the decision and began putting online maps into the public domain. Simon Greenman, co-founder of the online map service MapQuest.com (launched in 1996), argues that this was a significant moment 'when many of us from the GIS industry saw the power of the Internet to bring mapping to the masses for free'.[41] Other companies like Multimap (launched in 1995) began selling digital maps, while others marketed a proliferation of GPS navigational devices, including relatively cheap personal satellite navigational systems. Avi Bar-Zeev is in no doubt about the significance of Gore's Digital Earth and SA initiatives:

> Without the open Internet, Google Earth (and this blog and a bunch of other things we like) would not exist. And for that, we owe some thanks

to Al Gore. So regardless of what you think of his politics, one of the clear motivations behind Google Earth was a shared desire to give people a vision of the Earth as a seamless whole and give them the tools to do something with that vision.[42]

Both these developments added an extra impetus to the rise of geospatial applications in the first years of the twenty-first century. But in the febrile dot-com world of 2000–2001, it was the scramble for commercial survival that soon became paramount. In March 2000, the dot-com bubble suddenly burst, wiping out trillions of dollars of the value of IT companies across the globe. At Keyhole, work had begun on an application called Earthviewer that they saw as following Al Gore's idea of 'Digital Earth', and which people like Mark Aubin thought could be marketed 'as a consumer product, giving it away to the world' that would raise revenue through advertising. But then 'the dot-bomb hit and the company never received funding to support that model, so the company changed gears and focused on commercial applications'.[43] Sony Broadband was already investing, but Keyhole wanted a broader portfolio of investors, and initially targeted the real-estate market. Although data for North America was easily available, the new tool was still limited in its global reach, so its use as an application to zoom in on a property and search the local area seemed attractive.

In June 2001 Keyhole launched Earthviewer 1.0 to a fanfare of critical praise across the industry. The program cost $69.95, with a limited promotional version released for free. Buyers could fly through a 3D digital model of the earth, at unprecedented levels of resolution and speed, although the early versions still had their limitations as they were only able to draw on a database of five to six terabytes of information. The full-earth coverage was disappointingly low-resolution, and many major cities outside the United States were poorly represented and some not visible at all. Keyhole simply could not afford to license enough data from commercial satellite companies to cover the whole earth, so even the UK was only visible at a resolution of 1 kilometre, making it impossible to make out streets. The elevation was often unaligned, with blurry imagery, and the application's perceptible 'flatness' made its 3D claim questionable to many reviewers.

Nevertheless, its usefulness soon became clear to those well beyond the real-estate market. When American and coalition forces invaded

Iraq in March 2003, the US news networks repeatedly used Earthviewer to visualize bombing targets across Baghdad. Newspapers reported that the coverage was 'making a surprise star of a tiny tech company and its super-sophisticated 3D maps'. As users overwhelmed the website and crashed it, CEO John Hanke is reported to have said, '[t]here are worse problems to have.'[44] The CIA was already taking an interest in Keyhole, and just weeks earlier had invested in the company through In-Q-Tel, a private non-profit company funded by the agency. The investment was In-Q-Tel's first with a private company on behalf of the National Imagery and Mapping Agency (NIMA). Formed in 1996 and run out of the Department of Defense, NIMA's mission was to provide accurate geospatial information in support of military combat and intelligence. Announcing its investment in Keyhole, In-Q-Tel revealed that in 'demonstrating the value of Keyhole's technology to the national security community, NIMA used the technology to support United States troops in Iraq'.[45] What exactly Keyhole did for the CIA remains unclear, but the injection of capital meant that the company's short-term success was assured. By late 2004 it had launched six versions of Earthviewer.

Then along came Google. In October 2004 the Internet search engine announced it had acquired Keyhole for an undisclosed sum. Jonathan Rosenberg, Google's vice-president of Product Management, expressed his delight. 'This acquisition gives Google users a powerful new search tool, enabling users to view 3D images of any place on earth as well as tap a rich database of roads, businesses and many other points of interest. Keyhole is a valuable addition to Google's efforts to organize the world's information and make it universally accessible and useful.'[46] Looking back, Avi Bar-Zeev saw the acquisition of Keyhole as providing Google with the technology to design an application 'to work like a physical globe on steroids'.[47] But at the time nobody seemed aware of just how important the purchase would prove to be for Google's wider business model.

The story of Google's rise to global pre-eminence has been told elsewhere,[48] but a brief account of its emergence as one of the key players in the online world provides some explanation of why Keyhole was such an important addition to the company. Google's founders, Sergey Brin and Larry Page, met at Stanford University in 1995 working as Ph.D. students in computer science. The World Wide Web was still in its infancy, and both Brin and Page grasped the massive potential of developing a search engine that could navigate users around its myriad sites and links. Search engines like AltaVista lacked the ability to conduct 'intelligent' searches that could organize information in terms of reliability and relevance, and weed out the Web's more unsavoury elements (including pornography).

For Page and Brin, looking at the situation in the late 1990s, the challenge was obvious. 'The biggest problem facing users of web search engines today', they said in April 1998, 'is the quality of the results they get back. While the results are often amusing and expand users' horizons, they are often frustrating and consume precious time.' Their solution was PageRank (a punning reference to Page's name), which attempted to measure the importance of a particular webpage by assessing the number and quality of hyperlinks to it. The cartographic language used by Brin and Page from the outset in describing PageRank is striking. 'The citation (link) graph of the web is an important resource that has largely gone unused in existing web search engines,' they wrote in 1998. 'We have created maps containing as many as 518 million of these hyperlinks, a significant sample of the total. These maps allow rapid calculation of a web page's "PageRank", an objective measure of its citation importance that corresponds well with people's subjective idea of importance.'[49] The result was a system that still drives each Google search – estimated at over 34,000 each second (2 million searches per minute or 3 billion per day) in 2011.[50]

In September 1997 Brin and Page registered 'Google' as a domain name (the intention was to use the name 'googol', the mathematical term for a one followed by a hundred zeros, but it was misspelt during the online registration). Within a year they had indexed 30 million pages online, and by July 2000 the figure stood at a billion. In August 2004 Google went public at $85 a share, raising $2 billion in the largest technology flotation ever. Between 2001 and 2009 its profits soared from an estimated $6 million to over $6 billion, with revenue of over $23 billion, 97 per cent of which came from advertising. With assets currently estimated at over $40 billion, Google processes 20 petabytes of information a day, and all with a global workforce of just 20,000, including an estimated 400 working on their geospatial applications.[51] With such an extraordinary rise came an equally innovative business philosophy. As well as wanting to organize the world's information and making it universally accessible, Google is driven by a series of beliefs laid out in its mission statement: 'democracy on the web works'; 'the need for information crosses all borders', and most contentiously of all, 'you can make money without doing evil'.[52]

By 2004 Google had fulfilled Claude Shannon's theory of quantifying information digitally: the question was how could such information be commodified and translated into financial profit? Google's motives for acquiring Keyhole were inextricably linked to answering this question, and also showed its ability to grasp how the Internet was changing. Rather than just passively viewing information, the online community was looking for greater interaction with the production of content and an increased capacity to manipulate it – a shift known as Web 2.0, which is characterized by blogging, networking and uploading a variety of different media. Google knew that if its ambition was to 'organise the world's information', it needed some way of depicting its geographical distribution and enticing commercial and personal users to buy and then interact with it. What it in fact needed was the largest virtual GIS application available, and Keyhole's Earthviewer provided the answer. Google's first move following its acquisition was to slash the price of buying Earthviewer from $69.95 to $29.95. They then got 'under the hood', in Jonathan Rosenberg's words, as they planned to rebrand it. In June 2005, eight months after first acquiring Keyhole, the company announced the launch of its new, free downloadable program: Google Earth.

The first reviews were ecstatic. Harry McCracken, editor in chief of *PC World*'s magazine and website, tested the application days before its official launch and called it 'spellbinding'. It ranked, he wrote, 'among the best free downloads in the history of free downloads'. He went on to outline the application's benefits. It did not need a super-powerful PC to run, it enabled users to swoop and circle across the world, and cities and landscapes featured amazing 3D renderings that 'are, indeed, wondrous'. Moving on to the drawbacks, McCracken admitted, 'Google Earth is so spectacular, particularly for a free programme, that my first impulse was to feel guilty about criticising it.' But the image resolution varied enormously, and some places were still not locatable (McCracken had a lot of trouble finding Hong Kong and Parisian restaurants). Data for the rest of the world was way behind that for the United States, and McCracken complained of difficulties in establishing what the application 'does and doesn't know'. He wondered how MSN's forthcoming Virtual Earth software would compare to Google Earth, but, considering the latter had been released in a beta (or trial) version, McCracken assumed, rightly, that it would quickly evolve.[53]

McCracken understood that Google Earth was effectively an updated version of Keyhole's Earthviewer (they both used the same code base), and that very little separated the two. What was new was the sheer amount of information that stood behind Google Earth. Google had poured in hundreds of millions of dollars of investment to buy and upload commercial satellite and aerial imagery that virtually no other company had the resources or foresight to spend. When Earthviewer was first demonstrated to Sergey Brin, he thought at one level it was simply 'cool'.[54] But Google's activities prior to their acquisition of the company suggest other factors were already at work. As early as 2002, long before it acquired Keyhole, Google had started buying high-resolution satellite imagery from companies like DigitalGlobe, whose two orbital satellites now capture imagery of up to a million square kilometres of the earth's surface every day, at a resolution of less than half a metre. Google takes this data and captures it with scanners capable of 1,800 dots per inch, or 14 microns. The imagery is then colour-balanced and 'warped' to account for the curvature of the earth's surface. It is then ready to be accessed by users. But Google does not rely on satellite imagery alone. It also uses aerial photography taken at an elevation of between 4,500 and 9,000 metres, using aeroplanes, hot-air balloons – even kites.[55] The need to diversify its access to photographic data stems from its powerlessness in preventing the data it receives being blurred out. Press stories from early 2009 claiming that the company censored sensitive locations by blurring places such as the US Vice President's residence proved to be inaccurate: the censorship apparently lay with the initial data obtained direct from the US military, not Google.[56]

The company's diversification was also driving another initiative. Just weeks before its acquisition of Keyhole in October 2004, Google had also acquired Where2, a small Australian-based digital mapping company, which began work on a new Google-branded map application. In February 2005, four months before Google Earth hit the market, Google announced the launch of Google Maps.[57] Ultimately, the synergy between the two applications would enable the viewer to see a graphic, virtual map overlaid on a photo-real image of the earth's surface, and today users are able to move between the two, depending on what kind of information they wish to access.

Having evolved through seven different versions since its launch in 2005, Google Earth continues to improve its realism and resolution, now at an unparalleled level. In 2008, in an attempt to expand its data, Google extended its commercial agreement with DigitalGlobe, also signing another deal with their rivals GeoEye to use its satellite data with a resolution of just 50 centimetres (the satellites have an even higher resolution of just 41 centimetres, but the licensing terms with the US government forbid its commercial release).[58] The rocket used by GeoEye to launch the satellite into orbit even featured Google's logo. On Google Earth the most recent challenge has been the introduction of 3D terrain modelling. Initially the application effectively showed a satellite image draped on a three-dimensional representation of the landscape, but, as McCracken noticed in 2005, it could not represent features like buildings. The information required digitally to see into the distance when looking at the earth horizontally or from an oblique angle is even more complicated than that needed for the classical aerial perspective looking directly down from above. Google's solution was to use a technique called 'ray tracing', which uses geometry to mimic the human eye. The application identifies the direction the viewer is looking in, and fills that part of the screen first before accessing the surrounding data that defines the eye's peripheral vision.

But the recent technical innovations within Google Earth are not the only developments within the company's plans for its geospatial applications. Michael T. Jones, Google's current Chief Technology Advocate and co-founder of Keyhole, recently claimed that Google Maps API is now used by more than 350,000 websites across the globe.[59] In June 2008 Google launched a new product called Map maker, within its Maps application, that now enables anyone in more than 180 countries to add or edit features such as roads, businesses and schools in their local areas, which are then incorporated into Google Maps. The information submitted is then moderated by other users and checked by Google in a peer-review system that the company claims enables users to make their own maps, while also benefiting from what is effectively free geographical data.

One important consequence of this initiative is that the dream (or fear) of a universally standardized virtual map of the world will never be realized. Ed Parsons, Google's geospatial technologist based in London, admits that Google initially held 'a naive view that we could have one global representation of the world'.[60] But once it became clear that national and local users wanted to retain certain ways of representing physical and human geographical features, Google decided to model a basic representation of the earth over which users added their own culturally specific codes and symbols. Critics argue that the Map maker program lacks the professional moderation of an organization like the Ordnance Survey, and that Google is effectively gathering information for free. But there is no doubt that this innovation allows people to make an image of their immediate environments in a way that has no parallel in the history of mapmaking.

The current excitement about the capabilities offered by Google Earth is understandable. Although the application is still in its relative infancy, with further developments planned for its global coverage and three-dimensional modelling, it is now technologically possible to envisage Borges's fantasy of the map on a scale of $1:1$. Parsons claims that 'if you talk to most people involved in internet mapping and doing what we do we completely accept the fact that you could build a one to one map'. But unlike the traditional paper map envisaged by Borges, such a virtual map would, Parsons says, operate 'at multiple levels of reality'. Google are storing different kinds of information that can be retrieved at any moment and layered upon a $1:1$ scale geospatial image: data on people's social networks, capital flows, underground transportation links, and a variety of commercial information, all of which can be called up instantaneously. The world picture of 'real virtuality' that Google Earth will soon be able to offer us will be limited only by military and legal constraints: the US military infrastructure still restricts access to satellite imagery of the highest resolution, including commercially available data, and how concerns over personal privacy will play out in the new realm of spatial law remains unknown. At 10 centimetre resolution, satellite imagery can now identify an individual's face, but until the law establishes whether such data can be made freely available, Google must wait.

In the meantime, the company continues to develop its geospatial applications alongside its search engine to create what is in effect a gigantic map of the Web. Brin and Page anticipated this development by the early establishment of PageRank, but their embryonic understanding of virtual mapping was very different to traditional definitions of cartography as mapping physical terrain. If the Web is a 'network of hyperlinked documents accessible via the Internet',[61] then Google is creating an infinite virtual map that attempts to represent an ever-expanding world of information. Google Earth is a compelling adjunct to this process, which allows people to start by looking at physical terrain, and from there drill down through potentially limitless layers of digital information, most of which cannot be 'seen' by the human eye. In April 2010 the Earth application became even more central, as Google integrated it into its 'Maps' site, allowing users to move seamlessly from one to the other.[62] For Google, one justification of its geospatial applications is that the digital image of the earth becomes the medium through which all information is accessed; writing in 2007, Michael T. Jones claimed that Google 'inverts the roles of Web browser as application and map as content, resulting in an experience where the planet itself is the browser'.[63] The Earth application – according to Google – is the first place a viewer goes to access and view information. This seems, for the moment at least, to be a completely pure definition of a world map made up from its own cultural beliefs and assumptions, all of which are now potentially available at the click of a computer mouse.

The scale of the data that can be uploaded onto the virtual map shows no signs of reaching its limit. In 2010, Ed Parsons estimated that if all the data ever recorded by humanity up to 1997 were digitized, the next thirteen years of Internet usage would double that figure. After that, he predicts it will double again in just eighteen months. Estimates suggest that the Web's current size is anything up to a staggering 1,800 exabytes (one exabyte contains one quintillion bytes, or $10^{30}$), with nearly 12 billion pages.[64] But capacity is not the problem. Parsons claims that 'if every planet had an Internet the size of the current Internet we'd easily fill it'. The challenge, as always in the history of mapmaking, is how to keep up with the accumulation, even overload, of information. Google and its geospatial applications may have the ability to keep pace with this phenomenal increase in data, but to map it will be an ongoing and – as Ptolemy, al-Idrīsī and the Cassinis discovered – an interminable process.

In 1970, the American geographer Waldo Tobler famously invoked what he called 'the first law of geography: everything is related to everything else, but near things are more related than distant things'.[65] Tobler, an early pioneer of computerized mapmaking, coined his First Law while developing computer-based simulations of population growth in Detroit. With its implications of global interconnectedness and the importance of computer technology in mapping human geography, Tobler's First Law acts as a metaphor for the Internet, and has become a driving principle for the geospatial technologists at Google Earth. The First Law acknowledges the fact that, since Ptolemy, geography has always been egocentric. Its users begin by finding themselves or their community on a map, but then gradually lose interest in 'distant things' on its margins. When they first log on to Google Earth (or any other geospatial application) most people begin by inputting their own location (in terms of a region, city, town or even a street), rather than using the application to expand their geographical knowledge.

For Google, Tobler's First Law offers a way of not only mapping the world online, but also of making money from its information. Ed Parsons points out that 'for us Google Earth and Google Maps are the visual representation of geography. But geography is buried in almost everything we do, because almost all information has some geographical context to it.' He estimates that over 30 per cent of all Google searches have some explicit geographical element: Google is effectively organizing information geographically, as well as alphabetically and numerically. Geospatial applications are now firmly embedded in the Google search experience. Any search allows for immediate comparison with its maps application as a way of situating information in space. If I type 'Chinese restaurants' into Google, I will be confronted with a list of seven restaurants in my local town, each with a place page alongside a Google map showing me their location. It is an aspect of the company's investment in these applications that most geographers have failed to notice, as they fixate on the strictly cartographic aspects of Google Earth and Maps. Parsons argues, directly following Tobler, that the increasing mobility of both individuals and their access to geospatial applications (such as via mobile phones) means that information 'that is close to us is going to be more important than information that is further away'. His example is advertising. If a business can 'show their advertisements to people within 100 metres of their business who have in the past expressed a preference for buying their sort of goods, that's a real lead. People would pay good money for that kind of information.'[66] A glance at Google's annual profits suggests that businesses are indeed paying for such information. In the hands of Google, Claude Shannon's theory of countable information has finally found its market. The far-away is close at hand in virtual images of elsewhere which, for Google, are proving to be extremely profitable.

Michael T. Jones anticipated Parsons's point in an interview in May 2006, just a year after Google's acquisition of Keyhole. Jones argued that

> to say that Google launched Google Earth to not make any money really doesn't make any sense. Google is a real business that makes profit. Google Earth connects the world with the world's information in a way that was never before possible and has excited the imagination of tens of millions of people. That's a good thing for Google. Even if our business model was to attract attention to Google and let people use Google search to pay for it, it's worked pretty well. So people who feel we went into Google Earth without the intention to make money really don't understand our business. Our business is not about the GIS components of our work. Those are the tools we use to build our business.[67]

The result is a model of e-commerce that economists call 'Google-nomics'. By 2002, Google had developed a new method of raising revenue by selling online advertising space in 'what may be the most successful business idea in history': Adwords.[68] Adwords uses a compli-cated algorithm to analyse every single Google search and determine which advertisers get to display their business on the 'sponsored links' found on every results page. Businesses tender sealed bids on how much they are willing to pay Google each time a user clicks on their advertise-ment, in the world's biggest and fastest auction. In a fraction of a second Google determines who will offer to pay the most, ranking their adverts on its sponsored links accordingly. Every time anyone searches on Google, they are unwittingly participating in a ceaseless, multi-billion-dollar global auction. Google sells the scheme to advertisers as enabling them to 'connect with potential customers at the magic moment they're searching for your products or services, and only pay when people click your ads'.[69] It represents what Google's head of advertising calls 'the physics of clicks'. The pursuit of profit is simultaneously also a method of acquiring more data. 'Selling ads', writes Steven Levy,

> doesn't generate only profits; it also generates torrents of data about users' tastes and habits, data that Google then sifts and processes in order to

predict future consumer behaviour, find ways to improve its products, and sell more ads. This is the heart and soul of Googlenomics. It's a system of constant self-analysis: a data-fueled feedback loop that defines not only Google's future but the future of anyone who does business online.[70]

At the centre of Googlenomics are the company's geospatial applications. As Adwords allows companies to target their advertisements more effectively, so Google Earth and Maps locate their product in both physical and virtual space. The geospatial application has found its definitive use, as Michael T. Jones suggested in a recent lecture during which he grandly announced what he called 'The new meaning of maps'. Jones defines the online map as 'a place of business', an 'application platform' from where businesses trade what he calls 'actionable information'.[71] There is clearly an increasingly commercial motivation behind the company's development of geospatial applications, but neither Jones's 'new meaning of maps' nor their intimate relation to business are quite as new as they would like to think. Google Earth is part of a long and distinguished cartographic tradition of mapping geography onto commerce that stretches right back at least to al-Idrīsī's regional maps of the commodities of the Mediterranean. It underlies the world maps of Ribeiro, driven by access to the commercial wealth of the Indonesian archipelago, Mercator's projection for navigators, Blaeu's atlases for the wealthy merchants and burghers of Holland, and even Halford Mackinder's world map of imperial conflict over increasingly competitive markets. Where maps and their makers are motivated by the apparently disinterested pursuit of geographical information, its acquisition requires patronage, state funding or commercial capital to make it viable. Mapping and money have always gone hand in hand and have reflected the vested interests of particular rulers, states, businesses or multinational corporations, but this does not necessarily negate the innovations made by the mapmakers they have financed.

But there is a crucial difference between what Google is doing and what went before, which is not simply about scale: it concerns the computerized source code used to build its geospatial applications, as well as Adwords and PageRank, which in principle remains true to Claude Shannon's basic formulations of how to communicate fungible information. For obvious commercial reasons, Google does not disclose the specific details of its code,[72] which means that, for the first time in recorded history, a world view is being constructed according to information which is not publicly and freely available. All prior methods of mapmaking ultimately disclosed their techniques and sources, even if, as in the case of sixteenth- and seventeenth-century mapmaking, they tried – but failed – to withhold its detail from their competitors. And even such examples of mapmaking were not exclusively designed to extract financial profit, nor were they built with the sheer scale of data that allows Google to restrict the circulation of its code within the public domain. Google Maps API allows users to reproduce Google's maps, but not to understand its code; and, like Adwords, by tracking the circulation of its maps, Google can simply expand its database on users' tastes and habits. The licensing terms of Google Maps API also reserves the company's right to place advertisements on websites that use their maps at any point in the future: it would be an aggressive and controversial move, but Google will not rule it out. 'So control of code is power,' writes the architectural historian William J. Mitchell. 'Who shall write the software that increasingly structures our daily lives? What shall that software allow and proscribe? Who shall be privileged by it and who marginalized? How shall the writers of the rules be answerable?'[73]

In a similar vein, the company's ongoing project to digitize the world's libraries is an attempt to make knowledge freely and instantly accessible online – although Google's critics argue that it represents an attempt to create an effective monopoly on such data, and point to the prevailing restrictions on it (Google's books cannot be printed, nor viewed in their entirety: such restrictions will probably only be lifted if the user pays a fee).[74] In March 2011 the US Federal judge Denny Chin rejected Google's planned $125 million deal with author and publisher groups to put over 150 million books online because it would give the company 'a significant advantage over competitors, rewarding it for engaging in wholesale copying of copyrighted works without permission', and arguably give Google a monopoly on the book search market.[75] In September 2011 the company faced a US Senate committee to answer allegations of abusing its predominant position in universal online searches to give its own services better placement, charges which seem likely to grow in subsequent years.[76]

Google responds that it is not in its interests to lose customer trust by compromising its status as an impartial provider of online search information. It also argues that copyright holders would benefit financially from the digitization of books. Ultimately, it insists that its users (and registered online groups like Google Earth Community) would not tolerate any move towards a monopoly of information, political partiality or acceptance of censorship. Nevertheless, fears remain about Google's ambitions in the field of geography as well as books. Users of its applications may not be sufficiently motivated or organized to resist a monopolization of information. Only governments can put in place the necessary checks and balances through competition laws. In the meantime, it seems unlikely that the company can sustain the growing tension between its commercial imperatives and a more progressive, interactive ethos. In this respect, the geospatial technologists at Google Earth resemble the humanist mapmakers of the sixteenth century, like Diogo Ribeiro and Martin Waldseemüller, working athwart political and commercial pressures to broaden the horizons of geographical information. But in contrast to the sixteenth century, modern civil societies have governments, NGOs and online communities to monitor and, where they feel it appropriate, criticize companies like Google.

We should also remain alive to the limitations of these geospatial applications. Technical problems remain. Google still has some way to go to provide standard high-resolution data of the entire planet – although it officially relishes the challenge of improving its coverage. In a survey of the four main online mapping sites (Google Earth, MSN Maps, MapQuest and Multimap), the Finnish Information Technology Consultant Annu-Maaria Nivala and her colleagues carried out a series of controlled tests with a group of users, who identified 403 problems, from difficulties with search operations to issues with the user interface, map visualization and map tools. The online maps were often 'messy, confusing, restless and awful to look at on screen'. Projections often looked 'weird', imagery was 'overloaded with information', panning and zooming was erratic, layout was poor, and data was inconsistent, leading participants to ask the age-old question, 'Who decides what is included or not in the map?'[77] Some of these problems could be remedied if standardized maps were adopted across all the competing commercial sites, but the chances of this happening in the near future are extremely unlikely.

Notwithstanding the ongoing technical challenges facing geospatial applications, an enduring problem is the so-called 'digital divide'.

Although Google Earth has been downloaded by over half a billion people, far eclipsing the estimated 80 million copies of the Peters projection circulated since the 1970s, that figure should be set against a global population of 7 billion people, many of them not only unable to access the Internet, but unaware of its existence. By 2011, out of a world population of nearly 7 billion people and an estimated 2 billion online users, only North America, Australasia and Europe could boast an Internet penetration rate of over 50 per cent. With a world average of 30 per cent, Asia's rate was 23.8 per cent and Africa's just 11.4 per cent or 110 million Internet users.[78] This is a problem not just of access to technology, but of access to information (or what's known in development studies as A2K).[79] These figures make the meaningful usage of applications like Google Earth largely limited to predominantly Western, educated elites. This also means that such applications are mapping parts of the world where the population has little or no knowledge of what is happening.

Nevertheless, Google Earth is a remarkable piece of technology with enormous potential, which probably signals the death, or at least in time the eclipse, of paper maps, as users increasingly favour online GPS technology over traditional maps and atlases of countries, cities and towns. At the moment, it allows anyone using the Internet unprecedented access to geographical information, and has been used for individuals and non-governmental organizations in a variety of progressive environmental and political situations. Google has created a personalized way of using maps, and of allowing them to be discarded, which is unprecedented, and promises future innovations which will take us further than ever from traditional perceptions of maps, with what Parsons calls 'augmented reality applications [which use computer-generated input such as sound and graphics to modify real-world environments] that overlay information onto an image of the world that may once have been represented by a map'.[80]

Despite these developments, Google Earth retains continuities with more traditional methods of cartographic presentation. The layout that invites you to view the whole earth and then descend to view continents, countries and discrete regions draws on the atlas format popularized by Mercator and Blaeu. The belief that its technology somehow 'mirrors' the earth in a transparent act of representation has been central to global mapping practices since at least the Renaissance, as is the abiding belief in the power of mathematics to project the globe onto a flat surface. The whole earth on Google Earth's home page may look like photo-real satellite imagery, but it still represents a three-dimensional object projected onto a plane surface, or a screen. As with all such images, it selects a particular projection, in this case the General Perspective Projection.[81] In choosing this projection, Google Earth brings this book full circle, as its inventor was none other than Ptolemy. In his *Geography*, Ptolemy described this projection of the 'globe in a plane' which 'is assumed to occupy the position of the meridian through the tropic points'.[82] On this projection the globe is viewed from a finite point in space, from either a vertical perspective (as described in Ptolemy), or a tilted one. Initially the projection had little practical value, as it was unable to represent the earth in any detail. However, photography and space travel revived the projection, because photographs of the earth like the Apollo 17 example have a tilted vertical perspective which mimics how the eye would see the globe from a distance. For applications like Google Earth, the General Perspective Projection is an ideal way of representing the three-dimensional globe in two dimensions, because it shows a pictorially satisfying image of the earth, while also allowing the viewer to then zoom down and fly across its surface to 'see' enhanced surface details, thanks to clip-mapping and ray tracing. Nevertheless, Google Earth still makes decisions about how it represents the globe in this way at the expense of geographical features such as the accurate representation of the polar regions. Geospatial applications convert the earth into strings of Claude Shannon's ones and zeros, which are then rendered by algorithms into a recognizable image of the world around us. So in these ways, Google Earth's methods are as old as Ptolemy, with his rudimentary geometry of seeing the globe from above, and his digital representation of the world according to the numerical calculation of latitude and longitude.[83]

The anxieties created by the rise and evolution of geospatial applications like Google Earth are nothing new. Similar fears have accompanied major shifts in the mediums of mapmaking at various points in history, from stone through parchment and paper, to manuscript illumination, woodcut printing, copperplate engraving, lithography and computer graphics. At each point mapmakers and map users have managed to exploit the religious, political or commercial pressures that shape maps according to their own particular interests. Current debates surrounding Google and its geospatial applications, which question whether its free dissemination of information and clashes with governmental authority are the result of a long-term monopolistic business model or of an inherently democratic belief in the power of the Internet, in some respects simply reflect an intensification of these historical trends.

As with most multinational companies, there are undoubted tensions within Google as to its future direction, but it seems increasingly unlikely that it can balance its aspirations for enormous profitability alongside its ostensibly democratic ideals. Like Claude Shannon's theories of electronic communication, the initial impulses that drove Google were predicated on the communication of quantifiable, noise-free information that can be circulated on a hitherto unimaginable scale. But Google has gone a step further in developing a method for not only quantifying geographical information, but also giving it a monetary value. The history of maps has never previously known the possibility of a monopoly of valuable geographical information falling into the hands of one company, and, as Google's share of the global online search market reaches 70 per cent, those working in the Internet industry are worried. Simon Greenman believes that although Google 'have done a wonderful job with Earth, they also have the potential to dominate world mapping on a scale that is historically unprecedented. If we fast forward ten to twenty years Google will own global mapping and geospatial applications.'[84] The company likes to say that, thanks to the ability of its online maps to pinpoint our location anywhere on the planet, we are the last generation to know what it means to be lost. We may also be the last generation to know what it means to see mapmaking generated by a range of individuals, states and organizations. We are on the brink of a new geography, but it is one that risks being driven as never before by a single imperative: the accumulation of financial profit through the monopolization of quantifiable information.

# Conclusion
## The Eye of History?

Each map described in this book is a world unto itself. Nevertheless, as well as providing a unique image of its time and place, I hope I have shown that certain features are common to all twelve. Each one accepts the reality of an external world, whatever its shape and dimensions. Such a belief is shared by virtually all cultures, as is the desire to reproduce it graphically in the form of a map. But the perception of that terrestrial world and the graphic methods used to express it differ enormously, from Greek circles to Chinese squares and Enlightnment triangles. Each one also accepts (either implicitly or explicitly) that the earth cannot be comprehensively mapped onto a flat surface. Ptolemy conceded that his projections were unsatisfactory responses to the problem; al-Idrīsī acknowledged the dilemma, but elided it in favour of sectional maps; Mercator believed that he offered the best available compromise; and Peters simply underlined the problem, in the process anticipating the current proliferation of geospatial applications, which offer a range of whole earth images across a variety of cartographic imperfections.

I hope the book has also shown that no world map is, or can be, a definitive, transparent depiction of its subject that offers a disembodied eye onto the world. Each one is a continual negotiation between its makers and users, as their understanding of the world changes. World maps are in a perpetual state of becoming, ongoing processes that navigate between the competing interests of patrons, makers, consumers and the worlds from which they arise. For the same reason, it is impossible to ever define a map as finished: the Cassini survey is the most obvious example of an endlessly unfolding map, but Ribeiro's series of world maps from the 1520s provide a similar example, and Blaeu completed only the first volume of an atlas that could have gone on indefinitely. Much as maps might try to encompass the world according to a defining principle, it is a continuously evolving space, and one which does not stop and wait for the completion of the mapmaker's labour – a fact that Google has now grasped and turned to its advantage better than any of its competitors.

Maps offer a proposal about the world, rather than just a reflection of it, and every proposal emerges from a particular culture's prevailing assumptions and preoccupations. The relationship between a map and these assumptions and preoccuptions is always reciprocal, but not necessarily fixed or stable. The Hereford *mappamundi* proposes a Christian understanding of the creation and anticipated ending of the world; the Kangnido map offers an image of the world with an imperial power at its centre and in which the belief in geomantic 'shapes and forces' are central to earthly existence. Both are logically consistent with the cultures from which they emerge, but they also extrapolate from systems of belief to aspire to a comprehensive view of the whole world. This reciprocal relationship is characteristic of all of my twelve maps. Each one is not just *about* the world, but also *of* it. For the historian, they all create the conditions for understanding a prevalent idea – religion, politics, equality, toleration – through which we make sense of ourselves, at the same moment as we come to understand the world around us.

Despite attempts by cartographers like Arthur Robinson to explain the cognitive processes by which maps transform people's beliefs and imaginative geography, it remains difficult to establish how people internalize the ways in which a map represents spatial information about the world around them. In their multi-volume *History of Cartography*, J. B. Harley and David Woodward (the latter one of Robinson's students) admitted that '[e]vidence for the level of map consciousness in early societies' is 'virtually nonexistent'.[1] A map can successfully innovate, but still apparently fail to affect people's perception of the world. Al-Idrīsī's maps proposed an ideal of the world born of cultural exchange between Islam and Christianity, but the collapse of the syncretic culture that produced then in twelfth-century Sicily meant that very few people probably saw them, and even fewer had the opportunity to accept their view of the world. In contrast, surveys of how professionals used Arno Peters's world map revealed little grasp of its flawed detail, but widespread acceptance of its demands for geographical equality. At other times individuals may suddenly disclose how a map corresponds to a prevalent concern or anxiety, as when twelfth-century Chinese poets describe maps as representing a lost, mythic empire, or the moment at which Napoleon's soldiers explain the magical ability of Cassini's maps to show an astonished priest the extent of the French nation. A map can draw assumptions from its culture which are either accepted or rejected by its users, because such assumptions are constantly being tested and renegotiated.

One assumption of the objective, scientific mapmaking that emerged in eighteenth-century Europe and which motivated the Cassinis and their followers was that at some point it would be possible to propose a universally accepted, standardized map of the world. Even today, no such map exists, even among the profusion of online geospatial applications, evidence that we must always make compromises when we choose our partial maps of the world, and accept that they are 'never fully formed and their work is never complete'.[2] I therefore end with the story of one final initiative that tried, but inevitably failed, to map the whole world.

In 1891, the internationally respected German geomorphologist Albrecht Penck proposed a new cartographic initiative at the Fifth International Geographical Congress in Berne. Anticipating Halford Mackinder's views on the state of geography at the end of the nineteenth century, Penck argued that enough information regarding the mapping of the earth's surface was now available to justify the creation of an international map of the world. Penck's scheme involved what he called 'the execution of a map of the world on the scale of 1:1,000,000 (15.78 miles to 1 inch [10 km to 1 cm]).' Penck pointed out that current world maps 'are not uniform in either scale, projection or style of execution; they are published at different places all over the world, and are often difficult to obtain'.[3] His solution was the International Map of the World (IMW).

Based on international collaboration between the world's leading mapmaking agencies, the IMW would involve the creation of 2,500 maps covering the entire earth. Each one would cover four degrees of latitude and six degrees of longitude, using a single projection – the modified conic – as well as standard conventions and symbols. The projection did not need to represent the entire globe accurately because, in an argument reminiscent of al-Idrīsī's method, Penck emphasized that it would be impractical to ever join all 2,500 maps together: the maps of Asia alone would cover a space of 2.8 square metres. In an echo of the great cosmographies of Mercator and Blaeu, Penck suggested that his idea 'might be rather described as an "Atlas of the World"'.[4] The prime meridian would run through Greenwich and the Latin alphabet would be used for all place names. The representation of physical and human geography would be strictly uniform, down to the width of lines used to represent political boundaries, and the colours chosen to depict natural features such as forests and rivers.

Penck estimated that 'the cost of production may be set at about £9 per square foot for an edition of 1,000 copies'. He admitted that if 'the whole edition were sold at 2s. a sheet, there would be a deficit of over £100,000', but pointed out that governments had spent far greater sums on scientific and colonial expeditions, such as 'the expenditure on Arctic exploration in the forties and fifties and on African exploration more recently'. The great imperial powers – Britain, Russia, the United States, France and China – would be responsible for the creation of over half the maps. In a plea for international cooperation regardless of cultural and ideological differences, Penck believed that if 'these countries give their approval to the scheme, its success will be assured, even if in some cases the work has to be done by private individuals or at the expense of geographical societies instead of by Government'.[5]

It was an idealistic plan, a summation of the Enlightenment belief in scientifically accurate standardized realism, and a global fulfilment of the national method of mapping represented by the *Carte de Cassini*. But there were two particular problems. It was not at all clear how countries with little experience of surveying would complete such a task, especially if they lacked the necessary financial resources, and Penck was unable to provide a sufficiently compelling account of the map's potential benefits. He claimed that the 'circumstances and interests of our civilised life make good maps almost a necessity. Maps of our own country are absolutely indispensable; commercial interests, missionary undertakings, and colonial enterprise create a demand for maps of foreign countries, while of the maps required for educational purposes and as illustrations of contemporary history, the name is legion.'[6] This was not good enough for many of the project's critics, one of whom wrote in 1913, 'I do not know that any very definite statement has ever been made of the precise purpose of this map ... We may think of it, perhaps, as meant for the use of the systematic geographer, whenever it shall have been determined what is the function of that person.'[7] Penck's conviction in the utility of the map fell back into the prevailing values of his day: it would define the modern nation state, facilitate global capitalism, enable the spread of Christianity, and justify the colonial expansion of Europe's empires. If, as Penck claimed, a 'uniform map of the world would be at the same time a uniform map of the British Empire', that might benefit the British, but not necessarily anyone else.

The Berne Congress agreed to investigate the implementation of the IMW, and subsequent congresses continued to support the idea, but with little practical consequence. It was not until 1909 that an International Map Committee (including Penck) met in London's Foreign Office. The committee was convened by the British Government, which grasped the advantages of shaping the project according to its own interests. The committee agreed on the form of the map's detail, including an indexed diagram of the whole project, and plans to produce its first maps. But by 1913 only six sheets of Europe had been drafted, and most of the countries represented rejected them for their own national and political reasons. A second meeting was convened in Paris in 1913 to establish uniformity across each map, but its deliberations received a setback with the news that the United States had chosen to implement its own independent 1:1,000,000 maps of South America.

As the IMW foundered, the British delegation proposed that a central bureau should run the project from the offices of the Ordnance Survey, with the Royal Geographical Society providing private funding ostensibly independent of politics. Few people were fooled; the initiative was supported by the Geographical Section of the General Staff (GSGS), otherwise known as MO4, part of the British government's intelligence organization, responsible for gathering and producing military maps. As war was declared in 1914, the OS, with the support of the RGS and the GSGS, produced a series of 1:1,000,000 maps of Europe, the Middle East and North Africa in support of the Allied war effort.[8] The national and political differences that Penck hoped would be transcended in the map's creation ultimately turned it into an instrument of warfare.

After 1918, the project limped on, but Penck distanced himself from it, disillusioned by what he regarded as the political injustice of the

Treaty of Versailles (which imposed its own cartographic divisions upon a defeated Germany). By 1925 the project's central bureau reported that just 200 1:1,000,000 maps had been produced, and only 21 of them conformed to the original criteria agreed by the delegates at Paris in 1913.[9] By 1939, only another 150 maps had been completed. With the outbreak of the Second World War the OS's involvement in the IMW was effectively over. The RGS's secretary Arthur Hinks concluded that the project's international ethos was mistaken. 'The moral seems to be', he wrote, 'that if you want a general map covering a continent, consistent in style, and available in quantity, then you must make it yourself, and whether you call it international or not is a matter of choice, or expediency.'[10]

As the Second World War established the importance of military control of the skies, aeronautical charts were regarded as more important than the relatively large-scale maps produced under the auspices of the OS. The war also took its toll on the OS itself. In November 1940 the bombing of Southampton destroyed most of the OS's offices and much of the material relating to the International Map. In 1949 those still involved recommended its transferral to the recently formed United Nations. The UN's Charter already acknowledged that 'accurate maps are a prerequisite to the proper development of the world's resources . . . such maps facilitate international trade, promote safety of navigation . . . and provide information required for the study of measures of peaceful adjustment . . . and for the application of security measures'.[11] At the thirteenth session of the UN's Economic and Social Council (ECOSOC), on 20 September 1951, Resolution 412 AII (XIII) was passed authorizing the transfer of the IMW's Central Bureau to the Cartographic Office of the UN Secretariat,[12] and in September 1953 the UN officially took control of the IMW. It inherited a project in disarray, with only 400 finished maps, a fraction of the number required to complete the project. The UN's first published index map provided a global summary of what had been published, revised, republished, received, and what still needed to be mapped. It was chaotic, and what remained to be done was daunting.[13]

As the Cold War intensified throughout the 1950s, it was obvious that the spirit of international cooperation that originally inspired the International Map was dead. In 1956 the USSR put forward a proposal to the ECOSOC for a new world map, based on the scale of 1:2,500,000.

Fig. 38 'Index Map Showing Status of Publication of the International One-millionth Map of the World' (IMW), 1952.

Not surprisingly, considering the UN's investment in the IMW, the proposal was rejected, but, in a bleak irony, the Hungarian National Office of Lands and Mapping took up the project, with the support of the other Communist states behind the Iron Curtain and China. The first printed sheets were published in 1964, and in 1976 the entire map was exhibited for the first time in Moscow. It included 224 full and 39 overlapping sheets, and although it lacked the scale and detail of Penck's original idea and achieved only limited circulation in Eastern Europe, it represented a Russian-sponsored attempt to show that the Soviet Bloc could match anything produced in the capitalist West.[14]

The UN tried to resuscitate the IMW throughout the 1960s, but to little effect, and leading cartographers lined up to castigate its tarnished international aspirations – including Arthur Robinson, who dismissed it as merely 'cartographic wallpaper'.[15] In 1989, the UN finally gave up and terminated the project. Less than a thousand maps were finished, and most were already obsolete. The world had moved on. The US government had already launched the National Center for Geographic Information and Analysis, just one of many state-funded organizations that would signal the birth of online geospatial applications and the death of the dream of an international state-funded world map based on global cooperation.

The nineteenth-century values that inspired the IMW – scientific progress, imperial dominion, global trade and the authority of the nation state – ultimately destroyed this cartographic Tower of Babel. The contradiction of trying to create the map is that the time of its establishment got (or actually did *not* get) the world map it deserved. Its imperial and national demands of transparent international cooperation based on the assumed superiority of Western mapping methods were simply too high, and overwhelmed the project's intellectual aspirations and scientific abilities. All the technical resources and state financial support available throughout the twentieth century could still not produce a standardized world map – and this was even before addressing the irresolvable dilemma of adopting a global map projection. Not only were Lewis Carroll's and Borges's tales of the 1 : 1 scale map pure fantasy, but so it would seem was the 1 : 1,000,000 map of the world.

Today's online geospatial applications show little appetite for revisiting such a project, despite Al Gore's dream of a 'digital earth'.[16] In 2008 a Japanese-led initiative supported by the US and Japanese governments was launched in an attempt to fulfil the dream of the 1 : 1,000,000 map of the world digitally. It is called simply 'Global Map'. The project's website mission statement claims that the 'Global Map is a platform for people to learn about the present state of the earth and take a broad view of the earth for the future'.[17] The fact that most people reading this book have never even heard of 'Global Map' tells you all you need to know about its impact. Even the engineers at Google Earth concede that their dream of a uniform virtual online world map is impossible. Their reason is simple: they want to retain the globe's national, local and linguistic diversity that the International Map wanted to transcend, because in today's global economy, diversity and difference are potentially profitable. Nobody wants to buy a product linked to a map that shows their local area labelled in a foreign language and covered in unfamiliar symbols.

For more than 3,000 years humanity has dreamt of creating a universally accepted map of the world, ever since the anonymous maker of the Babylonian map of the world first shaped his tablet out of the earth's clay. Today, it still seems an idealistic fantasy, and will always be doomed by the impossibility of creating a commonly accepted, global projection of the earth. Despite the claims of Google Earth, will it ever be possible, or even desirable, to create what Abraham Ortelius desired, a comprehensive and universally accepted map the whole earth that can act as the omniscient eye of history?

From a practical point of view, surveyors and geodesists would probably answer yes, but they would need to provide convincing answers as to why such a project would be necessary, notwithstanding the technical problems of projection, scale and execution. Penck never provided one sufficient to withstand the intemperate politics of the twentieth century, and the ineffectiveness of the more recent 'Global Map' suggests that its vaguely environmental mission statement is not the answer either. What all the maps discussed in this book have shown is that their proposal about how to see the world emanates from a particular vision of it, something which both Penck and the 'Global Map' lack. When the sheer scale of implementing such a project still requires some form of state or corporate funding, it is difficult to imagine it could escape the perennial political or commercial manipulation that has so often tried to impose a single image upon the sheer variety of the earth and its people.

But to answer no appears to endorse a partial vision that turns its back on both the inevitability of globalization and the possibility of celebrating a common international humanity through geography. Virtually all twelve maps discussed in this book have successfully struggled with such a partial global vision of the world. Every culture has a specific way of seeing and representing its world through maps, and this is as true for Google Earth as it is for the Hereford *mappamundi* and the Kangnido world map. Perhaps the answer is less an unqualified no, and more a sceptical yes. There will always be maps of the world, and their technology and appearance at some point in the future will make the world map in a modern atlas, and even Google Earth's home page, seem as quaint and unfamiliar as the Babylonian world map. But they will also inevitably pursue a particular agenda, insist on a certain geographical interpretation at the expense of possible alternatives, and ultimately define the earth in one way rather than another. But they certainly will not show the world 'as it really is', because that cannot be represented. There is simply no such thing as an accurate map of the world, and there never will be. The paradox is that we can never know the world without a map, nor definitively represent it with one.

# Notes

## INTRODUCTION

1. J. E. Reade, 'Rassam's Excavations at Borsippa and Kutha, 1879–82', *Iraq*, 48 (1986), pp. 105–16, and 'Hormuzd Rassam and his Discoveries', *Iraq*, 55 (1993), pp. 39–62.
2. The map's transcriptions are quoted from Wayne Horowitz, 'The Babylonian Map of the world', *Iraq*, 50 (1988), pp. 147–65, his subsequent book, *Mesopotamian Cosmic Geography* (Winona Lake, Ind., 1998), pp. 20–42, and I. L. Finkel and M. J. Seymour (eds.), *Babylon: Myth and Reality* (London, 2008), p. 17.
3. Catherine Delano-Smith, 'Milieus of Mobility: Itineraries, Route Maps and Road Maps', in James R. Akerman (ed.), *Cartographies of Travel and Navigation* (Chicago, 2006), pp. 16–68.
4. Catherine Delano-Smith, 'Cartography in the Prehistoric Period in the Old World: Europe, the Middle East, and North Africa', in J. B. Harley and David Woodward (eds.), *The History of Cartography*, vol. 1: *Cartography in Prehistoric, Ancient, and Medieval Europe and the Mediterranean* (Chicago, 1987), pp. 54–101.
5. James Blaut, David Stea, Christopher Spencer and Mark Blades, 'Mapping as a Cultural and Cognitive Universal', *Annals of the Association of American Geographers*, 93/1 (2003), pp. 165–85.
6. Robert M. Kitchin, 'Cognitive Maps: What Are They and Why Study Them?', *Journal of Environmental Psychology*, 14 (1994), pp. 1–19.
7. G. Malcolm Lewis, 'Origins of Cartography', in Harley and Woodward, *History of Cartography*, vol. 1, pp. 50–53, at p. 51.
8. Denis Wood, 'The Fine Line between Mapping and Mapmaking', *Cartographica*, 30/4 (1993), pp. 50–60.
9. J. B. Harley and David Woodward, 'Preface', in Harley and Woodward, *History of Cartography*, vol. 1, p. xvi.
10. J. H. Andrews, 'Definitions of the Word "Map"', 'MapHist' discussion papers, 1998, accessed at: http://www.maphist.nl/discpapers.html.
11. Harley and Woodward, *History of Cartography*, vol. 1, p. xvi.

12. Denis Cosgrove, 'Mapping the World', in James R. Akerman and Robert W. Karrow (eds.), *Maps: Finding our Place in the World* (Chicago, 2007), pp. 65–115.

13. Denis Wood, 'How Maps Work', *Cartographica*, 29/3–4 (1992), pp. 66–74.

14. See Alfred Korzybski, 'General Semantics, Psychiatry, Psychotherapy and Prevention' (1941), in Korzybski, *Collected Writings, 1920–1950* (Fort Worth, Tex., 1990), p. 205.

15. Gregory Bateson, 'Form, Substance, and Difference', in Bateson, *Steps to an Ecology of Mind: Collected Essays in Anthropology, Psychiatry, Evolution, and Epistemology* (London, 1972), p. 460.

17. Lewis Carroll, *Sylvie and Bruno Concluded* (London, 1894), p. 169.

18. Jorge Luis Borges, 'On Rigour in Science', in Borges, *Dreamtigers*, trans. Mildred Boyer and Harold Morland (Austin, Tex., 1964), p. 90.

19. Mircea Eliade, *Images and Symbols: Studies in Religious Symbolism*, trans. Philip Mairet (Princeton, 1991), pp. 27–56. See Frank J. Korom, 'Of Navels and Mountains: A Further Inquiry into the History of an Idea', *Asian Folklore Studies*, 51/1 (1992), pp. 103–25.

20. Denis Cosgrove, *Apollo's Eye: A Cartographic Genealogy of the Earth in the Western Imagination* (Baltimore, 2001).

21. Christian Jacob, *The Sovereign Map: Theoretical Approaches to Cartography throughout History* (Chicago, 2006), pp. 337–8.

22. Abraham Ortelius, 'To the Courteous Reader', in Ortelius, *The Theatre of the Whole World*, English translation (London, 1606), unpaginated.

23. David Woodward, 'The Image of the Spherical Earth', *Perspecta*, 25 (1989), pp. 2–15.

24. Stefan Hildebrandt and Anthony Tromba, *The Parsimonious Universe: Shape and Form in the Natural World* (New York, 1995), pp. 115–16.

25. Leo Bagrow, *The History of Cartography*, 2nd edn. (Chicago, 1985).

26. Matthew H. Edney, 'Cartography without "Progress": Reinterpreting the Nature and Historical Development of Mapmaking', *Cartographica*, 30/2–3 (1993), pp. 54–68.

27. Quoted in James Welu, 'Vermeer: His Cartographic Sources', *Art Bulletin*, 57 (1975), pp. 529–47, at p. 547.

28. Oscar Wilde, 'The Soul of Man under Socialism' (1891), in Wilde, *The Soul of Man under Socialism and Selected Critical Prose*, ed. Linda C. Dowling (London, 2001), p. 141.

29. Denis Wood with John Fels, *The Power of Maps* (New York, 1992), p. 1.

## CHAPTER 1. SCIENCE: PTOLEMY'S GEOGRAPHY, C. AD 150

1. On the stone tower of Pharos, see Rory MacLeod (ed.), *The Library of Alexandria: Centre of Learning in the Ancient World* (London and New York, 2000).

2. See *The Cambridge Ancient History*, vol. 7, part 1: *The Hellenistic World*, 2nd edn., ed. F. W. Walbank *et al.* (Cambridge, 1984).

3. Quoted in James Raven (ed.), *Lost Libraries: The Destruction of Great Book Collections in Antiquity* (Basingstoke, 2004), p. 15.

4. See Bruno Latour, *Science in Action* (Cambridge, Mass., 1983), p. 227, and Christian Jacob, 'Mapping in the Mind', in Denis Cosgrove (ed.), *Mappings* (London, 1999), p. 33.

5. Quoted in J. Lennart Berggren and Alexander Jones (eds. and trans.), *Ptolemy's Geography: An Annotated Translation of the Theoretical Chapters* (Princeton, 2000), pp. 57–8.

6. Ibid., pp. 3–5.

7. Quoted ibid., p. 82.

8. On Ptolemy's life, see G. J. Toomer, 'Ptolemy', in Charles Coulston Gillispie (ed.), *Dictionary of Scientific Biography*, 16 vols. (New York, 1970–80), vol. 11, pp. 186–206.

9. See Germaine Aujac, 'The Foundations of Theoretical Cartography in Archaic and Classical Greece', in J. B. Harley and David Woodward (eds.), *The History of Cartography*, vol. 1: *Cartography in Prehistoric, Ancient and Medieval Europe and the Mediterranean* (Chicago, 1987), pp. 130–47; Christian Jacob, *The Sovereign Map: Theoretical Approaches to Cartography throughout History* (Chicago, 2006), pp. 18–19; James Romm, *The Edges of the Earth in Ancient Thought* (Princeton, 1992), pp. 9–10.

10. Strabo, *The Geography of Strabo*, 1. 1. 1, trans. Horace Leonard Jones, 8 vols. (Cambridge, Mass., 1917–32).

11. Crates of Mallos, quoted in Romm, *Edges of the Earth*, p. 14.

12. All quotations taken from Richmond Lattimore (ed. and trans.), *The Iliad of Homer* (Chicago, 1951).

13. P. R. Hardie, 'Imago Mundi: Cosmological and Ideological Aspects of the Shield of Achilles', *Journal of Hellenic Studies*, 105 (1985), pp. 11–31.

14. G. S. Kirk, *Myth: Its Meaning and Function in Ancient and Other Cultures* (Berkeley and Los Angeles, 1970), pp. 172–205; Andrew Gregory, *Ancient Greek Cosmogony* (London, 2008).

15. Quoted in Aujac, 'The Foundations of Theoretical Cartography', p. 134.

16. Quoted in Charles H. Kahn, *Anaximander and the Origins of Greek Cosmology* (New York, 1960), p. 87.

17. Quoted ibid., pp. 76, 81.

18. See Jacob, 'Mapping in the Mind', p. 28; on *omphalos* and *periploi*, see the entries in John Roberts (ed.), *The Oxford Dictionary of the Classical World* (Oxford, 2005).

19. Herodotus, *The Histories*, trans. Aubrey de Selincourt (London, 1954), p. 252.

20. Ibid., p. 253.

21. Ibid., p. 254.

22. Plato, *Phaedo*, trans. David Gallop (Oxford, 1975), 108c–109b.

23. Ibid., 109b–110b.

24. Ibid., 110c.

25. See Germaine Aujac, 'The Growth of an Empirical Cartography in Hellenistic Greece', in Harley and Woodward, *History of Cartography*, vol. 1, pp. 148–60, at p. 148.

26. Aristotle, *De caelo*, 2. 14.

27. Aristotle, *Meteorologica*, trans. H. D. P. Lee (Cambridge, Mass., 1952), 338b.

28. Ibid., 362b.

29. D. R. Dicks, 'The Klimata in Greek Geography', *Classical Quarterly*, 5/3–4 (1955), pp. 248–55.

30. Herodotus, *The Histories*, pp. 328–9.

31. See C. F. C. Hawkes, *Pytheas: Europe and the Greek Explorers* (Oxford, 1977).

32. Claude Nicolet, *Space, Geography, and Politics in the Early Roman Empire* (Ann Arbor, 1991), p. 73.

33. Jacob, *Sovereign Map*, p. 137.

34. Berggren and Jones, *Ptolemy's Geography*, p. 32.

35. Aujac, 'Growth of an Empirical Cartography', pp. 155–6.

36. Strabo, *Geography*, 1. 4. 6.

37. O. A. W. Dilke, *Greek and Roman Maps* (London, 1985), p. 35.

38. See chapters 12, 13 and 14 in Harley and Woodward, *History of Cartography*, vol. 1, and Richard J. A. Talbert, 'Greek and Roman Mapping: Twenty-First Century Perspectives', in Richard J. A. Talbert and Richard W. Unger (eds.), *Cartography in Antiquity and the Middle Ages: Fresh Perspectives, New Methods* (Leiden, 2008), pp. 9–28.

39. Strabo, *Geography*, 1. 2. 24.

40. Ibid., 1. 1. 12.

41. Ibid., 2. 5. 10.

42. Ibid., 1. 1. 18.

43. Quoted in Nicolet, *Space, Geography, and Politics*, p. 31.

44. See Toomer, 'Ptolemy'.

45. Quoted in D. R. Dicks, *The Geographical Fragments of Hipparchus* (London, 1960), p. 53.

46. Ptolemy, *Almagest*, 2. 13, quoted in Berggren and Jones, *Ptolemy's Geography*, p. 19.

47. Ptolemy, *Geography*, 1. 5–6.

48. Jacob, 'Mapping in the Mind', p. 36.

49. Ptolemy, *Geography*, 1. 1.

50. Ibid., 1. 9–12; O. A. W. Dilke, 'The Culmination of Greek Cartography in Ptolemy', Harley and Woodward, *History of Cartography*, vol. 1, p. 184.

51. Ptolemy, *Geography*, 1. 23.

52. Ibid., 1. 20.

53. Ibid., 1. 23.
54. Ibid.
55. David Woodward, 'The Image of the Spherical Earth', *Perspecta*, 25 (1989), p. 9.
56. See Leo Bagrow, 'The Origin of Ptolemy's *Geographia*', *Geografiska Annaler*, 27 (1943), pp. 318–87; for a more recent summary of the controversy, see O. A. W. Dilke, 'Cartography in the Byzantine Empire', in Harley and Woodward, *History of Cartography*, vol. 1, pp. 266–72.
57. Berggren and Jones, *Ptolemy's Geography*, p. 47.
58. T. C. Skeat, 'Two Notes on Papyrus', in Edda Bresciani *et al.* (eds.), *Scritti in onore di Orsolino Montevecchi* (Bologna, 1981), pp. 373–83.
59. Berggren and Jones, *Ptolemy's Geography*, p. 50.
60. See Raven, *Lost Libraries*.
61. Ptolemy, *Geography*, 1. 1.

## CHAPTER 2. EXCHANGE: AL-IDRĪSĪ, AD 1154

1. See Elisabeth van Houts, 'The Normans in the Mediterranean', in van Houts, *The Normans in Europe* (Manchester, 2000), pp. 223–78.
2. For the best account of al-Idrīsī's life and works in English, see S. Maqbul Ahmad, 'Cartography of al-Sharīf al-Idrīsī', in J. B. Harley and David Woodward (eds.), *The History of Cartography*, vol. 2, bk. 1: *Cartography in the Traditional Islamic and South Asian Societies* (Chicago, 1987), pp. 156–74.
3. Anthony Pagden, *Worlds at War: The 2,500-Year Struggle between East and West* (Oxford, 2008), pp. 140–42.
4. B. L. Gordon, 'Sacred Directions, Orientation, and the Top of the Map', *History of Religions*, 10/3 (1971), pp. 211–27.
5. Ibid., p. 221.
6. David A. King, *World-Maps for Finding the Direction and Distance of Mecca: Innovation and Tradition in Islamic Science* (Leiden, 1999).
7. Ahmet T. Karamustafa, 'Introduction to Islamic Maps', in Harley and Woodward (eds.), *History of Cartography*, vol. 2, bk. 1, p. 7.
8. Ahmet T. Karamustafa, 'Cosmographical Diagrams', in Harley and Woodward, *History of Cartography*, vol. 2, bk. 1, pp. 71–2; S. Maqbul Ahmad and F. Taeschnes, 'Djugrafiya', in *The Encyclopaedia of Islam*, 2nd edn., vol. 2 (Leiden, 1965), p. 577.
9. Ibid., p. 574.
10. On the early history of Islam, see Patricia Crone and Martin Hinds, *God's Caliph: Religious Authority in the First Centuries of Islam* (Cambridge, 1986).
11. Quoted in Gerald R. Tibbetts, 'The Beginnings of a Cartographic Tradition', in Harley and Woodward, *History of Cartography*, vol. 2, bk. 1, p. 95.

12. Ibid., pp. 94–5; André Miquel, 'Iḳlīm', in *The Encyclopaedia of Islam*, 2nd edn., vol. 3 (Leiden, 1971), pp. 1076–8.

13. Quoted ibid., p. 1077.

14. Quoted in Edward Kennedy, 'Suhrāb and the World Map of al-Ma'mūn', in J. L. Berggren *et al.* (eds.), *From Ancient Omens to Statistical Mechanics: Essays on the Exact Sciences Presented to Asger Aaboe* (Copenhagen, 1987), pp. 113–19.

15. Quoted in Raymond P. Mercer, 'Geodesy', in Harley and Woodward (eds.), *History of Cartography*, vol. 2, bk. 1, pp. 175–88, at p. 178.

16. On Ibn Khurradādhbih and the administrative tradition, see Paul Heck, *The Construction of Knowledge in Islamic Civilisation* (Leiden, 2002), pp. 94–146, and Tibbetts, 'Beginnings of a Cartographic Tradition', pp. 90–92.

17. Ralph W. Brauer, 'Boundaries and Frontiers in Medieval Muslim Geography', *Transactions of the American Philosophical Society*, new series, 85/6 (1995), pp. 1–73.

18. Quoted in Gerald R. Tibbetts, 'The Balkhī School of Geographers', in Harley and Woodward (eds.), *History of Cartography*, vol. 2, bk. 1, pp. 108–36, at p. 112.

19. Konrad Miller, *Mappae Arabicae: Arabische Welt- und Länderkasten des 9.-13. Jahrshunderts*, 6 vols. (Stuttgart, 1926–31), vol. 1, pt. 1.

20. On Córdoba, see Robert Hillenbrand, '"The Ornament of the World": Medieval Córdoba as a Cultural Centre', in Salma Khadra Jayyusi (ed.), *The Legacy of Muslim Spain* (Leiden, 1992), pp. 112–36, and Heather Ecker, 'The Great Mosque of Córdoba in the Twelfth and Thirteenth Centuries', *Muqarnas*, 20 (2003), pp. 113–41.

21. Quoted in Hillenbrand, '"The Ornament of the World"', p. 112.

22. Quoted ibid., p. 120.

23. Maqbul Ahmad, 'Cartography of al-Idrīsī', p. 156.

24. Jeremy Johns, *Arabic Administration in Norman Sicily: The Royal Dīwān* (Cambridge, 2002), p. 236.

25. Quoted in Hubert Houben, *Roger II of Sicily: A Ruler between East and West* (Cambridge, 2002), p. 106.

26. Helen Wieruszowski, 'Roger II of Sicily, Rex Tyrannus, in Twelfth-Century Political Thought', *Speculum*, 38/1 (1963), pp. 46–78.

27. Donald Matthew, *The Norman Kingdom of Sicily* (Cambridge, 1992).

28. Quoted in R. C. Broadhurst (ed. and trans.), *The Travels of Ibn Jubayr* (London, 1952), pp. 339–41.

29. Charles Haskins and Dean Putnam Lockwood, 'The Sicilian Translators of the Twelfth Century and the First Latin Version of Ptolemy's Almagest', *Harvard Studies in Classical Philology*, 21 (1910), pp. 75–102.

30. Houben, *Roger II*, p. 102.

31. Ibid., pp. 98–113; Matthew, *Norman Kingdom*, pp. 112–28.

32. Quoted in Ahmad, 'Cartography of al-Idrīsī', p. 159.

33. Ibid.
34. Ibid.
35. Ibid., p. 160.
36. Quoted in Pierre Jaubert (ed. and trans.), *Géographie d'Édrisi*, 2 vols. (Paris, 1836), vol. 1, p. 10. Jaubert's translation is somewhat erratic, and where possible corrected based on a comparison with the partial translation in Reinhart Dozy and Michael Jan de Goeje (eds. and trans.), *Description de l'Afrique et de l'Espagne par Edrîsî* (Leiden, 1866).
37. S. Maqbul Ahmad, *India and the Neighbouring Territories in the 'Kitāb nuzhat al-mushtāq fi khtirāq al-āfāq' of al-Sharīf al-Idrīsī* (Leiden, 1960), pp. 12–18.
38. Quoted in Jaubert, *Géographie d'Édrisi*, vol. 1, p. 140.
39. Quoted ibid., pp. 137–8.
40. Quoted ibid., vol. 2, p. 156.
41. Quoted ibid., p. 252.
42. Quoted ibid., pp. 342–3.
43. Quoted ibid., pp. 74–5.
44. Brauer, 'Boundaries and Frontiers', pp. 11–14.
45. J. F. P. Hopkins, 'Geographical and Navigational Literature', in M. J. L. Young, J. D. Latham and R. B. Serjeant (eds.), *Religion, Learning and Science in the 'Abbasid Period* (Cambridge, 1990), pp. 301–27, at pp. 307–11.
46. *The History of the Tyrants of Sicily by 'Hugo Falcandus' 1154–69*, trans. Graham A. Loud and Thomas Wiedemann (Manchester, 1998), p. 59.
47. Matthew, *Norman Kingdom*, p. 112; on Frederick's Sicilian reign, see David Abulafia, *Frederick II: A Medieval Emperor* (Oxford, 1988), pp. 340–74.
48. Ibn Kaldūn, *The Muqadimah: An Introduction to History*, trans. Franz Rosenthal (Princeton, 1969), p. 53.
49. Jeremy Johns and Emilie Savage-Smith, 'The Book of Curiosities: A Newly Discovered Series of Islamic Maps', *Imago Mundi*, 55 (2003), pp. 7–24, Yossef Rapoport and Emilie Savage-Smith, 'Medieval Islamic Views of the Cosmos: The Newly Discovered *Book of Curiosities*', *Cartographic Journal*, 41/3 (2004), pp. 253–9, and Rapoport and Savage-Smith, 'The Book of Curiosities and a Unique Map of the World', in Richard J. A. Talbert and Richard W. Unger (eds.), *Cartography in Antiquity and the Middle Ages: Fresh Perspectives, New Methods* (Leiden, 2008), pp. 121–38.

## CHAPTER 3. FAITH: HEREFORD MAPPAMUNDI, C. 1300

1. Colin Morris, 'Christian Civilization (1050–1400)', in John McManners (ed.), *The Oxford Illustrated History of Christianity* (Oxford, 1990), pp. 196–232.

2. On Cantilupe's career and conflict with Pecham, see the essays in Meryl Jancey (ed.), *St. Thomas Cantilupe, Bishop of Hereford: Essays in his Honour* (Hereford, 1982).

3. See Nicola Coldstream, 'The Medieval Tombs and the Shrine of Saint Thomas Cantilupe', in Gerald Aylmer and John Tiller (eds.), *Hereford Cathedral: A History* (London, 2000), pp. 322–30.

4. David Woodward, 'Medieval *Mappaemundi*', in J. B. Harley and David Woodward (eds.), *The History of Cartography*, vol. 1: *Cartography in Prehistoric, Ancient, and Medieval Europe and the Mediterranean* (Chicago, 1987), p. 287.

5. Scott D. Westrem, *The Hereford Map: A Transcription and Translation of the Legends with Commentary* (Turnhout, 2001), p. 21. Unless otherwise stated, all quotations from the map are taken from Westrem.

6. Ibid., p. 8.

7. Quoted in Woodward, 'Medieval *Mappaemundi*', p. 299.

8. Quoted in Natalia Lozovsky, '*The Earth is Our Book*': *Geographical Knowledge in the Latin West ca. 400–1000* (Ann Arbor, 2000), p. 11.

9. Quoted ibid., p. 12.

10. Quoted ibid., p. 49.

11. Sallust, *The Jugurthine War/The Conspiracy of Catiline*, trans. S. A. Handford (London, 1963), pp. 53–4.

12. Evelyn Edson, *Mapping Time and Space: How Medieval Mapmakers Viewed their World* (London, 1997), p. 20.

13. Alfred Hiatt, 'The Map of Macrobius before 1100', *Imago Mundi*, 59 (2007), pp. 149–76.

14. Quoted in William Harris Stahl (ed.), *Commentary on the Dream of Scipio by Macrobius* (Columbia, NY, 1952), pp. 201–3.

15. Ibid., p. 216.

16. Roy Deferrari (ed.), *Paulus Orosius: The Seven Books of History against the Pagans* (Washington, 1964), p. 7.

17. Quoted in Edson, *Mapping Time and Space*, p. 38.

18. Quoted ibid., p. 48.

19. Lozovsky, '*The Earth is Our Book*', p. 105; Edson, *Mapping Time and Space*, p. 49.

20. William Harris Stahl *et al.* (eds. and trans.), *Martianus Capella and the Seven Liberal Arts*, vol. 2: *The Marriage of Philology and Mercury* (New York, 1997), p. 220.

21. Lozovsky, '*The Earth is Our Book*', pp. 28–34.

22. Erich Auerbach, *Mimesis: The Representation of Reality in Western Literature* (Princeton, 1953), pp. 73–4, 195–6.

23. See Patrick Gautier Dalché, 'Maps in Words: The Descriptive Logic of Medieval Geography', in P. D. A. Harvey (ed.), *The Hereford World Map: Medieval World Maps and their Context* (London, 2006), pp. 223–42.

24. Conrad Rudolph, '"First, I Find the Center Point": Reading the Text of Hugh of Saint Victor's *The Mystic Ark*', *Transactions of the American Philosophical Society*, 94/4 (2004), pp. 1–110.

25. Quoted in Alessandro Scafi, *Mapping Paradise: A History of Heaven on Earth* (London, 2006), p. 123.

26. Quoted in Woodward, 'Medieval *Mappaemundi*', p. 335.

27. Quoted in Mary Carruthers, *The Book of Memory: A Study of Memory in Medieval Culture* (Cambridge, 2nd edn., 2007), p. 54.

28. Quoted in Scafi, *Mapping Paradise*, pp. 126–7.

29. Westrem, *The Hereford Map*, pp. 130, 398.

30. Peter Barber, 'Medieval Maps of the World', in Harvey, *The Hereford World Map*, pp. 1–44, at p. 13.

31. Westrem, *The Hereford Map*, p. 326; G. R. Crone, 'New Light on the Hereford Map', *Geographical Journal*, 131 (1965), pp. 447–62.

32. Ibid., p. 451; P. D. A. Harvey, 'The Holy Land on Medieval World Maps', in Harvey, *The Hereford World Map*, p. 248.

33. Brouria Bitton-Ashkelony, *Encountering the Sacred: The Debate on Christian Pilgrimage in Late Antiquity* (Berkeley and Los Angeles, 2006), pp. 110–15; Christian K. Zacher, *Curiosity and Pilgrimage: The Literature of Discovery in Fourteenth-Century England* (Baltimore, 1976).

34. Robert Norman Swanson, *Religion and Devotion in Europe, 1215–1515* (Cambridge, 1995), pp. 198–9.

35. Valerie J. Flint, 'The Hereford Map: Its Author(s), Two Scenes and a Border', *Transactions of the Royal Historical Society*, sixth series, 8 (1998), pp. 19–44.

36. Ibid., pp. 37–9.

37. Dan Terkla, 'The Original Placement of the Hereford Mappa Mundi', *Imago Mundi*, 56 (2004), pp. 131–51, and 'Informal Cathechesis and the Hereford Mappa Mundi', in Robert Bork and Andrea Kann (eds.), *The Art, Science and Technology of Medieval Travel* (Aldershot, 2008), pp. 127–42.

38. Martin Bailey, 'The Rediscovery of the Hereford Mappamundi: Early References, 1684–1873', in Harvey, *The Hereford World Map*, pp. 45–78.

39. Martin Bailey, 'The Discovery of the Lost Mappamundi Panel: Hereford's Map in a Medieval Altarpiece?', in Harvey, *The Hereford World Map*, pp. 79–93.

40. Quoted in Daniel K. Connolly, 'Imagined Pilgrimage in the Itinerary Maps of Matthew Paris', *Art Bulletin*, 81/4 (1999), pp. 598–622, at p. 598.

## CHAPTER 4. EMPIRE: KANGNIDO WORLD MAP, 1402

1. Martina Deuchlar, *The Confucian Transformation of Korea: A Study of Society and Ideology* (Cambridge, Mass., 1992).

2. John B. Duncan, *The Origins of the Chosŏn Dynasty* (Washington, 2000).

3. Tanaka Takeo, 'Japan's Relations with Overseas Countries', in John Whitney Hall and Takeshi Toyoda (eds.), *Japan in the Muromachi Age* (Berkeley and Los Angeles, 1977), pp. 159–78.

4. Joseph Needham *et al.*, *The Hall of Heavenly Records: Korean Astronomical Instruments and Clocks* (Cambridge, 1986), pp. 153–9, and F. Richard Stephenson, 'Chinese and Korean Star Maps and Catalogs', in J. B. Harley and David Woodward (eds.), *The History of Cartography*, vol. 2, bk. 2: *Cartography in the Traditional East and Southeast Asian Societies* (Chicago, 1987), pp. 560–68.

5. The Chinese map known as *Da Ming hunyi tu* ('Integrated Map of the Great Ming'), held in the First Historical Archives of China, Beijing, bears many similarities to the Kangnido map, and is dated by some scholars to 1389. However, others argue that there is no physical evidence to ascribe it to such an early date, and suggest that it is a reproduction from the late sixteenth or early seventeenth century. See Kenneth R. Robinson, 'Gavin Menzies, 1421, and the Ryūkoku *Kangnido* World Map', *Ming Studies*, 61 (2010), pp. 56–70, at p. 62. I am grateful to Cordell Yee for corresponding with me about this map.

6. The most recent detailed description of the map is Kenneth R. Robinson, 'Chosŏn Korea in the Ryūkoku *Kangnido*: Dating the Oldest Extant Korean Map of the World (15th Century)', *Imago Mundi*, 59/2 (2007), pp. 177–92.

7. Ibid., pp. 179–82.

8. Joseph Needham, with Wang Ling, *Science and Civilisation in China*, vol. 3: *Mathematics and the Sciences of the Heavens and the Earth* (Cambridge, 1959), pp. 555–6.

9. Ibid., p. 555.

10. C. Dale Walton, 'The Geography of Universal Empire: A Revolution in Strategic Perspective and its Lessons', *Comparative Strategy*, 24 (2005), pp. 223–35.

11. Quoted in Gari Ledyard, 'Cartography in Korea', in Harley and Woodward, *The History of Cartography*, vol. 2, bk. 2, pp. 235–345, at p. 245.

12. Timothy Brook, *The Troubled Empire: China in the Yuan and Ming Dynasties* (Cambridge, Mass., 2010), pp. 164, 220. I am deeply grateful to Professor Brook for drawing this illustration and further references to my attention, and enabling me to reproduce it here.

13. Kenneth R. Robinson, 'Yi Hoe and his Korean Ancestors in T'aean Yi Genealogies', *Seoul Journal of Korean Studies*, 21/2 (2008), pp. 221–50, at pp. 236–7.

14. Hok-lam Chan, 'Legitimating Usurpation: Historical Revisions under the Ming Yongle Emperor (r. 1402–1424)', in Philip Yuen-sang Leung (ed.), *The Legitimation of New Orders: Case Studies in World History* (Hong Kong, 2007), pp. 75–158.

15. Zheng Qiao (AD 1104–62), quoted in Francesca Bray, 'Introduction: The Powers of *Tu*', in Francesca Bray, Vera Dorofeeva-Lichtmann and Georges Métailié (eds.), *Graphics and Text in the Production of Technical Knowledge in China* (Leiden, 2007), pp. 1–78, at p. 1.

16. Nathan Sivin and Gari Ledyard, 'Introduction to East Asian Cartography', in Harley and Woodward, *The History of Cartography*, vol. 2, bk. 2, pp. 23–31, at p. 26.

17. Bray, 'The Powers of *tu*', p. 4.

18. Quoted in Needham, *Science and Civilisation*, vol. 3, p. 217.

19. Ibid., p. 219.

20. Quoted in John S. Major, *Heaven and Earth in Early Han Thought* (New York, 1993), p. 32.

21. John B. Henderson, 'Nonary Cosmography in Ancient China', in Kurt A. Raaflaub and Richard J. A. Talbert (eds.), *Geography and Ethnography: Perceptions of the World in Pre-Modern Societies* (Oxford, 2010), pp. 64–73, at p. 64.

22. Sarah Allan, *The Shape of the Turtle: Myth, Art and Cosmos in Early China* (Albany, NY, 1991).

23. Mark Edward Lewis, *The Flood Myths of Early China* (Albany, NY, 2006), pp. 28–30.

24. Quoted in Needham, *Science and Civilisation*, vol. 3, p. 501.

25. Vera Dorofeeva-Lichtmann, 'Ritual Practices for Constructing Terrestrial Space (Warring States – Early Han)', in John Lagerwey and Marc Kalinowski (eds.), *Early Chinese Religion*, pt. 1: *Shang through Han (1250 BC–220 AD)* (Leiden, 2009), pp. 595–644.

26. Needham, *Science and Civilisation*, vol. 3, pp. 501–3.

27. Quoted in William Theodore De Bary (ed.), *Sources of East Asian Tradition*, vol. 1: *Premodern Asia* (New York, 2008), p. 133.

28. Quoted in Mark Edward Lewis, *The Construction of Space in Early China* (Albany, NY, 2006), p. 248.

29. Quoted in Cordell D. K. Yee, 'Chinese Maps in Political Culture', in Harley and Woodward, *History of Cartography*, vol. 2, bk. 2, pp. 71–95, at p. 72.

30. Hung Wu, *The Wu Liang Shrine: The Ideology of Early Chinese Pictorial Art* (Stanford, Calif., 1989), p. 54.

31. Quoted in Yee, 'Chinese Maps', p. 74.

32. Ibid., p. 74.

33. Nancy Shatzman Steinhardt, 'Mapping the Chinese City', in David Buisseret (ed.), *Envisioning the City: Six Studies in Urban Cartography* (Chicago, 1998), pp. 1–33, at p. 11; Cordell D. K. Yee, 'Reinterpreting Traditional Chinese Geographical Maps', in Harley and Woodward, *History of Cartography*, vol. 2, bk. 2, pp. 35–70, at p. 37.

34. Craig Clunas, *Art in China* (Oxford, 1997), pp. 15–44.

35. Yee, 'Chinese Maps', pp. 75–6.

36. Quoted in Needham, *Science and Civilisation*, vol. 3, pp. 538–40.
37. Cordell D. K. Yee, 'Taking the World's Measure: Chinese Maps between Observation and Text', in Harley and Woodward, *History of Cartography*, vol. 2, bk. 2, pp. 96–127.
38. Quoted ibid., p. 113.
39. Quoted in Needham, *Science and Civilisation*, vol. 3, p. 540.
40. Ibid., p. 546.
41. Quoted in Alexander Akin, 'Georeferencing the Yujitu', accessed at: http://www.davidrumsey.com/china/Yujitu_Alexander_Akin.pdf.
42. Tsien Tsuen-Hsuin, 'Paper and Printing', in Joseph Needham, *Science and Civilisation in China*, vol. 5, pt. 1: *Chemistry and Chemical Technology: Paper and Printing* (Cambridge, 1985).
43. Patricia Buckley Ebrey, *The Cambridge Illustrated History of China* (Cambridge, 1996), pp. 136–63.
44. Vera Dorofeeva-Lichtmann, 'Mapping a "Spiritual" Landscape: Representation of Terrestrial Space in the *Shanhaijing*', in Nicola Di Cosmo and Don J. Wyatt (eds.), *Political Frontiers, Ethnic Boundaries, and Human Geographies in Chinese History* (Oxford, 2003), pp. 35–79.
45. Quoted in Hilde De Weerdt, 'Maps and Memory: Readings of Cartography in Twelfth- and Thirteenth-Century Song China', *Imago Mundi*, 61/2 (2009), pp. 145–67, at p. 156.
46. Ibid., p. 159.
47. Quoted in Ledyard, 'Cartography in Korea', p. 240.
48. Ibid., pp. 238–79.
49. Quoted in Steven J. Bennett, 'Patterns of the Sky and Earth: A Chinese Science of Applied Cosmology', *Chinese Science*, 3 (1978), pp. 1–26, at pp. 5–6.
50. David J. Nemeth, *The Architecture of Ideology: Neo-Confucian Imprinting on Cheju Island, Korea* (Berkeley and Los Angeles, 1987), p. 114.
51. Quoted in Ledyard, 'Cartography in Korea', p. 241.
52. Quoted in Nemeth, *Architecture of Ideology*, p. 115.
53. Ledyard, 'Cartography in Korea', pp. 276–9.
54. Ibid., pp. 291–2.
55. I am deeply grateful to Gari Ledyard for explaining this point to me.
56. Quoted in Dane Alston, 'Emperor and Emissary: The Hongwu Emperor, Kwŏn Kŭn, and the Poetry of Late Fourteenth Century Diplomacy', *Korean Studies*, 32 (2009), pp. 104–47, at p. 111.
57. Quoted ibid., p. 112.
58. Ibid., p. 120.
59. Ibid., p. 125.
60. Ibid., p. 129.
61. Ibid., p. 131.
62. Ibid., p. 134.
63. Etsuko Hae-Jin Kang, *Diplomacy and Ideology in Japanese-Korean*

*Relations: From the Fifteenth to the Eighteenth Century* (London, 1997), pp. 49-83.

64. Quoted in Ledyard, 'Cartography in Korea', p. 245.
65. Robinson, 'Chosŏn Korea in the Ryūkoku *Kangnido*', pp. 185-8.
66. Bray, 'The Powers of *Tu*', p. 8.

## CHAPTER 5. DISCOVERY: MARTIN WALDSEEMÜLLER, WORLD MAP, 1507

1. All subsequent quotations relating to the acquisition of the map are taken from the files held in the US Library of Congress Map Division collection. I am grateful to John Hessler and John Herbert of the Map Division for allowing me access to these files, and to Philip Burden for supplying emails and discussing with me his involvement in the acquisition.
2. Quoted in Seymour I. Schwartz, *Putting 'America' on the Map: The Story of the Most Important Graphic Document in the History of the United States* (New York, 2007), pp. 251-2.
3. *New York Times*, 20 June 2003.
4. See http://www.loc.gov/today/pr/2001/01-093.html.
5. Jacob Burckhardt, *The Civilization of the Renaissance in Italy*, trans. S. G. C. Middlemore (London, 1990), pp. 213-22.
6. Quoted in John Hessler, *The Naming of America: Martin Waldseemüller's 1507 World Map and the 'Cosmographiae Introductio'* (London, 2008), p. 34.
7. Ibid., p. 17.
8. Samuel Eliot Morison, *Portuguese Voyages to America in the Fifteenth Century* (Cambridge, Mass., 1940), pp. 5-10.
9. On the early history of printing and the volume of publications, see Elizabeth Eisenstein, *The Printing Press as an Agent of Change*, 2 vols. (Cambridge, 1979), and Lucien Febvre, *The Coming of the Book*, trans. David Gerard (London, 1976).
10. Quoted in Barbara Crawford Halporn (ed.), *The Correspondence of Johann Amerbach* (Ann Arbor, 2000), p. 1.
11. For a more sceptical approach to the 'revolutionary' thesis, see Adrian Johns, *The Nature of the Book: Print and Knowledge in the Making* (Chicago, 1998).
12. William Ivins, *Prints and Visual Communications* (Cambridge, Mass., 1953), pp. 1-50.
13. Robert Karrow, 'Centers of Map Publishing in Europe, 1472-1600', in David Woodward (ed.), *The History of Cartography*, vol. 3: *Cartography in the European Renaissance*, pt. 1 (Chicago, 2007), pp. 611-21.
14. Quoted in Schwartz, *Putting 'America' on the Map*, p. 36.

15. See Denis Cosgrove, 'Images of Renaissance Cosmography, 1450–1650', in Woodward, *History of Cartography*, vol. 3, pt. 1, pp. 55–98.

16. Patrick Gautier Dalché, 'The Reception of Ptolemy's *Geography* (End of the Fourteenth to Beginning of the Sixteenth Century)', in Woodward, *History of Cartography*, vol. 3, pt. 1 pp. 285–364.

17. Tony Campbell, *The Earliest Printed Maps, 1472–1500* (London, 1987), p. 1.

18. Quoted in Schwartz, *Putting 'America' on the Map*, pp. 39–40.

19. See Luciano Formisano (ed.), *Letters from a New World: Amerigo Vespucci's Discovery of America*, trans. David Jacobson (New York, 1992).

20. Quoted in Joseph Fischer SJ and Franz von Weiser, *The Cosmographiae Introductio of Martin Waldseemüller in Facsimile* (Freeport, NY, 1960), p. 88.

21. All quotes from the text are taken from Hessler, *The Naming of America*, although see also Charles George Herbermann (ed.), *The Cosmographia Introductio of Martin Waldseemüller* (New York, 1907).

22. Quoted in Hessler, *Naming of America*, p. 88.

23. Ibid., p. 94.

24. Ibid., pp. 100–101. See also Toby Lester, *The Fourth Part of the World: The Epic Story of History's Greatest Map* (New York, 2009).

25. Quoted in Christine R. Johnson, 'Renaissance German Cosmographers and the Naming of America', *Past and Present*, 191/1 (2006), pp. 3–43, at p. 21.

26. Miriam Usher Chrisman, *Lay Culture, Learned Culture: Books and Social Changes in Strasbourg, 1480–1599* (New Haven, 1982), p. 6.

27. R. A. Skelton, 'The Early Map Printer and his Problems', *Penrose Annual*, 57 (1964), pp. 171–87.

28. Quoted in Halporn (ed.), *Johann Amerbach*, p. 2.

29. See David Woodward (ed.), *Five Centuries of Map Printing* (Chicago, 1975), ch. 1.

30. Quoted in Schwartz, *Putting 'America' on the Map*, p. 188.

31. Quoted in E. P. Goldschmidt, 'Not in Harrisse', in *Essays Honoring Lawrence C. Wroth* (Portland, Me., 1951), pp. 135–6.

32. Quoted in J. Lennart Berggren and Alexander Jones (eds. and trans.), *Ptolemy's Geography: An Annotated Translation of the Theoretical Chapters* (Princeton, 2000), pp. 92–3.

33. On Ptolemy's projection, see ibid., and O. A. W. Dilke, 'The Culmination of Greek Cartography in Ptolemy', in J. B. Harley and David Woodward (eds.), *The History of Cartography*, vol. 1: *Cartography in Prehistoric, Ancient, and Medieval Europe and the Mediterranean* (Chicago, 1987), pp. 177–200.

34. Using computational modelling and a technique known as 'polynomial warping', Hessler has produced some controversial evidence that throws intriguing light on the creation of the *Universalis cosmographia*. Hessler describes polynomial warping as 'a mathematical transformation or mapping from a

distorted image, such as an early map or a map with an unknown scale or geometric grid, to a target image that is well known. The objective is to perform a spatial transformation, or *warp*, so that the corrected image can be measured or have a metric placed upon it relative to a known map or grid.' John Hessler, 'Warping Waldseemüller: A Phenomenological and Computational Study of the 1507 World Map', *Cartographica*, 41/2 (2006), pp. 101–13.

35. Quoted in Franz Laubenberger and Steven Rowan, 'The Naming of America', *Sixteenth Century Journal*, 13/4 (1982) , p. 101.

36. Quoted in Joseph Fischer SJ and Franz von Wieser (eds.), *The World Maps of Waldseemüller (Ilacomilus) 1507 and 1516* (Innsbruck, 1903), pp. 15–16.

37. Quoted in Johnson, 'Renaissance German Cosmographers', p. 32.

38. See Laubenberger and Rowan, 'The Naming of America'.

39. Johnson, 'Renaissance German Cosmographers', pp. 34–5.

40. Quoted in Schwartz, *Putting 'America' on the Map*, p. 212.

41. Elizabeth Harris, 'The Waldseemüller Map: A Typographic Appraisal', *Imago Mundi*, 37 (1985), pp. 30–53.

42. Michel Foucault, 'Nietzsche, Genealogy, History', in Foucault, *Language, Counter-Memory, Practice: Selected Essays and Interviews*, ed. and trans. Donald Bouchard (New York, 1977), pp. 140–64, at p. 142.

### CHAPTER 6. GLOBALISM: DIOGO RIBEIRO, WORLD MAP, 1529

1. Quoted in Frances Gardiner Davenport and Charles Oscar Paullin (eds.), *European Treaties Bearing on the History of the United States and its Dependencies*, 4 vols. (Washington, 1917), vol. 1, p. 44.

2. Quoted ibid., p. 95.

3. Quoted in Francis M. Rogers (ed.), *The Obedience of a King of Portugal* (Minneapolis, 1958), p. 48.

4. Quoted in Davenport and Paullin, *European Treaties*, vol. 1, p. 161.

5. Quoted in Donald Weinstein (ed.), *Ambassador from Venice: Pietro Pasqualigo in Lisbon, 1501* (Minneapolis, 1960), pp. 29–30.

6. See Sanjay Subrahmanyam and Luis Filipe F. R. Thomaz, 'Evolution of Empire: The Portuguese in the Indian Ocean during the Sixteenth Century', in James Tracey (ed.), *The Political Economy of Merchant Empires* (Cambridge, 1991), pp. 298–331.

7. Quoted in W. B. Greenlee (ed.), *The Voyage of Pedro Alvares Cabral to Brazil and India* (London, 1937), pp. 123–4.

8. Quoted in Carlos Quirino (ed.), *First Voyage around the World by Antonio Pigafetta and 'De Moluccis Insulis' by Maximilianus Transylvanus* (Manila, 1969), pp. 112–13.

9. See Richard Hennig, 'The Representation on Maps of the Magalhães Straits before their Discovery', *Imago Mundi*, 5 (1948), pp. 32–7.

10. See Edward Heawood, 'The World Map before and after Magellan's Voyage', *Geographical Journal*, 57 (1921), pp. 431–42.

11. Lord Stanley of Alderley (ed.), *The First Voyage around the World by Magellan* (London, 1874), p. 257.

12. Quoted in Marcel Destombes, 'The Chart of Magellan', *Imago Mundi*, 12 (1955), pp. 65–88, at p. 68.

13. Quoted in R. A. Skelton (ed.), *Magellan's Voyage: A Narrative Account of the First Circumnavigation*, 2 vols. (New Haven, 1969), vol. 1, p. 128.

14. Quoted in Samuel Eliot Morison, *The European Discovery of America: The Northern Voyages*, A.D. 500–16 (Oxford, 1974), p. 473.

15. Quoted in Quirino, *First Voyage around the World*, pp. 112–13; Julia Cartwright (ed.), *Isabella d'Este, Marchioness of Mantua 1474–1539: A Study of the Renaissance*, 2 vols. (London, 1903), vol. 2, pp. 225–6.

16. Quoted in Morison, *European Discovery*, p. 472.

17. Peter Martyr, *The Decades of the Newe Worlde*, trans. Richard Eden (London, 1555), p. 242.

18. Antonio Barrera-Osorio, *Experiencing Nature: The Spanish American Empire and the Early Scientific Revolution* (Austin, Tex., 2006), pp. 29–55; Maria M. Portuondo, *Secret Science: Spanish Cosmography and the New World* (Chicago, 2009).

19. Destombes, 'The Chart of Magellan', p. 78.

20. L. A. Vigneras, 'The Cartographer Diogo Ribeiro', *Imago Mundi*, 16 (1962), pp. 76–83.

21. Quoted in Destombes, 'The Chart of Magellan', p. 78.

22. Bartholomew Leonardo de Argensola, *The Discovery and Conquest of the Molucco Islands* (London, 1708).

23. Quoted in Emma H. Blair and James A. Robertson (eds.), *The Philippine Islands: 1493–1898*, 55 vols. (Cleveland, 1903–9), vol. 1, pp. 176–7.

24. Peter Martyr, *The Decades of the Newe Worlde*, p. 242.

25. Quoted in Blair and Robertson, *The Philippine Islands*, vol. 1, pp. 209–10.

26. Ibid., p. 201.

27. Ibid., p. 197.

28. Ibid., p. 205.

29. Quoted in Vigneras, 'Ribeiro', p. 77.

30. Quoted in Armado Cortesão and Avelino Teixeira da Mota, *Portugaliae Monumenta Cartographica*, 6 vols. (Lisbon, 1960–62), vol. 1, p. 97.

31. Vigneras, 'Ribeiro', pp. 78–9.

32. Surekha Davies, 'The Navigational Iconography of Diogo Ribeiro's 1529 Vatican Planisphere', *Imago Mundi*, 55 (2003), pp. 103–12.

33. Bailey W. Diffie and George D. Winius, *Foundations of the Portuguese Empire, 1415–1580* (Minneapolis, 1977), p. 283.

34. Robert Thorne, 'A Declaration of the Indies', in Richard Hakluyt, *Divers Voyages Touching America* (London, 1582), sig. C3.
35. Quoted in Cortesão and da Mota, *Portugaliae Monumenta Cartographica*, vol. 1, p. 100.
36. Davenport, *European Treaties*, p. 188.
37. Ibid., pp. 186-97.
38. Jerry Brotton, *Trading Territories: Mapping the Early Modern World* (London, 1997), pp. 143-4.
39. Quoted in Cortesão and da Mota, *Portugaliae Monumenta Cartographica*, vol. 1, p. 102.
40. Konrad Eisenbichler, 'Charles V in Bologna: The Self-Fashioning of a Man and a City', *Renaissance Studies*, 13/4 (2008), pp. 430-39.
41. Jerry Brotton and Lisa Jardine, *Global Interests: Renaissance Art between East and West* (London, 2000), pp. 49-62.

## CHAPTER 7. TOLERATION: GERARD MERCATOR, WORLD MAP, 1569

1. For the most comprehensive account of the heresy executions, see H. Averdunk and J. Müller-Reinhard, *Gerhard Mercator und die Geographen unter seinen Nachkommen* (Gotha, 1904). For the most recent English-language biography of Mercator, see Nicholas Crane, *Mercator: The Man who Mapped the Planet* (London, 2003).
2. Paul Arblaster, '"Totius Mundi Emporium": Antwerp as a Centre for Vernacular Bible Translations, 1523-1545', in Arie-Jan Gelderblom, Jan L. de Jong and Marc van Vaeck (eds.), *The Low Countries as a Crossroads of Religious Belief* (Leiden, 2004), pp. 14-15.
3. William Monter, 'Heresy Executions in Reformation Europe, 1520-1565', in Ole Peter Grell and Bob Scribner (eds.), *Tolerance and Intolerance in the European Reformation* (Cambridge, 1996), pp. 48-64.
4. Karl Marx, 'The Eighteenth Brumaire of Napoleon Bonaparte' (1852), in David McLellan (ed.), *Karl Marx: Selected Writings* (Oxford, 2nd edn. 2000), pp. 329-55.
5. The preceding lines and the concept of 'self-fashioning' are deeply indebted to Stephen Greenblatt, *Renaissance Self-Fashioning: From More to Shakespeare* (Chicago, 1980), pp. 1-2.
6. Quoted in Crane, *Mercator*, p. 193.
7. Ibid., p. 194.
8. Ibid., p. 44.
9. Quoted in A. S. Osley (ed.), *Mercator: A Monograph on the Lettering of Maps, etc. in the 16th Century Netherlands with a Facsimile and Translation of his Treatise on the Italic Hand and a Translation of Ghim's 'Vita Mercatoris'* (London, 1969), p. 185.

10. Quoted in Peter van der Krogt, *Globi Neerlandici: The Production of Globes in the Low Countries* (Utrecht, 1993), p. 42.

11. On the globe, see ibid., pp. 53–5; Robert Haardt, 'The Globe of Gemma Frisius', *Imago Mundi*, 9 (1952), pp. 109–10. On the cost of globes, see Steven Vanden Broeke, *The Limits of Influence: Pico, Louvain and the Crisis of Astrology* (Leiden, 2003).

12. Quoted in Robert W. Karrow, Jr., *Mapmakers of the Sixteenth Century and their Maps: Bio-Bibliographies of the Cartographers of Abraham Ortelius, 1570* (Chicago, 1993), p. 377.

13. Quoted in M. Büttner, 'The Significance of the Reformation for the Reorientation of Geography in Lutheran Germany', *History of Science*, 17 (1979), pp. 151–69, at p. 160.

14. The following passages are deeply indebted to Catherine Delano-Smith and Elizabeth Morley Ingram, *Maps in Bibles, 1500–1600: An Illustrated Catalogue* (Geneva, 1991), and Delano-Smith, 'Maps as Art and Science: Maps in Sixteenth Century Bibles', *Imago Mundi*, 42 (1990), pp. 65–83.

15. Quoted in Delano-Smith and Morley, *Maps in Bibles*, p. xxvi.

16. Delano-Smith, 'Maps as Art', p. 67.

17. Quoted in Delano-Smith and Morley, *Maps in Bibles*, p. xxv.

18. Robert Karrow, 'Centers of Map Publishing in Europe, 1472–1600', in David Woodward (ed.), *The History of Cartography*, vol. 3: *Cartography in the European Renaissance*, pt. 1 (Chicago, 2007), pp. 618–19.

19. On the history of Renaissance map projections, see Johannes Keuning, 'A History of Geographical Map Projections until 1600', *Imago Mundi*, 12 (1955), pp. 1–24; John P. Snyder, *Flattening the Earth: Two Thousand Years of Map Projections* (Chicago, 1993), and his 'Map Projections in the Renaissance', in David Woodward (ed.), *The History of Cartography*, vol.3: *Cartography in the European Renaissance*, pt.1(Chicago, 2007), pp. 365–81.

20. Rodney W. Shirley, *The Mapping of the World: Early Printed World Maps, 1472–1700* (London, 1983), p. 84.

21. See Robert L. Sharp, 'Donne's "Good-Morrow" and Cordiform Maps', *Modern Language Notes*, 69/7 (1954), pp. 493–5; Julia M. Walker, 'The Visual Paradigm of "The Good-Morrow": Donne's Cosmographical Glasse', *Review of English Studies*, 37/145 (1986), pp. 61–5.

22. Eric Jager, *The Book of the Heart* (Chicago, 2000), pp. 139, 143.

23. William Harris Stahl (ed.), *Commentary on the Dream of Scipio by Macrobius* (Columbia, NY, 1952), pp. 72, 216.

24. Quoted in Denis Cosgrove, *Apollo's Eye: A Cartographic Genealogy of the Earth in the Western Imagination* (Baltimore, 2001), p. 49.

25. Giorgio Mangani, 'Abraham Ortelius and the Hermetic Meaning of the Cordiform Projection', *Imago Mundi*, 50 (1998), pp. 59–83. On Melanchthon, see Crane, *Mercator*, p. 96.

26. Quoted in Osley, *Mercator*, p. 186.

27. See Geoffrey Parker, *The Dutch Revolt* (London, 1979), p. 33.

28. Rolf Kirmse, 'Die grosse Flandernkarte Gerhard Mercators (1540) – ein Politicum?', *Duisburger Forschungen*, l (1957), pp. 1-44; Crane, *Mercator*, pp. 102-10.

29. See Marc Boone, 'Urban Space and Political Conflict in Late Medieval Flanders', *Journal of Interdisciplinary History*, 32/4 (2002), pp. 621-40.

30. Diarmaid MacCulloch, *Reformation: Europe's House Divided, 1490-1700* (London, 2003), pp. 75, 207-8.

31. Quoted in Rienk Vermij, 'Mercator and the Reformation', in Manfred Büttner and René Dirven (eds.), *Mercator und Wandlungen der Wissenschaften im 16. und 17. Jahrhundert* (Bochum, 1993), pp. 77-90, at p. 85.

32. Alison Anderson, *On the Verge of War: International Relations and the Jülich-Kleve Succession Crises* (Boston, 1999) , pp. 18-21.

33. Andrew Taylor, *The World of Gerard Mercator: The Man who Revolutionised Geography* (London, 2005), pp. 128-9.

34. Quoted in Crane, *Mercator*, p. 160.

35. Karrow, *Mapmakers of the Sixteenth Century*, p. 386.

36. Quoted in Crane, *Mercator*, p. 194.

37. On the crisis of sixteenth-century cosmography, see Frank Lestringant, *Mapping the Renaissance World: The Geographical Imagination in the Age of Discovery*, trans. David Fausett (Oxford, 1994), and Denis Cosgrove, 'Images of Renaissance Cosmography, 1450-1650', in Woodward, *History of Cartography*, vol. 3, pt. 1; on chronology, see Anthony Grafton, 'Joseph Scaliger and Historical Chronology: The Rise and Fall of a Discipline', *History and Theory*, 14/2 (1975), pp. 156-85, 'Dating History: The Renaissance and the Reformation of Chronology', *Daedalus*, 132/2 (2003), pp. 74-85, and *Joseph Scaliger: A Study in the History of Classical Scholarship*, vol. 2: *Historical Chronology* (Oxford, 1993).

38. Quoted ibid., p. 13.

39. Ibid., p. 9.

40. Quoted in Vermij, 'Mercator and the Reformation', p. 86.

41. On Mercator's *Chronologia*, see Rienk Vermij, 'Gerard Mercator and the Science of Chronology', in Hans Blotevogel and Rienk Vermij (eds.), *Gerhard Mercator und die geistigen Strömungen des 16. und 17. Jahrhunderts* (Bochum, 1995), pp. 189-98.

42. Ibid., p. 192.

43. Grafton, 'Dating History', p. 75.

44. On this view of cosmography, see Cosgrove, *Apollo's Eye*; Lestringant, *Mapping the Renaissance World*.

45. All quotes from the map's legends are taken from the anonymous article, 'Text and Translation of the Legends of the Original Chart of the World by Gerhard Mercator, Issued in 1569', *Hydrographic Review*, 9 (1932), pp. 7-45.

46. On loxodromes, see James Alexander, 'Loxodromes: A Rhumb Way to Go',

*Mathematics Magazine*, 7/5 (2004), pp. 349–56; Mark Monmonier, *Rhumb Lines and Map Wars: A Social History of the Mercator Map Projection* (Chicago, 2004), pp. 1–24.

47. See Lloyd A. Brown, *The Story of Maps* (New York, 1949), p. 137.
48. Monmonier, *Rhumb Lines and Map Wars*, pp. 4–5.
49. William Borough, *A Discourse on the Variation of the Compass*, quoted in E. J. S. Parsons and W. F. Morris, 'Edward Wright and his Work', *Imago Mundi*, 3 (1939), pp. 61–71, at p. 63.
50. Eileen Reeves, 'Reading Maps', *Word and Image*, 9/1 (1993), pp. 51–65.
51. Gerardus Mercator, *Atlas sive cosmographicae meditationes de fabrica mundi et fabricate figura* (CD-ROM, Oakland, Calif., 2000), p. 106.
52. Ibid.
53. Ibid., p. 107.
54. Quoted in Lucia Nuti, 'The World Map as an Emblem: Abraham Ortelius and the Stoic Contemplation', *Imago Mundi*, 55 (2003), pp. 38–55, at p. 54.
55. See Lestringant, *Mapping the Renaissance World*, p. 130; Cosgrove, 'Images of Renaissance Cosmography', p. 98.
56. David Harvey, 'Cosmopolitanism and the Banality of Geographical Evils', *Public Culture*, 12/2 (2000), pp. 529–64, at p. 549.

## CHAPTER 8. MONEY: JOAN BLAEU, ATLAS MAIOR, 1662

1. Quoted in Maarten Prak, *The Dutch Republic in the Seventeenth Century* (Cambridge, 2005), p. 262.
2. On Blaeu's map, see Minako Debergh, 'A Comparative Study of Two Dutch Maps, Preserved in the Tokyo National Museum: Joan Blaeu's Wall Map of the World in Two Hemispheres, 1648 and its Revision ca. 1678 by N. Visscher', *Imago Mundi*, 35 (1983), pp. 20–36.
3. Derek Croxton, 'The Peace of Westphalia of 1648 and the Origins of Sovereignty', *International History Review*, 21/3 (1999), pp. 569–91.
4. Oscar Gelderblom and Joost Jonker, 'Completing a Financial Revolution: The Finance of the Dutch East India Trade and the Rise of the Amsterdam Capital Market, 1595–1612', *Journal of Economic History*, 64/3 (2004), pp. 641–72; Jan de Vries and Ad van der Woude, *The First Modern Economy: Success, Failure and Perseverance of the Dutch Economy, 1500–1815* (Cambridge, 1997).
5. Kees Zandvliet, *Mapping for Money: Maps, Plans and Topographic Paintings and their Role in Dutch Overseas Expansion during the 16th and 17th Centuries* (Amsterdam, 1998), pp. 33–51.
6. Cornelis Koeman, Günter Schilder, Marco van Egmond and Peter van der Krogt, 'Commercial Cartography and Map Production in the Low Countries,

1500–ca. 1672', in David Woodward (ed.), *The History of Cartography*, vol. 3: *Cartography in the European Renaissance*, pt. 1 (Chicago, 2007), pp. 1296–1383.

7. Herman de la Fontaine Verwey, 'Het werk van de Blaeus', *Maandblad Amstelodamum*, 39 (1952), p. 103.

8. Simon Schama, *The Embarrassment of Riches: An Interpretation of Dutch Culture in the Golden Age* (London, 1987).

9. Svetlana Alpers, *The Art of Describing: Dutch Art in the Seventeenth Century* (Chicago, 1983).

10. Herman de la Fontaine Verwey, 'Dr Joan Blaeu and his Sons', *Quaerendo*, 11/1 (1981), pp. 5–23.

11. C. Koeman, 'Life and Works of Willem Janszoon Blaeu: New Contributions to the Study of Blaeu, Made during the Last Hundred Years', *Imago Mundi*, 26 (1972), pp. 9–16, gives this date as 1617. I am grateful to Jan Werner for providing the correct date.

12. Herman Richter, 'Willem Jansz. Blaeu with Tycho Brahe on Hven, and his Map of the Island: Some New Facts', *Imago Mundi*, 3 (1939), pp. 53–60.

13. Quoted in Klaas van Berkel, 'Stevin and the Mathematical Practitioners', in Klaas van Berkel, Albert van Helden and Lodewijk Palm (eds.), *A History of Science in the Netherlands* (Leiden, 1999), pp. 13–36, at p. 19.

14. Peter Burke, *A Social History of Knowledge: From Gutenberg to Diderot* (Oxford, 2000), pp. 163–5.

15. Günter Schilder, 'Willem Jansz. Blaeu's Wall Map of the World, on Mercator's Projection, 1606–07 and its Influence', *Imago Mundi*, 31 (1979), pp. 36–54.

16. Quoted ibid., pp. 52–3.

17. James Welu, 'Vermeer: His Cartographic Sources', *Art Bulletin*, 57 (1975), p. 529.

18. Nadia Orenstein *et al.*, 'Print Publishers in the Netherlands 1580–1620', in *Dawn of the Golden Age*, exhibition catalogue, Rijksmuseum (Amsterdam, 1993), pp. 167–200.

19. Cornelis Koeman and Marco van Egmond, 'Surveying and Official Mapping in the Low Countries, 1500–ca. 1670', in Woodward, *History of Cartography*, vol. 3, pt. 1, pp. 1246–95, at p. 1270.

20. Zandvliet, *Mapping for Money*, pp. 97–8, and 'Mapping the Dutch World Overseas in the Seventeenth Century', in Woodward, *History of Cartography*, vol. 3, pt. 1, pp. 1433–62.

21. J. Keuning, 'The History of an Atlas: Mercator-Hondius', *Imago Mundi*, 4 (1947), pp. 37–62, Peter van der Krogt, *Koeman's Atlantes Neerlandici*, 3 vols. (Houten, 1997), vol. 1, pp. 145–208.

22. J. Keuning, 'Jodocus Hondius Jr', *Imago Mundi*, 5 (1948), pp. 63–71, Ir. C. Koeman, *Atlantes Neerlandici: Bibliography of Terrestrial, Maritime, and Celestial Atlases and Pilot Books, Published in the Netherlands up to 1800*, 6 vols. (Amsterdam, 1969), vol. 2, pp. 159–88.

23. Quoted in J. Keuning, 'Blaeu's *Atlas*', *Imago Mundi*, 14 (1959), pp. 74–89, at pp. 76–7; Koeman, *Atlantes Neerlandici*, vol. 1, pp. 73–85; van der Krogt, *Koeman's Atlantes*, vol. 1, pp. 31–231.

24. Edward Luther Stevenson, *Willem Janszoon Blaeu, 1571–1638* (New York, 1914), pp. 25–6.

25. Günter Schilder, *The Netherland Nautical Cartography from 1550 to 1650* (Coimbra, 1985), p. 107.

26. Koeman *et al.*, 'Commercial Cartography', pp. 1324–30.

27. Quoted in Keuning, 'Blaeu's *Atlas*', p. 77.

28. Jonathan Israel, 'Frederick Henry and the Dutch Political Factions, 1625–1642', *English Historical Review*, 98 (1983), pp. 1–27.

29. Zandvliet, *Mapping for Money*, p. 91.

30. Keuning, 'Blaeu's *Atlas*', pp. 78–9, Koeman, *Atlantes Neerlandici*, vol. 1, pp. 86–198, van der Krogt, *Koeman's Atlantes*, vol. 1, pp. 209–466.

31. Quoted in Keuning, 'Blaeu's *Atlas*', p. 80.

32. Rienk Vermij, *The Calvinist Copernicans: The Reception of the New Astronomy in the Dutch Republic, 1575–1750* (Cambridge, 2002), pp. 107–8.

33. De Vries and van der Woude, *The First Modern Economy*, pp. 490–91; J. R. Bruin *et al.* (eds.), *Dutch-Asiatic Shipping in the 17th and 18th Centuries*, 3 vols. (The Hague, 1987), vol. 1, pp. 170–88.

34. Günter Schilder, 'Organization and Evolution of the Dutch East India Company's Hydrographic Office in the Seventeenth Century', *Imago Mundi*, 28 (1976), pp. 61–78; Zandvliet, *Mapping for Money*, p. 120.

35. Ibid., pp. 122–4.

36. Ibid., p. 122.

37. Ibid., p. 124.

38. Ir. C. Koeman, *Joan Blaeu and his Grand Atlas* (Amsterdam, 1970), pp. 8–10.

39. Verwey, 'Blaeu and his Sons', p. 9.

40. Koeman, *Grand Atlas*, pp. 9–10.

41. Koeman, *Atlantes Neerlandici*, vol. 1, pp. 199–294, van der Krogt, *Koeman's Atlantes*, vol. 2, pp. 316–458.

42. Koeman, *Grand Atlas*, pp. 43–6, Peter van der Krogt, 'Introduction', in Joan Blaeu, *Atlas maior of 1665* (Cologne, 2005), pp. 36–7.

43. Koeman, *Grand Atlas*, pp. 53–91.

44. Joan Blaeu, *Atlas maior of 1665*, p. 12.

45. Ibid.

46. See e.g. Vermij, *The Calvinist Copernicans*, pp. 222–37.

47. Quoted in Alpers, *The Art of Describing*, p. 159.

48. Herman de la Fontaine Verwey, 'The Glory of the Blaeu Atlas and "the Master Colourist"', *Quaerendo*, 11/3 (1981), pp. 197–229.

49. Johannes Keuning, 'The *Novus Atlas* of Johannes Janssonius', *Imago Mundi*, 8 (1951), pp. 71–98.

50. Quoted in Koeman, *Grand Atlas*, p. 95.

51. Koeman, *Atlantes Neerlandici*, vol. 1, pp. 199–200.
52. Peter van der Krogt and Erlend de Groot (eds.), *The Atlas Blaeu-Van der Hem*, 7 vols. (Utrecht, 1996); Verwey, 'The Glory of the Blaeu Atlas', pp. 212–19.

## CHAPTER 9. NATION: THE CASSINI FAMILY MAP OF FRANCE, 1793

1. Quoted in Monique Pelletier, *Les Cartes des Cassini: la science au service de l'état et des régions* (Paris, 2002), p. 167.
2. Quoted ibid.
3. Quoted in Anne Godlewska, 'Geography and Cassini IV: Witness and Victim of Social and Disciplinary Change', *Cartographica*, 35/3–4 (1998), pp. 25–39, at p. 35.
4. To avoid confusion between the four generations of Cassinis, historians label the Cassinis I to IV.
5. Marcel Roncayolo, 'The Department', in Pierre Nora (ed.), *Rethinking France: Les Lieux de Mémoire*, vol. 2: *Space* (Chicago, 2006), pp. 183–231.
6. Montesquieu, quoted in David A. Bell, *The Cult of the Nation in France: Inventing Nationalism, 1680–1800* (Cambridge, Mass., 2001), p. 11.
7. Benedict Anderson, *Imagined Communities: Reflections on the Origin and Spread of Nationalism* (London, 1983, rev. edn. 1991).
8. James R. Akerman, 'The Structuring of Political Territory in Early Printed Atlases', *Imago Mundi*, 47 (1995), pp. 138–54, at p. 141; David Buisseret, 'Monarchs, Ministers, and Maps in France before the Accession of Louis XIV', in Buisseret (ed.), *Monarchs, Ministers, and Maps: The Emergence of Cartography as a Tool of Government in Early Modern Europe* (Chicago, 1992), pp. 99–124, at p. 119.
9. Jacob Soll, *The Information Master: Jean-Baptiste Colbert's Secret State Intelligence System* (Ann Arbor, 2009).
10. Quoted in David J. Sturdy, *Science and Social Status: The Members of the Académie des Sciences, 1666–1750* (Woodbridge, 1995), p. 69.
11. Ibid., pp. 151–6.
12. David Turnbull, 'Cartography and Science in Early Modern Europe: Mapping the Construction of Knowledge Spaces', *Imago Mundi*, 48 (1996), pp. 5–24.
13. Quoted in Pelletier, *Cassini*, p. 39.
14. Ibid., p. 40. On the changing role of the surveyor, see E. G. R. Taylor, 'The Surveyor', *Economic History Review*, 17/2 (1947), pp. 121–33.
15. John Leonard Greenberg, *The Problem of the Earth's Shape from Newton to Clairaut* (Cambridge, 1995), pp. 1–2.
16. Josef W. Konvitz, *Cartography in France, 1660–1848: Science, Engineering and Statecraft* (Chicago, 1987), pp. 5–6.
17. Ibid., p. 7.

18. Quoted in Pelletier, *Cassini*, p. 54.
19. Mary Terrall, 'Representing the Earth's Shape: The Polemics Surrounding Maupertuis's Expedition to Lapland', *Isis*, 83/2 (1992), pp. 218–37.
20. Pelletier, *Cassini*, p. 79.
21. Quoted in Terrall, 'Representing the Earth's Shape', p. 223.
22. Mary Terrall, *The Man who Flattened the Earth: Maupertuis and the Sciences in the Enlightenment* (Chicago, 2002), pp. 88–130.
23. Quoted in Michael Rand Hoare, *The Quest for the True Figure of the Earth* (Aldershot, 2005), p. 157.
24. Quoted in Pelletier, *Cassini*, p. 79.
25. Quoted in Monique Pelletier, 'Cartography and Power in France during the Seventeenth and Eighteenth Centuries', *Cartographica*, 35/3–4 (1998), pp. 41–53, at p. 49.
26. Konvitz, *Cartography in France*, p. 14, Graham Robb, *The Discovery of France* (London, 2007), pp. 4–5.
27. Charles Coulston Gillispie, *Science and Polity in France: The Revolutionary and Napoleonic Years* (Princeton, 1980), p. 115, Konvitz, *Cartography in France*, p. 16.
28. Quoted in Mary Sponberg Pedley, *The Commerce of Cartography: Making and Marketing Maps in Eighteenth-Century France and England* (Chicago, 2005), pp. 22–3.
29. Christine Marie Petto, *When France was King of Cartography: The Patronage and Production of Maps in Early Modern France* (Plymouth, 2007); Mary Sponberg Pedley, 'The Map Trade in Paris, 1650–1825', *Imago Mundi*, 33 (1981), pp. 33–45.
30. Josef W. Konvitz, 'Redating and Rethinking the Cassini Geodetic Surveys of France, 1730–1750', *Cartographica*, 19/1 (1982), pp. 1–15.
31. Quoted in Pelletier, *Cassini*, p. 95.
32. On Cassini III's estimates, see Konvitz, *Cartography in France*, pp. 22–4. On salaries, see Peter Jones, 'Introduction: Material and Popular Culture', in Martin Fitzpatrick, Peter Jones, Christa Knellwolf and Iain McCalman (eds.), *The Enlightenment World* (Oxford, 2004), pp. 347–8.
33. Quoted in Pelletier, *Cassini*, pp. 117–18.
34. Ibid., pp. 123–4.
35. Ibid., p. 128.
36. Ibid., p. 143.
37. Ibid., p. 144.
38. Ibid., pp. 232–3.
39. Pedley, *Commerce of Cartography*, pp. 85–6.
40. Quoted in Pelletier, *Cassini*, p. 135.
41. Ibid., p. 140.
42. Quoted in Bell, *The Cult of the Nation*, p. 70.
43. Ibid., p. 15.

44. Quoted in Anne Godlewska, *Geography Unbound: French Geographic Science from Cassini to Humboldt* (Chicago, 1999), p. 80.
45. Bell, *The Cult of the Nation*, p. 69.
46. Emmanuel-Joseph Sieyès, quoted in Linda and Marsha Frey, *The French Revolution* (Westport, Conn., 2004), p. 3.
47. Quoted in Bell, *The Cult of the Nation*, p. 76.
48. Ibid., pp. 14, 22, 13–14.
49. Quoted in Pelletier, *Cassini*, p. 165.
50. Ibid., p. 169.
51. Quoted in Godlewska, *Geography Unbound*, p. 84.
52. Quoted in Pelletier, *Cassini*, p. 170.
53. Quoted in Robb, *Discovery of France*, pp. 202–3.
54. *London Literary Gazette*, no. 340, Saturday, 26 July 1823, p. 471.
55. Quoted in Pelletier, *Cassini*, p. 244.
56. Ibid., pp. 246–7.
57. Ibid., p. 243.
58. Sven Widmalm, 'Accuracy, Rhetoric and Technology: The Paris–Greenwich Triangulation, 1748–88', in Tore Frängsmyr, J. L. Heilbron and Robin E. Rider (eds.), *The Quantifying Spirit in the Eighteenth Century* (Berkeley and Los Angeles, 1990), pp. 179–206.
59. Konvitz, *Cartography in France*, pp. 25–8; Gillispie, *Science and Polity*, pp. 122–30; Lloyd Brown, *The Story of Maps* (New York, 1949), pp. 255–65.
60. Ibid., p. 255.
61. Bernard de Fontenelle, quoted in Matthew Edney, 'Mathematical Cosmography and the Social Ideology of British Cartography, 1780–1820', *Imago Mundi*, 46 (1994), pp. 101–16, at p. 104.
62. Quoted in Godlewska, *Geography Unbound*, p. 83.
63. Pedley, *Commerce of Cartography*, p. 22.
64. Quoted in Pelletier, *Cassini*, p. 133.
65. Bell, *The Cult of the Nation*, p. 6.
66. Anderson, *Imagined Communities*, pp. 11, 19.
67. Quoted in Helmut Walser Smith, *The Continuities of German History: Nation, Religion and Race across the Long Nineteenth Century* (Cambridge, 2008), p. 47.
68. Anderson, *Imagined Communities*, p. 22. Anderson rectified the omission of maps in the second edition of his book, but confined his analysis to their usage by modern colonial states.

## CHAPTER 10. GEOPOLITICS: HALFORD MACKINDER, 'THE GEOGRAPHICAL PIVOT OF HISTORY', 1904

1. 'Prospectus of the Royal Geographical Society', *Journal of the Royal Geographical Society*, 1 (1831), pp. vii–xii.

2. Ibid., pp. vii–viii.

3. *Quarterly Review*, 46 (Nov. 1831), p. 55.

4. David Smith, *Victorian Maps of the British Isles* (London, 1985).

5. Walter Ristow, 'Lithography and Maps, 1796–1850', in David Woodward (ed.), *Five Centuries of Map Printing* (Chicago, 1975), pp. 77–112.

6. Arthur Robinson, 'Mapmaking and Map Printing: The Evolution of a Working Relationship', in Woodward, *Five Centuries of Map Printing*, pp. 14–21.

7. Matthew Edney, 'Putting "Cartography" into the History of Cartography: Arthur H. Robinson, David Woodward, and the Creation of a Discipline', *Cartographic Perspectives*, 51 (2005), pp. 14–29; Peter van der Krogt, '"Kartografie" or "Cartografie"?', *Caert-Thresoor*, 25/1 (2006), pp. 11–12; *Oxford English Dictionary*, entries on 'cartography' and 'cartographer'.

8. Matthew Edney, 'Mathematical Cosmography and the Social Ideology of British Cartography, 1780–1820', *Imago Mundi*, 46 (1994), pp. 101–16, at p. 112.

9. John P. Snyder, *Flattening the Earth: Two Thousand Years of Map Projections* (Chicago, 1993), pp. 98–9, 112–13, 150–54, 105.

10. Arthur Robinson, *Early Thematic Mapping in the History of Cartography* (Chicago, 1982), pp. 15–17.

11. Ibid., pp. 160–62.

12. Simon Winchester, *The Map that Changed the World* (London, 2001).

13. Karen Severud Cook, 'From False Starts to Firm Beginnings: Early Colour Printing of Geological Maps', *Imago Mundi*, 47 (1995), pp. 155–72, at pp. 160–62.

14. Quoted in Smith, *Victorian Maps*, p. 13.

15. Matthew Edney, *Mapping an Empire: The Geographical Construction of British India, 1765–1843* (Chicago, 1997), pp. 2–3.

16. Joseph Conrad, *Heart of Darkness*, ed. Robert Hampson (London, 1995), p. 25.

17. Halford Mackinder, *Britain and the British Seas* (London, 1902), p. 343.

18. Jeffrey C. Stone, 'Imperialism, Colonialism and Cartography', *Transactions of the Institute of British Geographers*, 13/1 (1988), pp. 57–64.

19. Quoted in William Roger Louis, 'The Berlin Congo Conference and the (Non-) Partition of Africa, 1884–85', in Louis, *Ends of British Imperialism: The Scramble for Empire, Suez and Decolonization* (London, 2006), pp. 75–126, at p. 102.

20. T. H. Holdich, 'How Are We to Get Maps of Africa', *Geographical Journal*, 18/6 (1901), pp. 590–601, at p. 590.

21. Halford Mackinder, 'The Round World and the Winning of the Peace', *Foreign Affairs*, 21/1 (1943), pp. 595–605, at p. 595.

22. Gerry Kearns, *Geopolitics and Empire: The Legacy of Halford Mackinder* (Oxford, 2009), p. 37; E. W. Gilbert, 'The Right Honourable Sir Halford J. Mackinder, P.C., 1861–1947', *Geographical Journal*, 110/1–3 (1947), pp. 94–9, at p. 99.

23. Halford Mackinder, 'Geography as a Pivotal Subject in Education', *Geographical Journal*, 27/5 (1921), pp. 376–84, at p. 377.

24. Brian Blouet, 'The Imperial Vision of Halford Mackinder', *Geographical Journal*, 170/4 (2004), pp. 322–9; Kearns, *Geopolitics and Empire*, pp. 39–50.

25. Francis Darwin (ed.), *The Life and Letters of Charles Darwin, including an Autobiographical Chapter*, 3 vols. (London, 1887), vol. 1, p. 336.

26. Quoted in Kearns, *Geopolitics and Empire*, p. 44.

27. Ibid., p. 47.

28. See Denis Cosgrove, 'Extra-terrestrial Geography', in Cosgrove, *Geography and Vision: Seeing, Imagining and Representing the World* (London, 2008), pp. 34–48.

29. Quoted in Charles Kruszewski, 'The Pivot of History', *Foreign Affairs*, 32 (1954), pp. 388–401, at p. 390.

30. Halford Mackinder, 'On the Scope and Methods of Geography', *Proceedings of the Royal Geographical Society*, 9/3 (1887), pp. 141–74, at p. 141.

31. Ibid., p. 145.

32. Ibid., pp. 159–60.

33. 'On the Scope and Methods of Geography – Discussion', *Proceedings of the Royal Geographical Society*, 9/3 (1887), pp. 160–74, at p. 166.

34. D. I. Scargill, 'The RGS and the Foundations of Geography at Oxford', *Geographical Journal*, 142/3 (1976), pp. 438–61.

35. Quoted in Kruszewski, 'Pivot of History', p. 390.

36. Halford Mackinder, 'Geographical Education: The Year's Progress at Oxford', *Proceedings of the Royal Geographical Society*, 10/8 (1888), pp. 531–3, at p. 532.

37. Halford Mackinder, 'Modern Geography, German and English', *Geographical Journal*, 6/4 (1895), pp. 367–79.

38. Ibid., pp. 374, 376.

39. Ibid., p. 379.

40. Quoted in Kearns, *Geopolitics and Empire*, p. 45.

41. Halford Mackinder, 'A Journey to the Summit of Mount Kenya, British East Africa', *Geographical Journal*, 15/5 (1900), pp. 453–76, at pp. 453–4.

42. Halford Mackinder, 'Mount Kenya in 1899', *Geographical Journal*, 76/6 (1930), pp. 529–34.

43. Mackinder, 'A Journey to the Summit', pp. 473, 475.

44. Ibid., p. 476.

45. Blouet, 'Imperial Vision', pp. 322–9.

46. Mackinder, *Britain and the British Seas*, p. 358.

47. Ibid., pp. 1–4.

48. Ibid., pp. 11–12.

49. Ibid., p. 358.

50. Max Jones, 'Measuring the World: Exploration, Empire and the Reform of

the Royal Geographical Society', in Martin Daunton (ed.), *The Organisation of Knowledge in Victorian Britain* (Oxford, 2005), pp. 313–36.

51. Paul Kennedy, *The Rise and Fall of British Naval Mastery* (London, 1976), p. 190.
52. Halford Mackinder, 'The Geographical Pivot of History', *Geographical Journal*, 23/4 (1904), pp. 421–37, at pp. 421–2.
53. Ibid., p. 422.
54. Ibid., p. 431.
55. Ibid., pp. 435–6.
56. Pascal Venier, 'The Geographical Pivot of History and Early Twentieth Century Geopolitical Culture', *Geographical Journal*, 170/4 (2004), pp. 330–36.
57. Gearóid Ó Tuathail, *Critical Geopolitics: The Politics of Writing Global Space* (Minneapolis, 1996), p. 24.
58. Mackinder, 'Geographical Pivot', p. 436.
59. Ibid., p. 437.
60. Spencer Wilkinson *et al.*, 'The Geographical Pivot of History: Discussion', *Geographical Journal*, 23/4 (1904), pp. 437–44, at p. 438.
61. Ibid., p. 438.
62. Halford Mackinder, *Democratic Ideals and Reality: A Study in the Politics of Reconstruction* (1919; Washington, 1996), pp. 64–5.
63. Ibid., p. 106.
64. Mackinder, 'The Round World', p. 601.
65. Ibid., pp. 604–5.
66. Colin S. Gray, 'The Continued Primacy of Geography', *Orbis*, 40/2 (1996), pp. 247–59, at p. 258.
67. Quoted in Kearns, *Geopolitics and Empire*, p. 8.
68. Ibid., p. 17.
69. Ibid., pp. 17–18.
70. Quotations from Geoffrey Parker, *Western Geopolitical Thought in the Twentieth Century* (Beckenham, 1985), pp. 16, 31.
71. Colin S. Gray and Geoffrey Sloan (eds.), *Geopolitics, Geography and Strategy* (Oxford, 1999), pp. 1–2; Parker, *Western Geopolitical Thought*, p. 6.
72. Quoted in Saul Bernard Cohen, *Geopolitics of the World System* (Lanham, Md., 2003), p. 11.
73. Alfred Thayer Mahan, *The Influence of Sea Power upon History, 1660–1783* (Boston, 1890), p. 42.
74. Kearns, *Geopolitics and Empire*, p. 4; Zachary Lockman, *Contending Visions of the Middle East: The History and Politics of Orientalism* (Cambridge, 2004), pp. 96–7.
75. Quoted in Ronald Johnston *et al.* (eds.), *The Dictionary of Human Geography*, 4th edn. (Oxford, 2000), p. 27.
76. Woodruff D. Smith, 'Friedrich Ratzel and the Origins of Lebensraum', *German Studies Review*, 3/1 (1980), pp. 51–68.

77. Kearns, *Geopolitics and Empire*; Brian Blouet (ed.), *Global Geostrategy: Mackinder and the Defence of the West* (Oxford, 2005); David N. Livingstone, *The Geographical Tradition: Episodes in the History of a Contested Enterprise* (Oxford, 1992), pp. 190–96; Colin S. Gray, *The Geopolitics of Super Power* (Lexington, Ky., 1988), pp. 4–12; Gray and Sloan, *Geopolitics*, pp. 15–62; and the special issue of *Geographical Journal*, 170 (2004).

78. Quoted in Kearns, *Geopolitics and Empire*, p. 62.

79. Livingstone, *Geographical Tradition*, p. 190.

80. Christopher J. Fettweis, 'Sir Halford Mackinder, Geopolitics and Policy-making in the 21st Century', *Parameters*, 30/2 (2000), pp. 58–72.

81. Paul Kennedy, 'The Pivot of History', *Guardian*, 19 June 2004, p. 23.

## CHAPTER 11. EQUALITY: THE PETERS PROJECTION, 1973

1. Quoted in Nicholas Mansergh (ed.), *The Transfer of Power, 1942–47*, 12 vols. (London, 1970), vol. 12, no. 488, appendix 1.

2. Quoted in Yasmin Khan, *The Great Partition: The Making of India and Pakistan* (New Haven, 2007), p. 125.

3. On the partition, see O. H. K. Spate, 'The Partition of the Punjab and of Bengal', *Geographical Journal*, 110/4 (1947), pp. 201–18, and Tan Tai Yong, '"Sir Cyril Goes to India": Partition, Boundary-Making and Disruptions in the Punjab', *Punjab Studies*, 4/1 (1997), pp. 1–20.

4. Quoted in John Pickles, 'Text, Hermeneutics and Propaganda Maps', in Trevor J. Barnes and James S. Duncan (eds.), *Writing Worlds: Discourse, Text and Metaphor in the Representation of Landscape* (London, 1992), pp. 193–230, at p. 197.

5. See Jeremy Black, *Maps and History: Constructing Images of the Past* (New Haven, 1997), pp. 123–8.

6. Denis Cosgrove, 'Contested Global Visions: One-World, Whole-Earth, and the Apollo Space Photographs', *Annals of the Association of American Geographers*, 84/2 (1994), pp. 270–94.

7. Quoted in Ursula Heise, *Sense of Place and Sense of Planet: The Environmental Imagination of the Global* (Oxford, 2008), p. 23.

8. Joe Alex Morris, 'Dr Peters' Brave New World', *Guardian*, 5 June 1973.

9. See Mark Monmonier, *Drawing the Line: Tales of Maps and Cartocontroversy* (New York, 1996), p. 10.

10. Arthur H. Robinson, 'Arno Peters and his New Cartography', *American Geographer*, 12/2 (1985), pp. 103–11, at p. 104.

11. Jeremy Crampton, 'Cartography's Defining Moment: The Peters Projection Controversy', *Cartographica*, 31/4 (1994), pp. 16–32.

12. *New Internationalist*, 124 (1983).

13. Jeremy Crampton, *Mapping: A Critical Introduction to Cartography and GIS* (Oxford, 2010), p. 92.

14. Derek Maling, 'A Minor Modification to the Cylindrical Equal-Area Projection', *Geographical Journal*, 140/3 (1974), pp. 509–10.

15. Norman Pye, review of the *Peters Atlas of the World* by Arno Peters, *Geographical Journal*, 155/2 (1989), pp. 295–7.

16. H. A. G. Lewis, review of *The New Cartography* by Arno Peters, *Geographical Journal*, 154/2 (1988), pp. 298–9.

17. Quoted in Stephen Hall, *Mapping the Next Millennium: The Discovery of New Geographies* (New York, 1992), p. 380.

18. Robinson, 'Arno Peters', pp. 103, 106.

19. Quoted in John Loxton, 'The Peters Phenomenon', *Cartographic Journal*, 22 (1985), pp. 106–10, at pp. 108, 110.

20. Quoted in Monmonier, *Drawing the Line*, pp. 30–32.

21. Maling, 'Minor Modification', p. 510.

22. Lewis, review of *The New Cartography*, pp. 298–9.

23. David Cooper, 'The World Map in Equal Area Presentation: Peters Projection', *Geographical Journal*, 150/3 (1984), pp. 415–16.

24. *The West Wing*, season 2, episode 16, first screened 28 February 2001.

25. Mark Monmonier, *Rhumb Lines and Map Wars: A Social History of the Mercator Map Projection* (Chicago, 2004), p. 15.

26. Arno Peters, 'Space and Time: Their Equal Representation as an Essential Basis for a Scientific View of the World', lecture presented at Cambridge University, 29 March 1982, trans. Ward L. Kaiser and H. Wohlers (New York, 1982), p. 1. On Peters's biography, see his obituary published in *The Times*, 10 December 2002, and the series of articles published to commemorate his life and work in the *Cartographic Journal*, 40/1 (2003).

27. Quoted in Stefan Muller, 'Equal Representation of Time and Space: Arno Peters's Universal History', *History Compass*, 8/7 (2010), pp. 718–29.

28. Quoted in Crampton, 'Cartography's Defining Moment', p. 23.

29. Peters, 'Space and Time', pp. 8–9.

30. Crampton, 'Cartography's Defining Moment', p. 22; see also *The Economist*'s review of the *Peters Atlas*, 25 March 1989.

31. *The Freeman: A Fortnightly for Individualists*, Monday, 15 December 1952, p. 188.

32. Arno Peters, *The New Cartography* [*Die Neue Kartographie*] (New York, 1983), p. 146.

33. Norman J. W. Thrower, *Maps and Civilization: Cartography in Culture and Society* (Chicago, 1996), p. 224.

34. Peters, *The New Cartography*, p. 102.

35. Ibid., pp. 102, 107–18.

36. See Monmonier, *Drawing the Line*, pp. 12–13; Robinson, 'Arno Peters',

p. 104; Norman Pye, review of 'Map of the World: Peters Projection', *Geographical Journal*, 157/1 (1991), p. 95.

37. Crampton, 'Cartography's Defining Moment', p. 24.
38. Pye, 'Map of the World', pp. 95–6.
39. Peters, *The New Cartography*, pp. 128, 148.
40. James Gall, *An Easy Guide to the Constellations* (Edinburgh, 1870), p. 3.
41. James Gall, 'Use of Cylindrical Projections for Geographical, Astronomical, and Scientific Purposes', *Scottish Geographical Journal*, 1/4 (1885), pp. 119–23, at p. 119.
42. James Gall, 'On Improved Monographic Projections of the World', *British Association of Advanced Science* (1856), p. 148.
43. Gall, 'Use of Cylindrical Projections', p. 121.
44. Monmonier, *Drawing the Line*, pp. 13–14.
45. Crampton, 'Cartography's Defining Moment', pp. 21–2.
46. Gall, 'Use of Cylindrical Projections', p. 122.
47. Quotations from *North-South: A Programme for Survival* (London, 1980). Figures taken from http://www.stwr.org/special-features/the-brandt-report.html#setting.
48. Paul Krugman, *The Conscience of a Liberal* (London, 2007), pp. 4–5, 124–9.
49. J. B. Harley, 'Deconstructing the Map', in Barnes and Duncan, *Writing Worlds*, pp. 231–47.
50. David N. Livingstone, *The Geographical Tradition: Episodes in the History of a Contested Enterprise* (Oxford, 1992).
51. Alfred Korzybski, 'General Semantics, Psychiatry, Psychotherapy and Prevention', in Korzybski, *Collected Writings* (Fort Worth, Tex., 1990), p. 205.
52. J. B. Harley, 'Can There Be a Cartographic Ethics?', *Cartographic Perspectives*, 10 (1991), pp. 9–16, at pp. 10–11.
53. Peter Vujakovic, 'The Extent of the Adoption of the Peters Projection by "Third World" Organizations in the UK', *Society of University Cartographers Bulletin* (*SUC*), 21/1 (1987), pp. 11–15, and 'Mapping for World Development', *Geography*, 74 (1989), pp. 97–105.

## CHAPTER 12. INFORMATION: GOOGLE EARTH, 2012

1. At least for those using the application based in Europe; by default it centres itself on the region in which the user logs on.
2. http://weblogs.hitwise.com/heather-dougherty/2009/04/google_maps_surpasses_mapquest.html. I am grateful to Simon Greenman for this reference.
3. http://www.comscore.com/Press_Events/Press_Releases/2011/11/comScore_Releases_October_2011_U.S._Search_Engine_Rankings.
4. http://www.thedomains.com/2010/07/26/googles-global-search-share-declines/.

5. Kenneth Field, 'Maps, Mashups and Smashups', *Cartographic Journal*, 45/4 (2008), pp. 241–5.

6. David Vise, *The Google Story: Inside the Hottest Business, Media and Technology Success of Our Time* (New York, 2006), pp. 1, 3.

7. http://www.nytimes.com/2005/12/20/technology/20image.html.

8. http://spatiallaw.blogspot.com/.

9. Jeremy W. Crampton, *Mapping: A Critical Introduction to Cartography and GIS* (Oxford, 2010), p. 129.

10. Field, 'Maps, Mashups', p. 242.

11. I am grateful to Patricia Seed for her views on this aspect of the application, and providing me with the phrase 'data aggregator' (personal email correspondence, November 2011).

12. David Y. Allen, 'A Mirror of our World: Google Earth and the History of Cartography', *Coordinates*, series b, 12 (2009), pp. 1–16, at p. 9.

13. Manuel Castells, *The Information Age: Economy, Society and Culture*, vol. 1: *The Rise of the Network Society* (Oxford, 1998; second edn., 2007), p. 509.

14. Manuel Castells, *The Information Age: Economy, Society and Culture*, vol. 3: *End of Millennium* (Oxford, 1998), p. 1.

15. Castells, *The Rise of the Network Society*, pp. 501, 52, 508.

16. Matthew A. Zook and Mark Graham, 'Mapping DigiPlace: Geocoded Internet Data and the Representation of Place', *Environment and Planning B: Planning and Design*, 34 (2007), pp. 466–82.

17. Eric Gordon, 'Mapping Digital Networks: From Cyberspace to Google', *Information, Communication and Society*, 10/6 (2007), pp. 885–901.

18. James Gleick, *The Information: A History, a Theory, a Flood* (London, 2011), pp. 8–10.

19. http://www.google.com/about/corporate/company/.

20. Norbert Wiener, *Cybernetics: Or, Control and Communication in the Animal and the Machine* (Cambridge, Mass., 1948), p. 11.

21. Ibid., p. 144.

22. Ronald E. Day, *The Modern Invention of Information: Discourse, History and Power* (Carbondale, Ill., 2008), pp. 38–43.

23. Claude Shannon, 'A Mathematical Theory of Communication', *Bell System Technical Journal*, 27 (1948), pp. 379–423, at p. 379.

24. Crampton, *Mapping*, pp. 49–52.

25. Quoted ibid., p. 58.

26. Castells, *The Rise of the Network Society*, p. 40.

27. Duane F. Marble, 'Geographic Information Systems: An Overview', in Donna J. Peuquet and Duane F. Marble (eds.), *Introductory Readings in Geographic Information Systems* (London, 1990), pp. 4–14.

28. Roger Tomlinson, 'Geographic Information Systems: A New Frontier', in Peuquet and Marble, *Introductory Readings*, pp. 15–27 at p. 17.

29. J. T. Coppock and D. W. Rhind, 'The History of GIS', in D. J. Maguire *et al.* (eds.), *Geographical Information Systems*, vol. 1 (New York, 1991), pp. 21–43.

30. Janet Abbate, *Inventing the Internet* (Cambridge, Mass., 2000).

31. Castells, *The Rise of the Network Society*, pp. 50–51.

32. Ibid., p. 61.

33. An earlier version of the film, entitled *Rough Sketch*, was made in 1968, and formed the basis of the slightly longer 1977 version, released under its current title. See http://powersof10.com/.

34. Christopher C. Tanner, Christopher J. Migdal and Michael T. Jones, 'The Clipmap: A Virtual Mipmap', Proceedings of the 25th Annual Conference on Computer Graphics and Interactive Techniques, July 1998, pp.151–8, at p. 151.

35. Avi Bar-Zeev, 'How Google Earth [Really] Works', accessed at: http://www.realityprime.com/articles/how-google-earth-really-works.

36. Mark Aubin, 'Google Earth: From Space to your Face ... and Beyond', accessed at http://www.google.com/librariancenter/articles/0604_01.html.

37. http://msrmaps.com/About.aspx?n=AboutWhatsNew&b=Newsite.

38. Mark Aubin, co-founder of Keyhole, Inc., in an article entitled 'Notes on the Origin of Google Earth', accessed at: http://www.realityprime.com/articles/notes-on-the-origin-of-google-earth.

39. Michael T. Jones, 'The New Meaning of Maps', talk delivered at the 'Where 2.0' conference, San Jose, California, 31 March 2010, accessed at: http://www.youtube.com/watch?v=UWj8qtIvkkg.

40. http://www.isde5.org/al_gore_speech.htm.

41. Simon Greenman, personal email communication, December 2010. I am very grateful to Simon for sharing his first-hand knowledge of the development of geospatial applications with me.

42. Avi Bar-Zeev, 'Notes on the Origin of Google Earth', accessed at: http://www.realityprime.com/articles/notes-on-the-origin-of-google-earth.

43. 'Google Earth Co-founder Speaks', accessed at: http://techbirmingham.wordpress.com/2007/04/26/googleearth-aita/.

44. For a visual example, see 'Tiny Tech Company Awes Viewers', *USA Today*, 21 March 2003, accessed at: http://www.usatoday.com/tech/news/techinnovations/2003-03-20-earthviewer_x.htm.

45. http://www.iqt.org/news-and-press/press-releases/2003/Keyhole_06-25-03.html.

46. http://www.google.com/press/pressrel/keyhole.html.

47. Quoted in Jeremy W. Crampton, 'Keyhole, Google Earth, and 3D Worlds: An Interview with Avi Bar-Zeev', *Cartographica*, 43/2 (2008), pp. 85–93, at p. 89.

48. Vise, *The Google Story*.

49. Sergey Brin and Larry Page, 'The Anatomy of a Large-Scale Hypertextual

Web Search Engine', Seventh International World-Wide Web Conference (WWW 1998), 14–18 April 1998, Brisbane, Australia, accessed at: http://ilpubs.stanford.edu:8090/361/.

50. http://ontargetwebsolutions.com/search-engine-blog/orlando-seo-statistics/. These figures are estimates and not verified by Google.

51. http://royal.pingdom.com/2010/02/24/google-facts-and-figures-massive-infographic/.

52. On Google's stated policy, see http://www.google.com/corporate/; http://www.google.com/corporate/tenthings.html.

53. Harry McCracken, 'First Impressions: Google's Amazing Earth', accessed at: http://blogs.pcworld.com/techlog/archives/000748.html.

54. Personal interviews with Ed Parsons, April 2009 and November 2010. All Parsons's subsequent quotes are based on these interviews. I am extremely grateful to Ed for taking the time to conduct these interviews.

55. Aubin, 'Google Earth', accessed at http://www.google.com/librariancenter/articles/0604_01.html.

56. http://www.techdigest.tv/2009/01/dick_cheneys_ho.html.

57. http://googleblog.blogspot.com/2005/02/mapping-your-way.html.

58. http://media.digitalglobe.com/index.php?s=43&item=147, http://news.cnet.com/8301-1023_3-10028842-93.html.

59. Jones, 'The New Meaning of Maps'.

60. Ed Parsons, personal interview, April 2010.

61. Crampton, *Mapping*, p. 133.

62. http://googleblog.blogspot.com/2010/04/earthly-pleasures-come-to-maps.html.

63. Michael T. Jones, 'Google's Geospatial Organizing Principle', *IEEE Computer Graphics and Applications* (2007), pp. 8–13, at p. 11.

64. http://www.emc.com/collateral/analyst-reports/diverse-exploding-digital-universe; http://www.worldwidewebsize.com/.

65. Waldo Tobler, 'A Computer Movie Simulating Urban Growth in the Detroit Region', *Economic Geography*, 46 (1970), pp. 234–40, at p. 236.

66. Personal interview with Ed Parsons, April 2010.

67. http://www.gpsworld.com/gis/integration-and-standards/the-view-google-earth-7434.

68. Steven Levy, 'Secret of Googlenomics: Data-Fueled Recipe Brews Profitability', *Wired Magazine*, 17.06, accessed at: http://www.wired.com/culture/culturereviews/magazine/17-06/nep_googlenomics?currentPage=all.

69. https://www.google.com/accounts/ServiceLogin?service=adwords&hl=en_GB&ltmpl=adwords&passive=true&ifr=false&alwf=true&continue=https://adwords.google.com/um/gaiaauth?apt%3DNone%26ugl%3Dtrue&gsessionid=2-eFqzo_CDGDCfqiSMq9sQ.

70. Levy, 'Secret of Googlenomics'.

71. Jones, 'The New Meaning of Maps'.

72. Matthew A. Zook and Mark Graham, 'The Creative Reconstruction of the Internet: Google and the Privatization of Cyberspace and DigiPlace', *Geoforum*, 38 (2007), pp. 1322–43.

73. William J. Mitchell, *City of Bits: Space, Place and the Infobahn* (Cambridge, Mass., 1996), p. 112.

74. http://www.nybooks.com/articles/archives/2009/feb/12/google-the-future-of-books/?pagination=false#fn2-496790631.

75. http://online.wsj.com/article/SB10001424052748704461304576216923562033348.html?mod=WSJ_hp_LEFTTopStories.

76. http://www.heritage.org/research/reports/2011/10/google-antitrust-and-not-being-evil.

77. Annu-Maaria Nivala, Stephen Brewster and L. Tiina Sarjakoski, 'Usability Evaluation of Web Mapping Sites', *Cartographic Journal*, 45/2 (2008), pp. 129–38.

78. http://www.internetworldstats.com/stats.htm.

79. Crampton, *Mapping*, pp. 139–40.

80. Ed Parsons, personal interview, November 2009.

81. Vittoria de Palma, 'Zoom: Google Earth and Global Intimacy', in Vittoria de Palma, Diana Periton and Marina Lathouri (eds.), *Intimate Metropolis: Urban Subjects in the Modern City* (Oxford, 2009), pp. 239–70, at pp. 241–2; Douglas Vandegraft, 'Using Google Earth for Fun and Functionality', *ACSM Bulletin*, (June 2007), pp. 28–32.

82. J. Lennart Berggren and Alexander Jones (eds. and trans.), *Ptolemy's Geography: An Annotated Translation of the Theoretical Chapters* (Princeton, 2000), p. 117.

83. Allen, 'A Mirror of our World', pp. 3–8.

84. Simon Greenman, personal email communication, December 2010.

## CONCLUSION: THE EYE OF HISTORY

1. J. B. Harley and David Woodward (eds.), *The History of Cartography*, vol. 1: *Cartography in Prehistoric, Ancient, and Medieval Europe and the Mediterranean* (Chicago, 1987), p. 508.

2. Rob Kitchin and Martin Dodge, 'Rethinking Maps', *Progress in Human Geography*, 31/3 (2007), pp. 331–44, at p. 343.

3. Albrecht Penck, 'The Construction of a Map of the World on a Scale of 1:1,000,000', *Geographical Journal*, 1/3 (1893), pp. 253–61, at p. 254.

4. Ibid., p. 256.

5. Ibid., p. 259.

6. Ibid., p. 254.

7. A. R. Hinks, quoted in G. R. Crone, 'The Future of the International Million Map of the World', *Geographical Journal*, 128/1 (1962), pp. 36–8, at p. 38.

8. Michael Heffernan, 'Geography, Cartography and Military Intelligence: The Royal Geographical Society and the First World War', *Transactions of the Institute of British Geographers*, new series, 21/3 (1996), pp. 504–33.

9. M. N. MacLeod, 'The International Map', *Geographical Journal*, 66/5 (1925), pp. 445–9.

10. Quoted in Alastair Pearson, D. R. Fraser Taylor, Karen Kline and Michael Heffernan, 'Cartographic Ideals and Geopolitical Realities: International Maps of the World from the 1890s to the Present', *Canadian Geographer*, 50/2 (2006), pp. 149–75, at p. 157.

11. Trygve Lie, 'Statement by the Secretary-General', *World Cartography*, 1 (1951), p. v.

12. 'Summary of International Meetings of Interest to Cartography (1951–1952)', *World Cartography*, 2 (1952), p. 103.

13. 'The International Map of the World on the Millionth Scale and the International Co-operation in the Field of Cartography', *World Cartography*, 3 (1953), pp. 1–13.

14. Sandor Radó, 'The World Map at the Scale of 1:2 500 000', *Geographical Journal*, 143/3 (1977), pp. 489–90.

15. Quoted in Pearson *et al.*, 'Cartographic Ideals', p. 163.

16. David Rhind, 'Current Shortcomings of Global Mapping and the Creation of a New Geographical Framework for the World', *Geographical Journal*, 166/4 (2000), pp. 295–305.

17. http://www.globalmap.org/english/index.html. See Pearson *et al.*, 'Cartographic Ideals', pp. 165–72.

# Acknowledgements

Readers of this book's title might be surprised to see its affinity with Neil MacGregor's *A History of the World in* 100 *Objects* (2010). Should anyone think that I am taking my admiration for MacGregor's wonderful book a little too far, I should perhaps point out that my own title was agreed (with the same publisher) back in 2006, and that I am not at all put out that he has used the formulation before me. Such is the nature of trying to capture the zeitgeist! Although the idea behind the book was conceived six years ago, it is the culmination of nearly twenty years of thinking about and publishing on maps. In that time I have been fortunate to have learnt from many friends and colleagues in the history of cartography, who have generously taken the time to read portions of the book, and provide invaluable criticism. At the British Museum Irving Finkel shared his voluminous knowledge of the Babylonian world map, and was kind enough to send me material on the subject. Mike Edwards very helpfully read the chapter on Ptolemy. Emilie Savage-Smith discussed al-Idrīsī with me, although I suspect she will not necessarily agree with all of my conclusions. Paul Harvey probably knows more about medieval *mappaemundi* than anyone else, and was extremely generous in his comments on the Hereford map, while Julia Boffey and Dan Terkla also offered helpful ideas for further reading. Gari Ledyard is the world's leading expert on the Korean Kangnido map, and steered me through the complexities of early Korean cartography. Kenneth R. Robinson generously provided me with a series of indispensable articles on the Kangnido and Korean history, and Cordell Yee offered insightful suggestions on Chinese materials. The wonderful Timothy Brook provided help on the Kangnido's Chinese sources and was gracious enough to enable me to reproduce a copy of Qingjun's map, which was his find, not mine.

At the US Library of Congress John Hessler allowed me access to papers relating to the acquisition of the Waldseemüller map, and also offered incisive comments on my chapter. Philip D. Burden shared his great love of antique maps as well as the remarkable story of evaluating Waldseemüller's map. Joaquim Alves Gaspar provided important research on sixteenth-century projections which helped on Ribeiro. Nick Crane gave me the benefit of his extensive knowledge on Mercator. Jan Werner commented extensively on the Blaeu chapter. David A. Bell offered shrewd ideas on the Cassini material and Josef Konvitz clarified some of its more arcane dimensions. Mark Monmonier read both the Mercator and Peters chapters with his typically penetrating eye. Dave Vest of Mythicsoft helped me on the technical aspects of Google Earth: his expertise rescued me on many occasions, for which I am extremely grateful. Simon Greenman also offered an insider's view of the rise of online mapping, and Patricia Seed provided a shrewd critique. At Google, Ed Parsons was enormously supportive of the entire project; he found time to conduct several interviews with me, provided access to a range of people, and also read the Google chapter. Even though the book has many reservations about Google's methods, Ed was exemplary in listening to criticism in my version of the Google Earth story. Many others have answered questions and provided references, including Angelo Cattaneo, Matthew Edney, John Paul Jones III, Eddy Maes, Nick Millea and Hilde De Weerdt. Christopher Nugee QC and Jim Smith spotted several errors, all of which were mine.

The completion of this book was supported by a generous research leave grant from the Arts and Humanities Research Council (ahrc. co.uk). The AHRC supports research that furthers our understanding of human culture and creativity, and I am very grateful that a book on the history of world mapmaking should be part of this endeavour. As a trustee of the J. B. Harley Trust, I have the great fortune to work with some of the world's leading experts in the history of cartography, and I would like to thank Peter Barber, Sarah Bendall, Catherine Delano-Smith, Felix Driver, David Fletcher, Paul Harvey, Roger Kain, Rose Mitchell, Sarah Tyacke and Charles Withers for helping me more than they probably know. Catherine supported the project from the outset and answered innumerable queries, as did Peter and Tony Campbell. I am deeply grateful to Peter and Catherine in particular for clarifying what the book was trying to achieve in its early stages, and all their help and friendship over the years. I am particularly fortunate that Peter took the time to read the entire manuscript, offering me the benefit of his unparalled expertise.

While writing the book, I was delighted to be asked to present a three-part BBC television series, 'Maps: Power, Plunder and Possession', which helped me not only to consolidate my relationship to many of the extraordinary maps that appear in this book but also to understand the importance of the story I have tried to tell. I am deeply grateful to the wonderful team responsible for making the series, in particular Louis Caulfield, Tom Cebula, Annabel Hobley, Helen Nixon and Ali Pares, and to Anne Laking and Richard Klein for commissioning the series.

Nearly every book I have written acknowledges the institutional support of Queen Mary, and this one is no exception. I am grateful to the English Department for allowing me a period of sabbatical leave to complete the research for the book, and in particular to Michèle Barrett, Julia Boffey, Markman Ellis, Alfred Hiatt, my surrogate Jewish mother Lisa Jardine, Philip Ogden, Chris Reid, Peggy Reynolds, Bill Schwarz and Morag Shiach. I only wish that the late Kevin Sharpe had the opportunity to read it; he is greatly missed, but never forgotten. As ever, David Colclough has been the greatest of friends, and it is a pleasure once again to thank him for sustaining me through a shared love of everything from Milton and Mercator to 1980s indie music.

When I was young, my limited book collection mainly consisted of Picador and Penguin titles, so I often have to pinch myself to appreciate that I ended up having Peter Straus as my agent and Stuart Proffitt as my editor. Peter is a legend and I want to thank him for all he has done for me over the last five years. Stuart has been an exemplary editor whose tireless labour on the book has been quite extraordinary (even as I write this I think of him, and worry over my sentence construction). I would like to acknowledge all his hard work and that of everyone at Allen Lane, especially Stuart's assistant, Shan Vahidy, for making this book possible. Elizabeth Stratford provided exemplary copy editing, and Cecilia Mackay was the best picture researcher I have ever worked with, extracting a series of seemingly impossible images with effortless ease.

Throughout the writing of this book I have needed the patience, humour, diversion and support of my friends and family. I would like to thank all the Brottons – Alan, Bernice, Peter, Susan, Diane and Tariq – for their faith in me, as well as Sophie and Dominik Beissel, Emma and James Lambe for Castle Farm, the 'Shed', and grandparenting above and beyond the call of duty. Simon Curtis, Matthew Dimmock, Rachel Garistina, Tim Marlow and Tanya Hudson, Rob Nixon, Grayson and Philippa Perry, Richard Scholar and Ita McCarthy, James Scott, Guy Richards Smit and Rebecca Chamberlain, and Dave and Emily Vest have all been great friends and helped me in particular and vital ways. Dafydd Roberts provided crucial help in translating key materials, and Michael Wheare was an indefatigable research assistant. Peter Florence provided me with 'The West Wing' and an unforgettable fortieth birthday in Granada, as well as the intellectual space to develop my own cultural geography. One of the book's inspirations was the work of my late friend Denis Cosgrove, who taught me so much about the global and transcendent possibilities of maps, and whose presence still pervades much of what I write.

I am fortunate to call Adam Lowe my greatest friend, and want to salute him as the presiding genius behind this book. Whenever I despair of the value of the arts, I look at what Adam does, and it fills me with wonder and inspiration. My world is an infinitely better place for his presence in it, for which I bless him on most days. I hope we will create more worlds within worlds together in the future.

Six years ago, I met my wife, Charlotte, for the second time. Since then she has filled my life with her love, and that of our two young children, Ruby and Hardie. Without Charlotte there would be no book, and very possibly no author. She has kept me going with passion, care, intelligence and sweeties, and gives me more in life than I ever thought possible. I love her beyond any measure expressed throughout the course of this book, which is why I dedicate it to her.

# Index

Page references in *italic* indicate Figures and illustrations.

ALLEN LANE
*an imprint of*
PENGUIN BOOKS

# Recently Published

Michael Axworthy, *Revolutionary Iran: A History of the Islamic Republic*

Jaron Lanier, *Who Owns the Future?*

John Gray, *The Silence of Animals: On Progress and Other Modern Myths*

Paul Kildea, *Benjamin Britten: A Life in the Twentieth Century*

Jared Diamond, *The World Until Yesterday: What Can We Learn from Traditional Societies?*

Nassim Nicholas Taleb, *Antifragile: How to Live in a World We Don't Understand*

Alan Ryan, *On Politics: A History of Political Thought from Herodotus to the Present*

Roberto Calasso, *La Folie Baudelaire*

Carolyn Abbate and Roger Parker, *A History of Opera: The Last Four Hundred Years*

Yang Jisheng, *Tombstone: The Untold Story of Mao's Great Famine*

Caleb Scharf, *Gravity's Engines: The Other Side of Black Holes*

Jancis Robinson, Julia Harding and José Vouillamoz, *Wine Grapes: A Complete Guide to 1,368 Vine Varieties, including their Origins and Flavours*

David Bownes, Oliver Green and Sam Mullins, *Underground: How the Tube Shaped London*

Niall Ferguson, *The Great Degeneration: How Institutions Decay and Economies Die*

Chrystia Freeland, *Plutocrats: The Rise of the New Global Super-Rich*

David Thomson, *The Big Screen: The Story of the Movies and What They Did to Us*

Halik Kochanski, *The Eagle Unbowed: Poland and the Poles in the Second World War*

Kofi Annan with Nader Mousavizadeh, *Interventions: A Life in War and Peace*

Mark Mazower, *Governing the World: The History of an Idea*

Anne Applebaum, *Iron Curtain: The Crushing of Eastern Europe 1944-56*

Steven Johnson, *Future Perfect: The Case for Progress in a Networked Age*

Christopher Clark, *The Sleepwalkers: How Europe Went to War in 1914*

Neil MacGregor, *Shakespeare's Restless World*

Nate Silver, *The Signal and the Noise: The Art and Science of Prediction*

Chinua Achebe, *There Was a Country: A Personal History of Biafra*

John Darwin, *Unfinished Empire: The Global Expansion of Britain*

Jerry Brotton, *A History of the World in Twelve Maps*

Patrick Hennessey, *KANDAK: Fighting with Afghans*

Katherine Angel, *Unmastered: A Book on Desire, Most Difficult to Tell*

David Priestland, *Merchant, Soldier, Sage: A New History of Power*

Stephen Alford, *The Watchers: A Secret History of the Reign of Elizabeth I*

Tom Feiling, *Short Walks from Bogotá: Journeys in the New Colombia*

Pankaj Mishra, *From the Ruins of Empire: The Revolt Against the West and the Remaking of Asia*

Geza Vermes, *Christian Beginnings: From Nazareth to Nicaea,*
*AD 30-325*

Steve Coll, *Private Empire: ExxonMobil and American Power*

Joseph Stiglitz, *The Price of Inequality*

Dambisa Moyo, *Winner Take All: China's Race for Resources and*
*What it Means for Us*

Robert Skidelsky and Edward Skidelsky, *How Much is Enough? The*
*Love of Money, and the Case for the Good Life*

Frances Ashcroft, *The Spark of Life: Electricity in the Human Body*

Sebastian Seung, *Connectome: How the Brain's Wiring Makes Us*
*Who We Are*

Callum Roberts, *Ocean of Life*

Orlando Figes, *Just Send Me Word: A True Story of Love and*
*Survival in the Gulag*

Leonard Mlodinow, *Subliminal: The Revolution of the New*
*Unconscious and What it Teaches Us about Ourselves*

John Romer, *A History of Ancient Egypt: From the First Farmers to*
*the Great Pyramid*

Ruchir Sharma, *Breakout Nations: In Pursuit of the Next Economic*
*Miracle*

Michael J. Sandel, *What Money Can't Buy: The Moral Limits of*
*Markets*

Dominic Sandbrook, *Seasons in the Sun: The Battle for Britain,*
*1974-1979*

Tariq Ramadan, *The Arab Awakening: Islam and the New Middle*
*East*

Jonathan Haidt, *The Righteous Mind: Why Good People are*
*Divided by Politics and Religion*

Colin McEvedy, *Cities of the Classical World: An Atlas and Gazetteer of 120 Centres of Ancient Civilization*

Heike B. Görtemaker, *Eva Braun: Life with Hitler*

Brian Cox and Jeff Forshaw, *The Quantum Universe: Everything that Can Happen Does Happen*

Nathan D. Wolfe, *The Viral Storm: The Dawn of a New Pandemic Age*

Norman Davies, *Vanished Kingdoms: The History of Half-Forgotten Europe*

Michael Lewis, *Boomerang: The Meltdown Tour*

Steven Pinker, *The Better Angels of Our Nature: The Decline of Violence in History and Its Causes*

Robert Trivers, *Deceit and Self-Deception: Fooling Yourself the Better to Fool Others*

Thomas Penn, *Winter King: The Dawn of Tudor England*

Daniel Yergin, *The Quest: Energy, Security and the Remaking of the Modern World*

Michael Moore, *Here Comes Trouble: Stories from My Life*

Ali Soufan, *The Black Banners: Inside the Hunt for Al Qaeda*

Jason Burke, *The 9/11 Wars*

Timothy D. Wilson, *Redirect: The Surprising New Science of Psychological Change*

Ian Kershaw, *The End: Hitler's Germany, 1944-45*

T M Devine, *To the Ends of the Earth: Scotland's Global Diaspora, 1750-2010*

Catherine Hakim, *Honey Money: The Power of Erotic Capital*

Douglas Edwards, *I'm Feeling Lucky: The Confessions of Google Employee Number 59*

John Bradshaw, *In Defence of Dogs*

Chris Stringer, *The Origin of Our Species*

Lila Azam Zanganeh, *The Enchanter: Nabokov and Happiness*

David Stevenson, *With Our Backs to the Wall: Victory and Defeat in 1918*

Evelyn Juers, *House of Exile: War, Love and Literature, from Berlin to Los Angeles*

Henry Kissinger, *On China*

Michio Kaku, *Physics of the Future: How Science Will Shape Human Destiny and Our Daily Lives by the Year 2100*

David Abulafia, *The Great Sea: A Human History of the Mediterranean*

John Gribbin, *The Reason Why: The Miracle of Life on Earth*

Anatol Lieven, *Pakistan: A Hard Country*

William Cohen, *Money and Power: How Goldman Sachs Came to Rule the World*

Joshua Foer, *Moonwalking with Einstein: The Art and Science of Remembering Everything*

Simon Baron-Cohen, *Zero Degrees of Empathy: A New Theory of Human Cruelty*

Manning Marable, *Malcolm X: A Life of Reinvention*

David Deutsch, *The Beginning of Infinity: Explanations that Transform the World*

David Edgerton, *Britain's War Machine: Weapons, Resources and Experts in the Second World War*

John Kasarda and Greg Lindsay, *Aerotropolis: The Way We'll Live Next*

David Gilmour, *The Pursuit of Italy: A History of a Land, Its Regions and Their Peoples*

Niall Ferguson, *Civilization: The West and the Rest*

Tim Flannery, *Here on Earth: A New Beginning*

Robert Bickers, *The Scramble for China: Foreign Devils in the Qing Empire, 1832-1914*

Mark Malloch-Brown, *The Unfinished Global Revolution: The Limits of Nations and the Pursuit of a New Politics*

King Abdullah of Jordan, *Our Last Best Chance: The Pursuit of Peace in a Time of Peril*

Eliza Griswold, *The Tenth Parallel: Dispatches from the Faultline between Christianity and Islam*

Brian Greene, *The Hidden Reality: Parallel Universes and the Deep Laws of the Cosmos*

John Gray, *The Immortalization Commission: The Strange Quest to Cheat Death*

Patrick French, *India: A Portrait*

Lizzie Collingham, *The Taste of War: World War Two and the Battle for Food*

Hooman Majd, *The Ayatollahs' Democracy: An Iranian Challenge*

Dambisa Moyo, *How The West Was Lost: Fifty Years of Economic Folly - and the Stark Choices Ahead*

Evgeny Morozov, *The Net Delusion: How Not to Liberate the World*

Ron Chernow, *Washington: A Life*

Nassim Nicholas Taleb, *The Bed of Procrustes: Philosophical and Practical Aphorisms*

Hugh Thomas, *The Golden Age: The Spanish Empire of Charles V*

Amanda Foreman, *A World on Fire: An Epic History of Two Nations Divided*

Nicholas Ostler, *The Last Lingua Franca: English until the Return of Babel*

Richard Miles, *Ancient Worlds: The Search for the Origins of Western Civilization*

Neil MacGregor, *A History of the World in 100 Objects*

Steven Johnson, *Where Good Ideas Come From: The Natural History of Innovation*

Dominic Sandbrook, *State of Emergency: The Way We Were: Britain, 1970-1974*

Jim Al-Khalili, *Pathfinders: The Golden Age of Arabic Science*

Ha-Joon Chang, *23 Things They Don't Tell You About Capitalism*

Robin Fleming, *Britain After Rome: The Fall and Rise, 400 to 1070*

Tariq Ramadan, *The Quest for Meaning: Developing a Philosophy of Pluralism*

Joyce Tyldesley, *The Penguin Book of Myths and Legends of Ancient Egypt*

Nicholas Phillipson, *Adam Smith: An Enlightened Life*

Paul Greenberg, *Four Fish: A Journey from the Ocean to Your Plate*

Clay Shirky, *Cognitive Surplus: Creativity and Generosity in a Connected Age*

Andrew Graham-Dixon, *Caravaggio: A Life Sacred and Profane*

Niall Ferguson, *High Financier: The Lives and Time of Siegmund Warburg*

Sean McMeekin, *The Berlin-Baghdad Express: The Ottoman Empire and Germany's Bid for World Power, 1898-1918*

Richard McGregor, *The Party: The Secret World of China's Communist Rulers*

Spencer Wells, *Pandora's Seed: The Unforeseen Cost of Civilization*

Francis Pryor, *The Making of the British Landscape: How We Have Transformed the Land, from Prehistory to Today*

Ruth Harris, *The Man on Devil's Island: Alfred Dreyfus and the Affair that Divided France*

Paul Collier, *The Plundered Planet: How to Reconcile Prosperity With Nature*

Norman Stone, *The Atlantic and Its Enemies: A History of the Cold War*

Simon Price and Peter Thonemann, *The Birth of Classical Europe: A History from Troy to Augustine*

Hampton Sides, *Hellhound on his Trail: The Stalking of Martin Luther King, Jr. and the International Hunt for His Assassin*

Jackie Wullschlager, *Chagall: Love and Exile*

Richard Miles, *Carthage Must Be Destroyed: The Rise and Fall of an Ancient Civilization*

Tony Judt, *Ill Fares The Land: A Treatise On Our Present Discontents*

Michael Lewis, *The Big Short: Inside the Doomsday Machine*

Oliver Bullough, *Let Our Fame Be Great: Journeys among the Defiant People of the Caucasus*

Paul Davies, *The Eerie Silence: Searching for Ourselves in the Universe*

Richard Wilkinson and Kate Pickett, *The Spirit Level: Why Equality is Better for Everyone*

Tom Bingham, *The Rule of Law*

Joseph Stiglitz, *Freefall: Free Markets and the Sinking of the Global Economy*

John Lanchester, *Whoops! Why Everyone Owes Everyone and No One Can Pay*

Chinua Achebe, *The Education of a British-Protected Child*

Jaron Lanier, *You Are Not A Gadget: A Manifesto*

John Cassidy, *How Markets Fail: The Logic of Economic Calamities*

Robert Ferguson, *The Hammer and the Cross: A New History of the Vikings*

Eugene Rogan, *The Arabs: A History*

Steven Johnson, *The Invention of Air: An experiment, a Journey, a New Country and the Amazing Force of Scientific Discovery*

Andrew Ross Sorkin, *Too Big to Fail: Inside the Battle to Save Wall Street*

Malcolm Gladwell, *What the Dog Saw and Other Adventures*

Steven D. Levitt, Stephen J. Dubner, *Superfreakonomics: Global Cooling, Patriotic Prostitutes and Why Suicide Bombers Should Buy Life Insurance*

Christopher Andrew, *The Defence of the Realm: The Authorized History of MI5*